ABSORPTION AND DRUG DEVELOPMENT

ABSORPTION AND DRUG DEVELOPMENT
Solubility, Permeability, and Charge State

ALEX AVDEEF
*p*ION, Inc.

WILEY-INTERSCIENCE
A JOHN WILEY & SONS, INC., PUBLICATION

Library of Congress Cataloging-in-Publication Data:

Avdeef, Alex.
   Absorption and drug development : solubility, permeability, and
   charge state / Alex Avdeef.
     p. cm.
Includes index.
   ISBN 0-471-42365-3 (Cloth)
   1. Drugs–Design. 2. Drugs–Metabolism. 3. Drug development.
4. Absorption. I. Title.
   RS420 .A935 2003
   615' .19–dc21   2003011397

Printed in the United States of America

10 9 8 7 6 5 4 3 2 1

Carla
Natalie
Michael

CONTENTS

PREFACE   xiii

ACKNOWLEDGMENTS   xvii

DEFINITIONS   ixx

1 INTRODUCTION   1

   1.1 Shotgun Searching for Drugs? / 1
   1.2 Screen for the Target or ADME First? / 2
   1.3 ADME and Multimechanism Screens / 3
   1.4 ADME and Medicinal Chemists / 4
   1.5 The "A" in ADME / 5
   1.6 It is Not Just a Number—It is a Multimechanism / 6

2 TRANSPORT MODEL   7

   2.1 Permeability-Solubility-Charge State and the
        pH Partition Hypothesis / 7
   2.2 Properties of the Gastrointestinal Tract (GIT) / 11
   2.3 pH Microclimate / 17
   2.4 Intracellular pH Environment / 18
   2.5 Tight-Junction Complex / 18
   2.6 Structure of Octanol / 19
   2.7 Biopharmaceutics Classification System / 20

3 CHARGE STATE   22

3.1   Constant Ionic Medium Reference State / 23

3.2   pK_a Databases / 24

3.3   Potentiometric Measurements / 25

    3.3.1   Bjerrum Plots / 25

    3.3.2   pH Definitions and Electrode Standardization / 27

    3.3.3   The "Solubility Problem" and Cosolvent Methods / 29

    3.3.4   Use of Cosolvents for Water-Soluble Molecules / 30

3.4   Spectrophotometric Measurements / 31

3.5   Capillary Electrophoresis Measurements / 32

3.6   Chromatographic pK_a Measurement / 33

3.7   pK_a Microconstants / 33

3.8   pK_a "Gold Standard" for Drug Molecules / 35

4 PARTITIONING INTO OCTANOL   42

4.1   Tetrad of Equilibria / 43

4.2   Conditional Constants / 45

4.3   log P Databases / 45

4.4   log D / 45

4.5   Partitioning of Quaternary Ammonium Drugs / 50

4.6   log D of Multiprotic Drugs and the Common-Ion Effect / 50

4.7   Summary of Charged-Species Partitioning in Octanol–Water / 53

4.8   Ion Pair Absorption of Ionized Drugs—Fact or Fiction? / 53

4.9   Micro-log P / 54

4.10 HPLC Methods / 54

4.11 IAM Chromatography / 54

4.12 Liposome Chromatography / 55

4.13 Other Chromatographic Methods / 55

4.14 pH-Metric log P Method / 55

4.15 High-Throughput log P Methods / 59

4.16 Octanol–Water log P^N, log P^I, and log $D_{7.4}$
     "Gold Standard" for Drug Molecules / 59

5 PARTITIONING INTO LIPOSOMES   67

5.1   Tetrad of Equilibria and Surface Ion Pairing (SIP) / 67

5.2   Databases / 69

5.3   Location of Drugs Partitioned into Bilayers / 69

5.4 Thermodynamics of Partitioning: Entropy- or Enthalpy-Driven? / 70

5.5 Electrostatic and Hydrogen Bonding in a Low-Dielectric Medium / 71

5.6 Water Wires, H^+/OH^- Currents, and the Permeability of Amino Acids and Peptides / 73

5.7 Preparation Methods: MLV, SUV, FAT, LUV, ET / 74

5.8 Experimental Methods / 75

5.9 Prediction of $\log P_{mem}$ from $\log P$ / 76

5.10 $\log D_{mem}$, $diff_{mem}$, and the Prediction of $\log P_{mem}^{SIP}$ from $\log P^I$ / 79

5.11 Three Indices of Lipophilicity: Liposomes, IAM, and Octanol / 83

5.12 Getting it Wrong from One-Point $\log D_{mem}$ Measurement / 84

5.13 Partitioning into Charged Liposomes / 85

5.14 pK_a^{mem} Shifts in Charged Liposomes and Micelles / 86

5.15 Prediction of Absorption from Liposome Partition Studies? / 90

5.16 $\log P_{mem}^N$, $\log P_{mem}^{SIP}$ "Gold Standard" for Drug Molecules / 90

6 SOLUBILITY   **91**

6.1 Solubility–pH Profiles / 92

  6.1.1 Monoprotic Weak Acid, HA (or Base, B) / 92

  6.1.2 Diprotic Ampholyte, XH_2^+ / 93

  6.1.3 Gibbs pK_a / 93

6.2 Complications May Thwart Reliable Measurement of Aqueous Solubility / 99

6.3 Databases and the "Ionizable Molecule Problem" / 100

6.4 Experimental Methods / 100

  6.4.1 Saturation Shake-Flask Methods / 101

  6.4.2 Turbidimetric Ranking Assays / 101

  6.4.3 HPLC-Based Assays / 101

  6.4.4 Potentiometric Methods / 101

  6.4.5 Fast UV Plate Spectrophotometer Method / 107

    6.4.5.1 Aqueous Dilution Method / 107

    6.4.5.2 Cosolvent Method / 108

6.5 Correction for the DMSO Effect by the Δ-Shift Method / 111

  6.5.1 DMSO Binding to the Uncharged Form of a Compound / 111

  6.5.2 Uncharged Forms of Compound–Compound Aggregation / 112

  6.5.3 Compound–Compound Aggregation of Charged Weak Bases / 112

  6.5.4 Ionizable Compound Binding by Nonionizable Excipients / 113

  6.5.5 Results of Aqueous Solubility Determined from Δ Shifts / 113

6.6   Limits of Detection / 115

6.7 $\log S_0$ "Gold Standard" for Drug Molecules / 115

7 PERMEABILITY   **116**

7.1   Permeability in the Gastrointestinal Tract and at the
      Blood–Brain Barrier / 116

7.2   Historical Developments in Artificial-Membrane Permeability
      Measurement / 118

      7.2.1   Lipid Bilayer Concept / 118

      7.2.2   Black Lipid Membranes (BLMs) / 123

      7.2.3   Microfilters as Supports / 124

      7.2.4   Octanol-Impregnated Filters with Controlled
              Water Pores / 128

7.3   Parallel Artificial-Membrane Permeability Assay (PAMPA) / 128

      7.3.1   Egg Lecithin PAMPA Model (Roche Model) / 128

      7.3.2   Hexadecane PAMPA Model (Novartis Model) / 129

      7.3.3   Brush-Border Lipid Membrane (BBLM) PAMPA Model
              (Chugai Model) / 130

      7.3.4   Hydrophilic Filter Membrane PAMPA Model
              (Aventis Model) / 131

      7.3.5   Permeability–Retention–Gradient–Sink PAMPA Models
              (pION Models) / 131

      7.3.6   Structure of Phospholipid Membranes / 131

7.4   The Case for the Ideal In Vitro Artificial Membrane
      Permeability Model / 132

      7.4.1   Lipid Compositions in Biological Membranes / 132

      7.4.2   Permeability–pH Considerations / 132

      7.4.3   Role of Serum Proteins / 135

      7.4.4   Effects of Cosolvents, Bile Acids, and
              Other Surfactants / 135

      7.4.5   Ideal Model Summary / 137

7.5   Derivation of Membrane-Retention Permeability Equations
      (One-Point Measurements, Physical Sinks, Ionization
      Sinks, Binding Sinks, Double Sinks) / 137

      7.5.1   Thin-Membrane Model (without Retention) / 139

      7.5.2   Iso-pH Equations with Membrane Retention / 142

              7.5.2.1   Without Precipitate in Donor Wells and without
                        Sink Condition in Acceptor Wells / 143

              7.5.2.2   Sink Condition in Acceptor Wells / 147

7.5.2.3 Precipitated Sample in the Donor Compartment / 147

7.5.3 Gradient pH Equations with Membrane Retention: Single and Double Sinks / 148

7.5.3.1 Single Sink: Eq. (7.34) in the Absence of Serum Protein or Sink in Acceptor Wells / 150

7.5.3.2 Double Sink: Eq. (7.34) in the Presence of Serum Protein or Sink in Acceptor Wells / 151

7.5.3.3 Simulation Examples / 152

7.5.3.4 Gradient pH Summary / 153

7.6 Permeability–Lipophilicity Relations / 153

7.6.1 Nonlinearity / 153

7.7 PAMPA: 50+ Model Lipid Systems Demonstrated with 32 Structurally Unrelated Drug Molecules / 156

7.7.1 Neutral Lipid Models at pH 7.4 / 160

7.7.1.1 DOPC / 166

7.7.1.2 Olive Oil / 167

7.7.1.3 Octanol / 168

7.7.1.4 Dodecane / 168

7.7.2 Membrane Retention (under Iso-pH and in the Absence of Sink Condition) / 169

7.7.3 Two-Component Anionic Lipid Models with Sink Condition in the Acceptor Compartment / 171

7.7.3.1 DOPC under Sink Conditions / 177

7.7.3.2 DOPC with Dodecylcarboxylic Acid under Sink Conditions / 179

7.7.3.3 DOPC with Phosphatidic Acid under Sink Conditions / 179

7.7.3.4 DOPC with Phosphatidylglycerol under Sink Conditions / 181

7.7.3.5 DOPC with Negative Lipids without Sink / 181

7.7.4 Five-Component Anionic Lipid Model (Chugai Model) / 181

7.7.5 Lipid Models Based on Lecithin Extracts from Egg and Soy / 183

7.7.5.1 Egg Lecithin from Different Sources / 183

7.7.5.2 Soy Lecithin and the Effects of Phospholipid Concentrations / 187

7.7.5.3 Lipophilicity and Decrease in Permeability with Increased Phospholipid Content in Dodecane / 194

7.7.5.4 Sink Condition to Offset the Attenuation of Permeability / 196

7.7.5.5 Comparing Egg and Soy Lecithin Models / 198

7.7.5.6 Titrating a Suspension of Soy Lecithin / 198

7.7.6 Intrinsic Permeability, Permeability–pH Profiles, Unstirred Water Layers (UWL), and the pH Partition Hypothesis / 199

7.7.6.1 Unstirred Water Layer Effect (Transport across Barriers in Series and in Parallel) / 199

7.7.6.2 Determination of UWL Permeability using pH Dependence (pK_a^{flux}) Method / 200

7.7.6.3 Determination of UWL Permeabilities using Stirring Speed Dependence / 205

7.7.6.4 Determination of UWL Permeabilities from Transport across Lipid-Free Microfilters / 207

7.7.6.5 Estimation of UWL Thickness from pH Measurements Near the Membrane Surface / 207

7.7.6.6 Prediction of Aqueous Diffusivities D_{aq} / 207

7.7.6.7 Intrinsic Permeability–log K_p Octanol–Water Relationship / 208

7.7.6.8 Iso-pH Permeability Measurements using Soy Lecithin–Dodecane–Impregnated Filters / 209

7.7.6.9 Gradient pH Effects / 211

7.7.6.10 Collander Relationship between 2% DOPC and 20% Soy Intrinsic Permeabilities / 215

7.7.7 Evidence of Transport of Charged Species / 215

7.7.7.1 The Case for Charged-Species Transport from Cellular and Liposomal Models / 218

7.7.7.2 PAMPA Evidence for the Transport of Charged Drugs / 221

7.7.8 Δ log P_e–Hydrogen Bonding and Ionic Equilibrium Effects / 222

7.7.9 Effects of Cosolvent in Donor Wells / 226

7.7.10 Effects of Bile Salts in Donor Wells / 228

7.7.11 Effects of Cyclodextrin in Acceptor Wells / 228

7.7.12 Effects of Buffer / 229

7.7.13 Effects of Stirring / 231

7.7.14 Errors in PAMPA: Intraplate and Interplate Reproducibility / 232

7.7.15 UV Spectral Data / 233

7.8 The Optimized PAMPA Model for the Gut / 236

7.8.1 Components of the Ideal GIT Model / 236

7.8.2   How Well Do Caco-2 Permeability Measurements Predict Human Jejunal Permeabilities? / 238

7.8.3   How Well Do PAMPA Measurements Predict the Human Jejunal Permeabilities? / 239

7.8.4   Caco-2 Models for Prediction of Human Intestinal Absorption (HIA) / 242

7.8.5   Novartis max-P_e PAMPA Model for Prediction of Human Intestinal Absorption (HIA) / 244

7.8.6 pION Sum-P_e PAMPA Model for Prediction of Human Intestinal Absorption (HIA) / 244

8   SUMMARY AND SOME SIMPLE RULES   **247**

REFERENCES   **250**

INDEX   **285**

PREFACE

This book is written for the practicing pharmaceutical scientist involved in absorption–distribution–metabolism–excretion (ADME) measurements who needs to communicate with medicinal chemists persuasively, so that newly synthesized molecules will be more "drug-like." ADME is all about "a day in the life of a drug molecule" (absorption, distribution, metabolism, and excretion). Specifically, this book attempts to describe the state of the art in measurement of ionization constants (pK_a), oil–water partition coefficients (log P/log D), solubility, and permeability (artificial phospholipid membrane barriers). Permeability is covered in considerable detail, based on a newly developed methodology known as *parallel artificial membrane permeability assay* (PAMPA).

These physical parameters form the major components of physicochemical profiling (the "A" in ADME) in the pharmaceutical industry, from drug discovery through drug development. But, there are opportunities to apply the methodologies in other fields, particularly the agrochemical and environmental industries. Also, new applications to augment animal-based models in the cosmetics industry may be interesting to explore.

The author has observed that graduate programs in pharmaceutical sciences often neglect to adequately train students in these classical solution chemistry topics. Often young scientists in pharmaceutical companies are assigned the task of measuring some of these parameters in their projects. Most find the learning curve somewhat steep. Also, experienced scientists in midcareers encounter the topic of physicochemical profiling for the first time, and find few resources to draw on, outside the primary literature.

The idea for a book on the topic has morphed through various forms, beginning with focus on the subject of metal binding to biological ligands, when the author was a postdoc (postdoctoral fellow) in Professor Ken Raymond's group at the University of California, Berkeley. When the author was an assistant professor of chemistry at Syracuse University, every time the special topics course on speciation analysis was taught, more notes were added to the "book." After 5 years, more than 300 pages of hand-scribbled notes and derivations accumulated, but no book emerged. Some years later, a section of the original notes acquired a binding and saw light in the form of *Applications and Theory Guide to pH-Metric pK$_a$ and log P Measurement* [112] out of the early effort in the startup of Sirius Analytical Instruments Ltd., in Forest Row, a charming four-pub village at the edge of Ashdown Forest, south of London. At Sirius, the author was involved in teaching a comprehensive 3-day training course to advanced users of pK$_a$ and log P measurement equipment manufactured by Sirius. The trainees were from pharmaceutical and agrochemical companies, and shared many new ideas during the courses. Since the early 1990s, Sirius has standardized the measurement of pK$_a$ values in the pharmaceutical and agrochemical industries. Some 50 courses later, the practice continues at another young company, *p*ION, located along hightech highway 128, north of Boston, Massachusetts. The list of topics has expanded since 1990 to cover solubility, dissolution, and permeability, as new instruments were developed. In 2002, an opportunity to write a review article came up, and a bulky piece appeared in *Current Topics in Medicinal Chemistry*, entitled "Physicochemical profiling (solubility, permeability and charge State)." [25] In reviewing that manuscript, Cynthia Berger (*p*ION) said that with a little extra effort, "this could be a book." Further encouragement came from Bob Esposito, of John Wiley & Sons. My colleagues at *p*ION were kind about my taking a sabbatical in England, to focus on the writing. For 3 months, I was privileged to join Professor Joan Abbott's neuroscience laboratory at King's College, London, where I conducted an informal 10-week graduate short course on the topics of this book, as the material was freshly written. After hours, it was my pleasure to jog with my West London Hash House Harrier friends. As the chapter on permeability was being written, my very capable colleagues at *p*ION were quickly measuring permeability of membrane models freshly inspired by the book writing. It is due to their efforts that Chapter 7 is loaded with so much original data, out of which emerged the *double-sink sum-P$_e$ PAMPA GIT* model for predicting human permeability. Per Nielsen (*p*ION) reviewed the manuscript as it slowly emerged, with a keen eye. Many late-evening discussions with him led to freshly inspired insights, now embedded in various parts of the book.

The book is organized into eight chapters. Chapter 1 describes the physico-chemical needs of pharmaceutical research and development. Chapter 2 defines the flux model, based on Fick's laws of diffusion, in terms of solubility, permeability, and charge state (pH), and lays the foundation for the rest of the book. Chapter 3 covers the topic of ionization constants—how to measure pK$_a$ values accurately and quickly, and which methods to use. Bjerrum analysis is revealed as the "secret weapon" behind the most effective approaches. Chapter 4 discusses experimental methods of measuring partition coefficients, log P and log D. It contains a description of the Dyrssen dual-phase potentiometric method, which truly is the "gold standard" method for measuring log P of ionizable molecules, having the unique 10-orders-of-magnitude range (log P from -2 to $+8$). High-throughput methods are also described. Chapter 5 considers the special topic of partition coefficients where the lipid phase is made of liposomes formed from vesicles made of bilayers of phospholipids. Chapter 6 dives into solubility measurements. A unique approach, based on the dissolution template titration method [473], has demonstrated capabilities to measure solubilities as low as 1 nanogram per milliliter (ng/mL). Also, high-throughput microtiter plate UV methods for determining "thermodynamic" solubility constants are described. At the ends of Chapters 3–6, an effort has been made to collect tables of critically-selected values of the constants of drug molecules, the best available values. Chapter 7 describes PAMPA (parallel artificial membrane permeability assay), the high-throughput method introduced by Manfred Kansy et al. of Hoffmann-La Roche [547]. Chapter 7 is the first thorough account of the topic and takes up almost half of the book. Nearly 4000 original measurements are tabulated in the chapter. Chapter 8 concludes with simple rules. Over 600 references and well over 100 drawings substantiate the book.

A. AVDEEF

ACKNOWLEDGMENTS

Professor Norman Ho (University of Utah) was very kind to critically read the Chapter 7 and comment on the various derivations and concepts of permeability. His unique expertise on the topic spans many decades. His thoughts and advice (30 pages of handwritten notes) inspired me to rewrite some of the sections in that chapter. I am very grateful to him. Special thanks go to Per Nielsen and Cynthia Berger of pION for critically reading and commenting on the manuscript. I am grateful to other colleagues at pION who expertly performed many of the measurements of solubility and permeability presented in the book: Chau Du, Jeffrey Ruell, Melissa Strafford, Suzanne Tilton, and Oksana Tsinman. Also, I thank Dmytro Voloboy and Konstantin Tsinman for their help in database, computational, and theoretical matters. The helpful discussion with many colleagues, particularly Manfred Kansy and Holger Fisher at Hoffmann La-Roche, Ed Kerns and Li Di at Wyeth Pharmaceuticals, and those at Sirius Analytical Instruments, especially John Comer and Karl Box, are gratefully acknowledged. Helpful comments from Professors John Dearden (Liverpool John Moores University) and Hugo Kubinyi (Heidelberg University) are greatly appreciated. I also thank Professor Anatoly Belyustin (St. Peterburgh University) for pointing out some very relevant Russian literature. Chris Lipinski (Pfizer) has given me a lot of good advice since 1992 on instrumentation and pharmaceutical research, for which I am grateful. Collaborations with Professors Krisztina Takács-Novák (Semmelweis University, Budapest) and Per Artursson (Uppsala University) have been very rewarding. James McFarland (Reckon.Dat) and Alanas Petrauskas (Pharma Algorithms) have been my teachers of in silico methods. I am in debt to Professor Joan Abbott and Dr. David Begley for allowing me to spend 3 months in their laboratory at King's College London, where I learned a lot about the blood–brain barrier. Omar at Cafe Minon, Warwick Street in Pimlico, London, was kind to let me spend many hours in his small place, as I wrote several papers and drank a lot of coffee. Lasting thanks go to David Dyrssen and the late Jannik Bjerrum for planting the seeds of most interesting and resilient pH-metric methodologies, and to Professor Bernard Testa of Lausanne University for tirelessly fostering the white light of physicochemical profiling. My congratulations to him on the occasion of his retirement.

DEFINITIONS

ACRONYMS[*]

AC	aminocoumarin
ADME	absorption, distribution, metabolism, excretion
ANS	anilinonaphthalenesulfonic acid
AUC	area under the curve
BA/BE	bioavailability–bioequivalence
BBB	blood–brain barrier
BBM	brush-border membrane
BBLM	brush-border lipid membrane
BCS	biopharmaceutics classification system
BLM	black lipid membrane
BSA	bovine serum albumin
CE	capillary electrophoresis
CHO	caroboxaldehyde
CMC	critical micelle concentration
CPC	centrifugal partition chromatography
CPZ	chlorpromazine
CTAB	cetyltrimethylammonium bromide

---

*Very common terms (IR, UV, etc.) and proprietary (agency, etc.) names (e.g., FDA) omitted from this list.

CV	cyclic votammetry
DA	dodecylcarboxylic acid
DOPC	dioleylphosphatidylcholine
DPPC	dipalmitoylphosphatidylcholine
DPPH	diphenylpicrylhydrazyl
DSHA	dansylhexadecylamine
DTT	dissolution template titration
EFA	evolving factor analysis
ET	extrusion technique (for preparing LUV)
FAT	freeze and thaw (step in LUV preparation)
FFA	free fatty acid
GIT	gastrointestinal tract
GMO	glycerol monooleate
HC	hydrocoumarin
HIA	human intestinal absorption
HJP	human jejunal permeability
HMW	high molecular weight
HTS	high-throughput screening
IAM	immobilized artificial membrane
IVIV	in vitro–in vivo
LUV	large unilamellar vesicle
MAD	maximum absorbable dose
MDCK	Madin–Darby canine kidney
MLV	multilamellar vesicle
M6G	morphine-6-glucuronide
NCE	new chemical entity
OD	optical density
PAMPA	parallel artifical membrane permeabillity assay
PC	phosphatidylcholine
PCA	principal-component analysis
PK	pharmacokinetic
QSPR	quantitative structure–property relationship
SCFA	short-chain fatty acid
SDES	sodium decyl sulfate
SDS	sodium dodecyl sulfate
SGA	spectral gradient analysis
SLS	sodium laurel sulfate
STS	sodium tetradecyl sulfate
SUV	small unilamellar vesicle
TFA	target factor analysis
TJ	tight junction
TMADPH	trimethylaminodiphylhexatriene chloride
UWL	unstirred water layer (adjacent to membrane surface)

NOMENCLATURE

C_A, C_D	aqueous solute concentrations on the acceptor and donor sides of a membrane, respectively (mol/cm^3)
C_0	aqueous concentration of the uncharged species (mol/cm^3)
C_m^x	solute concentration inside a membrane, at position x (mol/cm^3)
δ	difference between the liposome–water and octanol–water log P for the uncharged species
diff	difference between the partition coefficient of the uncharged and the charged species
Δ shift	the difference between the true pK_a and the apparent pK_a observed in a solubility–pH profile, due to DMSO–drug binding, or drug–drug aggregation binding
D_{aq}	diffusivity of a solute in aqueous solution (cm^2/s)
D_m	diffusivity of a solute inside a membrane (cm^2/s)
eggPC	egg phosphatidylcholine
h	membrane thickness (cm)
hit	a molecule with confirmed activity from a primary assay, a good profile in secondary assays, and with a confirmed structure
J	flux across a membrane (mol cm^{-2} s^{-1})
K_{sp}	solubility product (e.g., [Na$^+$][A$^-$] or [BH$^+$][Cl$^-$])
lead	a hit series for which the structure–activity relationship is shown and activity demonstrated in vivo
K_d or D	lipid–water distribution pH-dependent function (also called the "apparent" partition coefficient)
K_p or P	lipid–water pH-independent partition coefficient
K_e	extraction constant
$\bar{n}_H$	Bjerrum function: average number of bound protons on a molecule at a particular pH
P_a	apparent artificial-membrane permeability (cm/s)—similar to P_e, but with some limiting assumption
P_e	effective artificial-membrane permeability (cm/s)
P_m	artificial-membrane permeability (cm/s)—similar to P_e, but corrected for the UWL
P_0	intrinsic artificial-membrane permeability (cm/s), that of the uncharged form of the drug
pH	operational pH scale
p$_c$H	pH scale based on hydrogen ion concentration
pK_a	ionization constant (negative log form), based on the concentration scale
p$_o$$K_a$	apparent ionization constant in an octanol–water titration
pK_a^{oct}	octanol pK_a (the limiting p$_o$$K_a$ in titrations with very high octanol–water volume ratios)
pK_a^{mem}	membrane pK_a

pK_a^{gibbs}	ionization constant corresponding to the pH at which both the uncharged and the salt form of a substance coprecipitate
pK_a^{flux}	apparent ionization constant in a log P_e–pH profile, shifted from the thermodynamic value as a consequence of the unstirred water layer; the pH where 50% of the resistance to transport is due to the UWL and 50% is due to the lipid membrane
sink	any process that can significantly lower the concentration of the neutral form of the sample molecule in the acceptor compartment; examples include physical sink (where the buffer solution in the acceptor compartment is frequently refreshed), ionization sink (where the concentration of the neutral form of the drug is diminished as a result of ionization), and binding sink (where the concentration of the neutral form of the drug is diminished because of binding with serum protein, cyclodextrin, or surfactants in the acceptor compartment)
double-sink	two sink conditions present: ionization and binding
S	solubility in molar, µg/mL, or mg/mL units
S_i	solubility of the ionized species (salt), a conditional constant, depending on the concentration of the counterion in solution
S_0	intrinsic solubility, that is, the solubility of the uncharged species
τ_{LAG}	the time for steady state to be reached in a permeation cell, after sample is introduced into the donor compartment; in the PAMPA model described in the book, this is approximated as the time that sample first appears detected in the acceptor well

CHAPTER 1

INTRODUCTION

1.1 SHOTGUN SEARCHING FOR DRUGS?

The search for new drugs is daunting, expensive, and risky.

If chemicals were confined to molecular weights of less than 600 Da and consisted of common atoms, the chemistry space is estimated to contain 10^{40} to 10^{100} molecules, an impossibly large space to search for potential drugs [1]. To address this limitation of vastness, "maximal chemical diversity" [2] was applied in constructing large experimental screening libraries. Such libraries have been directed at biological "targets" (proteins) to identify active molecules, with the hope that some of these "hits" may someday become drugs. The current target space is very small—less than 500 targets have been used to discover the known drugs [3]. This number may expand to several thousand in the near future as genomics-based technologies uncover new target opportunities [4]. For example, the human genome mapping has identified over 3000 transcription factors, 580 protein kinases, 560 G-protein coupled receptors, 200 proteases, 130 ion transporters, 120 phosphatases, over 80 cation channels, and 60 nuclear hormone receptors [5].

Although screening throughputs have massively increased since the early 1990s, lead discovery productivity has not necessarily increased accordingly [6–8]. Lipinski has concluded that maximal chemical diversity is an *inefficient* library design strategy, given the enormous size of the chemistry space, and especially that clinically useful drugs appear to exist as small tight clusters in chemistry space:

Absorption and Drug Development: Solubility, Permeability, and Charge State. By Alex Avdeef
ISBN 0-471-423653.   Copyright © 2003 John Wiley & Sons, Inc.

"one can make the argument that screening truly diverse libraries for drug activity is the fastest way for a company to go bankrupt because the screening yield will be so low" [1]. Hits *are* made in pharmaceutical companies, but this is because the most effective (not necessarily the largest) screening libraries are highly focused, to reflect the putative tight clustering. Looking for ways to reduce the number of tests, to make the screens "smarter," has an enormous cost reduction implication.

The emergence of combinatorial methods in the 1990s has lead to enormous numbers of new chemical entities (NCEs) [9]. These are the molecules of the newest screening libraries. A large pharmaceutical company may screen 3 million molecules for biological activity each year. Some 30,000 hits are made. Most of these molecules, however potent, do not have the right physical, metabolic, and safety properties. Large pharmaceutical companies can cope with about 30 molecules taken into development each year. A good year sees three molecules reach the product stage. Some years see none. These are just rough numbers, recited at various conferences.

A drug product may cost as much as $880 M (million) to bring out. It has been estimated that about 30% of the molecules that reach development are eventually rejected due to ADME (absorption, distribution, metabolism, excretion) problems. Much more money is spent on compounds that fail than on those that succeed [10,11]. The industry has started to respond by attempting to screen out those molecules with inappropriate ADME properties during discovery, before the molecules reach development. However, that has led to another challenge: how to do the additional screening quickly enough, while keeping costs down [6,12].

1.2 SCREEN FOR THE TARGET OR ADME FIRST?

Most commercial combinatorial libraries, some of which are very large and may be diverse, have a very small proportion of drug-like molecules [1]. Should only the small drug-like fraction be used to test against the targets? The industry's current answer is "No." The existing practice is to screen for the receptor activity before "drug-likeness." The reasoning is that structural features in molecules rejected for poor ADME properties may be critical to biological activity related to the target. It is believed that active molecules with liabilities can be modified later by medicinal chemists, with minimal compromise to potency. Lipinski [1] suggests that the order of testing may change in the near future, for economic reasons. When a truly new biological therapeutic target is examined, nothing may be known about the structural requirements for ligand binding to the target. Screening may start as more or less a random process. A library of compounds is tested for activity. Computational models are constructed on the basis of the results, and the process is repeated with newly synthesized molecules, perhaps many times, before satisfactory hits are revealed. With large numbers of molecules, the process can be very costly. If the company's library is first screened for ADME properties, that screening is done only once. The same molecules may be recycled against existing or future targets many times, with knowledge of drug-likeness to fine-tune the optimization process. If some of the molecules with very poor ADME properties are judiciously filtered out, the biological activity testing process would be less costly. But the order of testing (activity vs. ADME) is likely to continue to be the subject of future debates [1].

1.3 ADME AND MULTIMECHANISM SCREENS

In silico property prediction is needed more than ever to cope with the screening overload. Improved prediction technologies are continuing to emerge [13,14]. However, reliably measured physicochemical properties to use as "training sets" for new target applications have not kept pace with the in silico methodologies.

Prediction of ADME properties should be simple, since the number of descriptors underlying the properties is relatively small, compared to the number associated with effective drug–receptor binding space. In fact, prediction of ADME is difficult! The current ADME experimental data reflect a multiplicity of mechanisms, making prediction uncertain. Screening systems for biological activity are typically single mechanisms, where computational models are easier to develop [1].

For example, aqueous solubility is a multimechanism system. It is affected by lipophilicity, H bonding between solute and solvent, intramolecular H bonding, intermolecular hydrogen and electrostatic bonding (crystal lattice forces), and charge state of the molecule. When the molecule is charged, the counterions in solution may affect the measured solubility of the compound. Solution microequilibria occur in parallel, affecting the solubility. Few of these physicochemical factors are well understood by medicinal chemists, who are charged with making new molecules that overcome ADME liabilities without losing potency.

Another example of a multi-mechanistic probe is the Caco-2 permeability assay (a topic covered in various sections of the book). Molecules can be transported across the Caco-2 monolayer by several mechanisms operating simultaneously, but to varying degrees, such as transcellular passive diffusion, paracellular passive diffusion, lateral passive diffusion, active influx or efflux mediated by transporters, passive transport mediated by membrane-bound proteins, receptor-mediated endocytosis, pH gradient, and electrostatic-gradient driven mechanisms. The P-glycoprotein (P-gp) efflux transporter can be saturated if the solute concentration is high enough during the assay. If the substance concentration is very low (perhaps because not enough of the compound is available during discovery), the importance of efflux transporters in gastrointestinal tract (GIT) absorption can be overestimated, providing the medicinal chemist with an overly pessimistic prediction of intestinal permeability [8,15,16]. Metabolism by the Caco-2 system can further complicate the assay outcome.

Compounds from traditional drug space ("common drugs"—readily available from chemical suppliers), often chosen for studies by academic laboratories for assay validation and computational model-building purposes, can lead to misleading conclusions when the results of such models are applied to 'real' discovery compounds, which most often have extremely low solubilities [16].

Computational models for single mechanism assays (e.g., biological receptor affinity) improve as more data are accumulated [1]. In contrast, computational models for multimechanism assays (e.g., solubility, permeability, charge state) worsen as more measurements are accumulated [1]. Predictions of human oral absorption using Caco-2 permeabilities can look very impressive when only a small number of molecules is considered. However, good correlations deteriorate as more molecules are included in the plot, and predictivity soon becomes meaningless. Lipinski states that "The solution to this dilemma is to carry out single mechanism ADME experimental assays and to construct single mechanism ADME computational models. The ADME area is at least 5 or more years behind the biology therapeutic target area in this respect" [1].

The subject of this book is to examine the components of the multimechanistic processes related to solubility, permeability, and charge state, with the aim of advancing improved strategies for in vitro assays related to drug absorption.

1.4 ADME AND MEDICINAL CHEMISTS

Although ADME assays are usually performed by analytical chemists, medicinal chemists—the molecule makers—need to have some understanding of the physicochemical processes in which the molecules participate. Peter Taylor [17] states:

> It is now almost a century since Overton and Meyer first demonstrated the existence of a relationship between the biological activity of a series of compounds and some simple physical property common to its members. In the intervening years the germ of their discovery has grown into an understanding whose ramifications extend into medicinal chemistry, agrochemical and pesticide research, environmental pollution and even, by a curious re-invention of familiar territory, some areas basic to the science of chemistry itself. Yet its further exploitation was long delayed. It was 40 years later that Ferguson at ICI applied similar principles to a rationalization of the comparative activity of gaseous anaesthetics, and 20 more were to pass before the next crucial step was formulated in the mind of Hansch. ... Without any doubt, one major factor [for delay] was compartmentalism. The various branches of science were much more separate then than now. It has become almost trite to claim that the major advances in science take place along the borders between its disciplines, but in truth this happened in the case of what we now call Hansch analysis, combining as it did aspects of pharmacy, pharmacology, statistics and physical organic chemistry. Yet there was another feature that is not so often remarked, and one with a much more direct contemporary implication. The physical and physical organic chemistry of equilibrium processes—solubility, partitioning, hydrogen bonding, etc.—is not a glamorous subject. It seems too simple. Even though the specialist may detect an enormous information content in an assemblage of such numbers, to synthetic chemists used to thinking in three-dimensional terms they appear structureless, with no immediate meaning that they can *visually* grasp. Fifty years ago it was the siren call of Ehrlich's lock-and-key theory that deflected medicinal chemists from a physical understanding that might otherwise have been attained much earlier. Today it is glamour of the television screen. No matter that what is on display may sometimes possess all the profundity of a

five-finger exercise. It is visual and therefore more comfortable and easier to assimilate. Similarly, MO theory in its resurgent phase combines the exotic appeal of a mystery religion with a new-found instinct for three-dimensional colour projection which really can give the ingenue the impression that he understands what it is all about. There are great advances and great opportunities in all this, but nevertheless a concomitant danger that medicinal chemists may forget or pay insufficient attention to hurdles the drug molecule will face if it is actually to perform the clever docking routine they have just tried out: hurdles of solubilization, penetration, distribution, metabolism and finally of its non-specific interactions in the vicinity of the active site, all of them the result of physical principles on which computer graphics has nothing to say. Such a tendency has been sharply exacerbated by the recent trend, for reasons of cost as much as of humanity, to throw the emphasis upon *in vitro* testing. All too often, chemists are disconcerted to discover that the activity they are so pleased with *in vitro* entirely fails to translate to the *in vivo* situation. Very often, a simple appreciation of basic physical principles would have spared them this disappointment; better, could have suggested in advance how they might avoid it. We are still not so far down the path of this enlightenment as we ought to be. What is more, there seems a risk that some of it may fade if the balance between a burgeoning receptor science and these more down-to-earth physical principles is not properly kept.

Taylor [17] described physicochemical profiling in a comprehensive and compelling way, but enough has happened since 1990 to warrant a thorough reexamination. Then, combichem, high-throughput screening (HTS), Caco-2, IAM, CE were in a preingenuic state; studies of drug-partitioning into liposomes were arcane; instrument companies took no visible interest in making pK_a, log P, or solubility analyzers; there was no biopharmaceutics classification system (BCS); it did not occur to anyone to do PAMPA. With all that is new, it is a good time to take stock of what we can learn from the work since 1990. In this book, measurement of solubility, permeability, lipophilicity, and charge state of drug molecules will be critically reexamined (with considerable coverage given to permeability, the property least explored). Fick's law of diffusion [18] in predicting drug absorption will be reexplored.

1.5   THE "A" IN ADME

In this book we will focus on physicochemical profiling in support of improved prediction methods for *absorption*, the "A" in ADME. Metabolism and other components of ADME will be beyond the scope of this book. Furthermore, we will focus on properties related to *passive* absorption, and not directly consider active transport mechanisms. The most important physicochemical parameters associated with passive absorption are *acid–base* character (which determines the charge state of a molecule in a solution of a particular pH), *lipophilicity* (which determines distribution of a molecule between the aqueous and the lipid environments), *solubility* (which limits the concentration that a dosage form of a molecule can present to the solution and the rate at which the molecule dissolves from the solid form), and membrane *permeability* (which determines how quickly molecules can cross membrane barriers). Current state of the art in measurement of these properties, as the ever important function of pH, will be surveyed, and in some cases (permeability), described in detail.

1.6   IT IS NOT JUST A NUMBER—IT IS A MULTIMECHANISM

Drugs exert their therapeutic effects through reactions with specific receptors. Drug–receptor binding depends on the concentration of the drug near the receptor. Its form and concentration near the receptor depend on its physical properties. Orally administered drugs need to be dissolved at the site of absorption in the gastrointestinal tract (GIT), and need to traverse several membrane barriers before receptor interactions can commence. As the drug distributes into the various compartments of the body, a certain (small) portion finds itself at the receptor site. Transport and distribution of most drugs are affected by passive diffusion, which depends on lipophilicity, since lipid barriers need to be crossed [19–24]. Passive transport is well described by the principles of physical chemistry [25–33].

The pK_a of a molecule, a charge-state-related parameter, is a descriptor of an acid–base equilibrium reaction [34,35]. Lipophilicity, often represented by the octanol–water partition coefficient K_p is a descriptor of a two-phase distribution equilibrium reaction [36]. So is solubility [37–39]. These three parameters are thermodynamic constants. On the other hand, permeability P_e is a rate coefficient, a kinetics parameter, most often posed in a first-order distribution reaction [40–42].

In high-throughput screening (HTS) these parameters are sometimes viewed simply as numbers, quickly and roughly determined, to be used to rank molecules into "good" and "bad" classes. An attempt will be made to examine this important aspect. In addition, how fundamental, molecular-level interpretations of the physical measurements can help to improve the design of the profiling assays will be examined, with the aim of promoting the data fodder of HTS to a higher level of quality, without compromising the need for high speed. Quality measurements in large quantities will lead to improved in silico methods. Simple rules (presented in visually appealing ways), in the spirit of Lipinski's rule of fives, will be sought, of use not only to medicinal chemists but also to preformulators [12,43]. This book attempts to make easier the dialog between the medicinal chemists charged with modifying test compounds and the pharmaceutical scientists charged with physicochemical profiling, who need to communicate the results of their assays in an optimally effective manner.

CHAPTER 2

TRANSPORT MODEL

2.1 PERMEABILITY-SOLUBILITY-CHARGE STATE AND THE pH PARTITION HYPOTHESIS

Fick's first law applied to a membrane [18,40–42] shows that passive diffusion of a solute is the product of the diffusivity and the concentration gradient of the solute *inside* the membrane. The membrane/water apparent partition coefficient relates the latter internal gradient to the external bulk water concentration difference between the two solutions separated by the membrane. For an ionizable molecule to permeate by passive diffusion most efficiently, the molecule needs to be in its uncharged form at the membrane surface. This is the essence of the pH partition hypothesis [44]. The amount of the uncharged form present at a given pH, which directly contributes to the flux, depends on several important factors, such as pH, binding to indigenous carriers (proteins and bile acids), self-binding (aggregate or micelle formation), and solubility (a solid-state form of self-binding). Low solubility enters the transport consideration as a thermodynamic "speed attenuator," as a condition that lowers the opportunity for transport. In this way, permeability and solubility are the linked kinetic and thermodynamic parts of transport across a membrane.

Consider a vessel divided into two chambers, separated by a homogeneous lipid membrane. Figure 2.1 is a cartoon of such an arrangement. The left side is the donor compartment, where the sample molecules are first introduced; the right side is the acceptor compartment, which at the start has no sample molecules.

Absorption and Drug Development: Solubility, Permeability, and Charge State. By Alex Avdeef
ISBN 0-471-423653. Copyright © 2003 John Wiley & Sons, Inc.

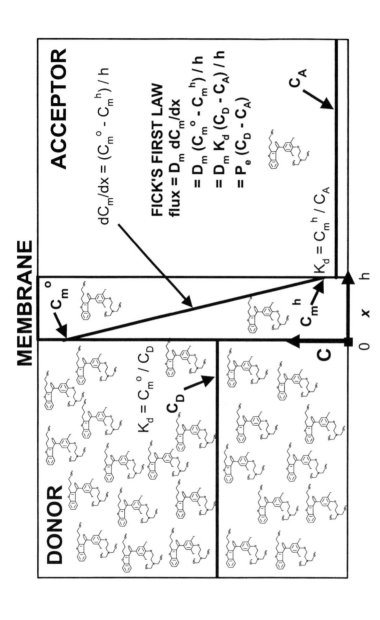

Figure 2.1 Transport model diagram, depicting two aqueous cells separated by a membrane barrier. The drug molecules are introduced in the donor cell. The concentration gradient in the membrane drives the molecules in the direction of the acceptor compartment. The apparent partition coefficient, $K_d = 2$. [Avdeef, A., *Curr. Topics Med. Chem.*, **1**, 277–351 (2001). Reproduced with permission from Bentham Science Publishers, Ltd.]

Fick's first law applied to homogeneous membranes at steady state is a transport equation

$$J = \frac{D_m dC_m}{dx} = \frac{D_m[C_m^0 - C_m^h]}{h} \tag{2.1}$$

where J is the flux, in units of mol cm^{-2} s^{-1}, where C_m^0 and C_m^h are the concentrations, in mol/cm^3 units, of the *uncharged* form of the solute within the membrane at the two water–membrane boundaries (at positions $x = 0$ and $x = h$ in Fig. 2.1, where h is the thickness of the membrane in centimeters) and where D_m is the diffusivity of the solute within the *membrane*, in units of cm^2/s. At steady state, the concentration gradient, dC_m/dx, within the membrane is linear, so the difference may be used in the right side of Eq. (2.1). Steady state takes about 3 min to be established in a membrane of thickness 125 µm [19,20], assuming that the solution is very well stirred.

The limitation of Eq. (2.1) is that measurement of concentrations of solute within different parts of the membrane is very inconvenient. However, since we can estimate (or possibly measure) the distribution coefficients between bulk water and the membrane, log K_d (the pH-dependent apparent partition coefficient), we can convert Eq. (2.1) into a more accessible form

$$J = \frac{D_m K_d (C_D - C_A)}{h} \tag{2.2}$$

where the substitution of K_d allows us to use bulk water concentrations in the donor and acceptor compartments, C_D and C_A, respectively. (With ionizable molecules, C_A and C_D refer to the concentrations of the solute summed over all forms of charge state.) These concentrations may be readily measured by standard techniques. Equation (2.2) is still not sufficiently convenient, since we need to estimate D_m and K_d. It is common practice to lump these parameters and the thickness of the membrane into a composite parameter, called *membrane permeability P_m*:

$$P_m = \frac{D_m K_d}{h} \tag{2.3}$$

The relevance of Eq. (2.2) (which predicts how quickly molecules pass through simple membranes) to solubility comes in the concentration terms. Consider "sink" conditions, where C_A is essentially zero. Equation (2.2) reduces to the following flux equation

$$J = P_m C_D \tag{2.4}$$

Flux depends on the product of membrane permeability of the solute times the concentration of the solute (summed over all charge state forms) at the water side of the donor surface of the membrane. This concentration ideally may be equal to the dose of the drug, unless the dose exceeds the solubility limit at the pH considered, in

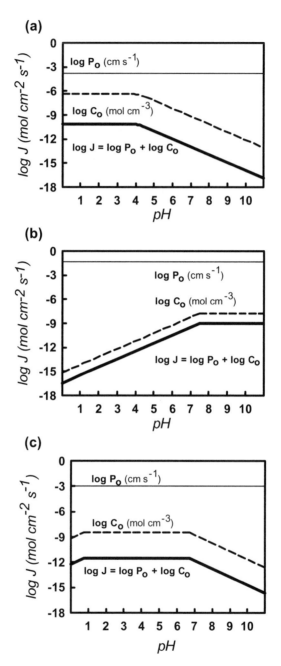

Figure 2.2 Log flux–pH profiles at dosing concentrations: (a) ketoprofen (acid, pK_a 3.98), dose 75 mg; (b) verapamil (base, pK_a 9.07), dose 180 mg; (c) piroxicam (ampholyte, pK_a 5.07, 2.33), dose 20 mg. The permeability and the concentration of the uncharged species are denoted P_0 and C_0, respectively. [Avdeef, A., *Curr. Topics Med. Chem.*, **1**, 277–351 (2001). Reproduced with permission from Bentham Science Publishers, Ltd.]

which case it is equal to the solubility. Since the uncharged molecular species is the permeant, Eq. (2.4) may be restated as

$$J = P_0 C_0 \leq P_0 S_0 \tag{2.5}$$

where P_0 and C_0 are the intrinsic permeability and concentration of the uncharged species, respectively. The intrinsic permeability does not depend on pH, but its cofactor in the flux equation C_0 does. The concentration of the uncharged species is always equal to or less than the intrinsic solubility of the specie, S_0, which never depends on pH. Note that for the uncharged species, Eq. (2.3) takes on the form

$$P_0 = \frac{D_m K_p}{h} \tag{2.6}$$

where $K_p = C_m(0)/C_{D0}$; also, $K_p = C_m(h)/C_{A0}$; C_{D0} and C_{A0} are the aqueous solution concentrations of the *uncharged* species in the donor and acceptor sides, respectively.

In solutions saturated (i.e., excess solid present) at some pH, the plot of log C_0 versus pH for an ionizable molecule is extraordinarily simple in form; it is a combination of straight segments, joined at points of discontinuity indicating the boundary between the saturated state and the state of complete dissolution. The pH of these junction points is dependent on the dose used in the calculation, and the maximum value of log C_0 is always equal to log S_0 in a saturated solution. [26] Figure 2.2 illustrates this idea using ketoprofen as an example of an acid, verapamil as a base, and piroxicam as an ampholyte. In the three cases, the assumed concentrations in the calculation were set to the respective doses [26]. For an acid, log C_0 (dashed curve in Fig. 2.2a) is a horizontal line (log $C_0 = $ log S_0) in the saturated solution (at low pH), and decreases with a slope of -1 in the pH domain where the solute is dissolved completely. For a base (Fig. 2.2b) the plot of log C_0 versus pH is also a horizontal line at high pH in a saturated solution and is a line with a slope of $+1$ for pH values less than the pH of the onset of precipitation.

We have called the plot of log C_0 versus pH the "flux factor" profile, with the idea that such a plot when combined with intrinsic permeability, can be the basis of an in vitro classification scheme to predict passive oral absorption as a function of pH. This will be discussed later.

Figures 2.1 and 2.2 represent the basic model that will be used to discuss the literature related to the measurement of the physicochemical parameters and the interpretation of their role in the oral absorption process [19,20,23,45–61].

2.2 PROPERTIES OF THE GASTROINTESTINAL TRACT (GIT)

The properties of the human GIT that are relevant to the absorption of drug products have been collected from several sources [62–69]. Figure 2.3 shows a cartoon of the GIT, indicating surface area and pH (fasted and fed state) in the various

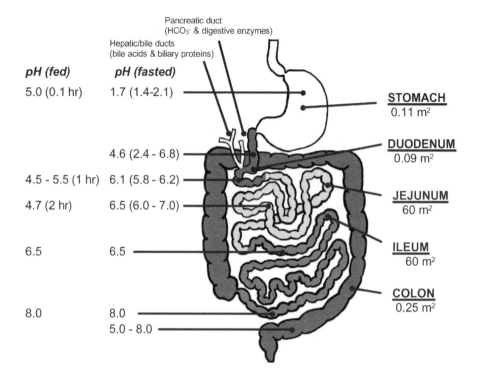

Figure 2.3 Physical properties of the GIT, with approximate values compiled from several sources [62–69]. The pH values refer mostly to median quantities and the range in parentheses generally refers to interquartile values [67,68]. The quoted surface areas are taken from Ref. 66. [Avdeef, A., *Curr. Topics Med. Chem.*, **1**, 277–351 (2001). Reproduced with permission from Bentham Science Publishers, Ltd.]

segments. The surface area available for absorption is highest in the jejunum and the ileum, accounting for more than 99% of the total. In the fasted state, the pH in the stomach is $\sim$1.7. The acidified contents of the stomach are neutralized in the duodenum by the infusion of bicarbonate ions through the pancreatic duct. Past the pyloric sphincter separating the stomach and the duodenum, the pH steeply rises to $\sim$4.6. Between the proximal jejunum and the distal ileum, the pH gradually rises from $\sim$6 to 8. The pH can drop to values as low as 5 in the colon, due to the microbial digestion of certain carbohydrates, producing short-chain fatty acids (SCFAs) in concentration as high as 60–120 mM. [70] The GIT exhibits a considerable pH gradient, and the pH partition hypothesis predicts that the absorption of ionizable drugs may be location-specific.

When food is ingested, the pH in the stomach can rise briefly to 7, but after 0.1 h drops to pH 5, after 1 h to pH 3, and after 3 h to the fasted value. The movement of food down the small intestine causes the pH in the proximal jejunum to drop to as low as 4.5 in 1–2 h after food intake, but the distal portions of the small intestine and the colon are not dramatically changed in pH due to the transit of food. The

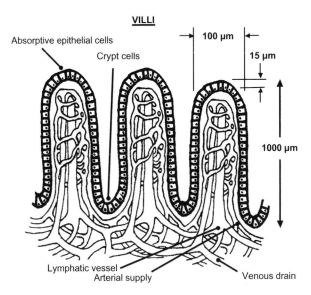

Figure 2.4 Schematic of the villi "fingers" covered by a monolayer of epithelial cells, separating the lumen from the blood capillary network [63,69]. [Avdeef, A., *Curr. Topics Med. Chem.*, **1**, 277–351 (2001). Reproduced with permission from Bentham Science Publishers, Ltd.]

stomach releases its contents periodically, and the rate depends on the contents. On an empty stomach, 200 mL of water have a transit half-life of 0.1–0.4 h, but solids (such as tablets) may reside for 0.5–3 h, with larger particles held back the longest. Food is retained for 0.5–13 h; fatty food and large particles are held the longest time. Transit time through the jejunum and ileum is about 3–5 h. Digesting food may stay in the colon for 7–20 h, depending on the sleep phase. Fatty foods trigger the release of bile acids, phospholipids, and biliary proteins via the hepatic/bile ducts into the duodenum. Bile acids and lecithin combine to form mixed micelles, which help solubilize lipid molecules, such as cholesterol (or highly lipophilic drugs). Under fasted conditions, the bile : lecithin concentrations in the small intestine are approximately 4 : 1 mM, but a fatty meal can raise the level to about 15 : 4 mM [68,71]. Thus, maximal absorption of drug products takes place in the jejunum and ileum over a period of 3–5 h, in a pH range of 4.5–8.0. This suggests that weak acids ought to be better absorbed in the jejunum, and weak bases in the ileum.

The surface area in the luminal side of the small intestine per unit length of the serosal (blood) side is enormous in the proximal jejunum, and steadily decreases (to about 20% of the starting value [62]) in the distal portions of the small intestine. The surface area is increased threefold [69] by ridges oriented circumferentially around the lumen. Similar folds are found in all segments of the GIT, except the mouth and esophagus [66]. Further 4–10-fold expansion [62,69] of the surface is produced by the villi structures, shown schematically in Fig. 2.4. The layer of epithelial cells lining the villi structures separate the lumen from the circulatory system. Epithelial cells are made in the crypt folds of the villi, and take about

14

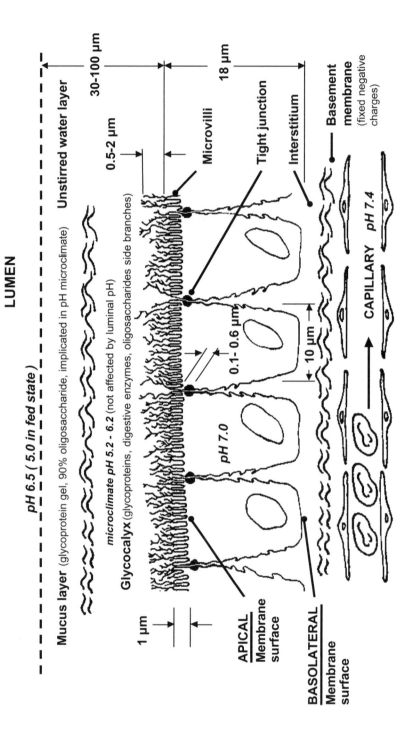

Figure 2.5 Schematic of the structure of epithelial cells, based on several literature sources [55,63,69,73,74,76,78,79]. The tight junctions and the basement membrane appear to be slightly ion-selective (lined with some negatively charged groups) [75,76,79]. [Avdeef, A., *Curr. Topics Med. Chem.*, **1**, 277–351 (2001). Reproduced with permission from

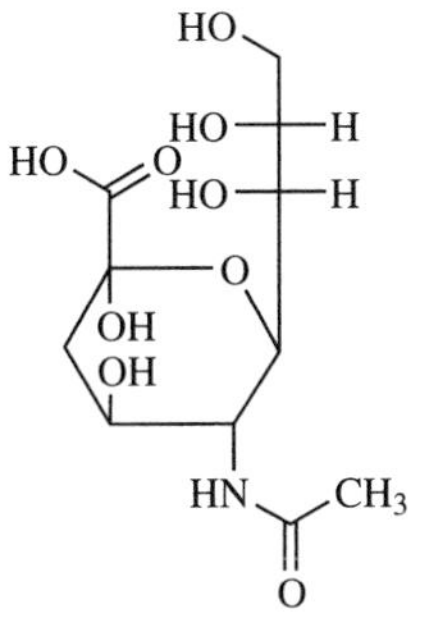

Figure 2.6   Sialic acid.

2 days to move to the region of the tips of the villi, where they are then shed into the lumen. A schematic view of the surface of the epithelial cells shows a further 10–30-fold surface expansion [62,63,69] structures, in the form of microvilli on the luminal side of the cell layer, as shown in Fig. 2.5.

The villi and microvilli structures are found in highest density in the duodenum, jejunum, and ileum, and in lower density in a short section of the proximal colon [66]. The microvilli have glycoproteins (the glycocalyx) protruding into the luminal fluid. There is residual negative charge in the glycoproteins. Some cells in the monolayer are known as goblet cells (not shown in Figs. 2.4 and 2.5), whose function is to produce the mucus layer that blankets the glycocalyx. The mucus layer is composed of a high-molecular-weight (HMW) (2×10^6 Da) glycoprotein, which is 90% oligosaccharide, rich in sialic acid (Fig. 2.6) residues, imparting negative charge to the layer [63]. Studies of the diffusion of drug molecules through the mucus layer suggest that lipophilic molecules are slowed by it [72].

The glycocalyx and the mucus layer make up the structure of the unstirred water layer (UWL) [73]. The thickness of the UWL is estimated to be 30–100 μm in vivo, consistent with very efficient stirring effects [74]. In isolated tissue (in the absence of stirring), the mucus layer is 300–700 μm thick [73]. The pH in the unstirred water layer is ~5.2–6.2, and might be regulated independently of the luminal pH (Section 2.3). The mucus layer may play a role in regulating the epithelial cell surface pH [73].

The membrane surface facing the lumen is called the *apical surface*, and the membrane surface on the side facing blood is called the *basolateral surface*. The intestinal cells are joined at the tight junctions [63,75]. These junctions have pores that can allow small molecules (MW < 200 Da) to diffuse through in aqueous solution. In the jejunum, the pores are ~7–9 Å in size. In the ileum the junctions are tighter, and pores are ~3–4 Å in size (i.e., dimensions of mannitol) [63].

The apical surface is loaded with more than 20 different digestive enzymes and proteins; the protein : lipid ratio is high: 1.7 : 1 [63]. The half-life of these proteins is ~6–12 h, whereas the epithelial cells last 2–3 days. So the cell must replace these constituents without depolarizing itself. The cytoskeleton may play a role

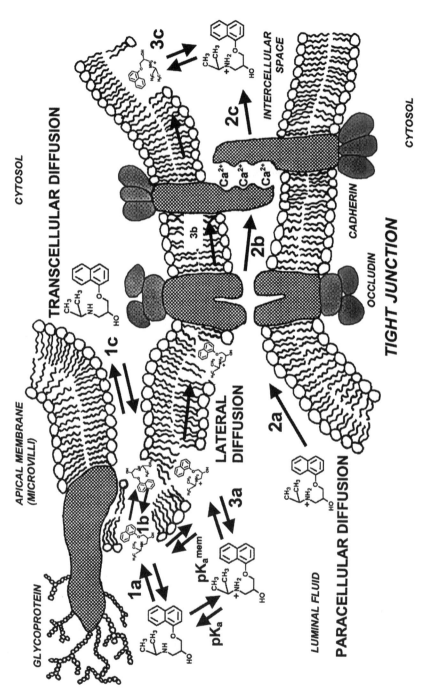

Figure 2.7 Schematic of the apical phospholipid bilayer surface of the epithelial cells, indicating three types of passive diffusion: transcellular (1a → 1b → 1c), paracellular (2a → 2b → 2c), and the hypothesized lateral, "under the skin of the tight junction" (3a → 3b → 3c) modes. Tight-junction matrix of proteins highly stylized, based on Ref. 75. [Avdeef, A., *Curr. Topics Med. Chem.*, **1**, 277–351 (2001). Reproduced with permission from Bentham Science Publishers, Ltd.]

16

in maintaining the polar distribution of the surface constituents [63]. After a permeant passes through the cell barrier, it encounters a charge-selective barrier in the basement membrane (Fig. 2.5) [76]. Positively charged drugs have a slightly higher permeability through it. After this barrier, drug molecules may enter the blood capillary network through openings in the highly fenestrated capillaries. Epithelial cell surfaces are composed of bilayers made with phospholipids, as shown in the highly stylized drawing in Fig. 2.7.

Two principal routes of passive diffusion are recognized: transcellular (1a → 1b → 1c in Fig. 2.7) and paracellular (2a → 2b → 2c). Lateral exchange of phospholipid components of the *inner* leaflet of the epithelial bilayer seems possible, mixing simple lipids between the apical and basolateral side. However, whether the membrane lipids in the *outer* leaflet can diffuse across the tight junction is a point of controversy, and there may be some evidence in favor of it (for some lipids) [63]. In this book, a third passive mechanism, based on lateral diffusion of drug molecules in the outer leaflet of the bilayer (3a → 3b → 3c), will be hypothesized as a possible mode of transport for polar or charged amphiphilic molecules.

In the transport across a phospholipid bilayer by passive diffusion, the permeability of the neutral form of a molecule is $\sim 10^8$ times greater than that of the charged form. For the epithelium, the discrimination factor is 10^5. The basement membrane (Fig. 2.5) allows passage of uncharged molecules more readily than charged species by a factor of 10 [76].

2.3 pH MICROCLIMATE

The absorption of short-chain weak acids in the rat intestine, as a function of pH, does not appear to conform to the pH partition hypothesis [44]. Similar anomalies were found with weak bases [77]. The apparent pK_a values observed in the absorption–pH curve were shifted to higher values for acids and to lower values for bases, compared with the true pK_a values. Such deviations could be explained by the effect of an acid layer on the apical side of cells, the so-called acid pH microclimate [44,70,73,76–84].

Shiau et al. [73] directly measured the microclimate pH, pH_m, to be 5.2–6.7 in different sections of the intestine (very reproducible values in a given segment) covered with the normal mucus layer, as the luminal (bulk) pH, pH_b, was maintained at 7.2. Good controls ruled out pH electrode artifacts. With the mucus layer washed off, pH_m rose from 5.4 to 7.2. Values of pH_b as low as 3 and as high as 10 remarkably did not affect values of pH_m. Glucose did not affect pH_m when the microclimate was established. However, when the mucus layer had been washed off and pH_m was allowed to rise to pH_b, the addition of 28 mM glucose caused the original low pH_m to be reestablished after 5 min. Shiau et al. [73] hypothesized that the mucus layer was an ampholyte (of considerable pH buffer capacity) that created the pH acid microclimate.

Said et al. [78] measured pH_m in rat intestine under in vitro and in vivo conditions. As pH_b was kept constant at 7.4, pH_m values varied within 6.4–6.3 (proximal to distal duodenum), 6.0–6.4 (proximal to distal jejunum), 6.6–6.9 (proximal to distal ileum), and were 6.9 in the colon. Serosal surface had normal pH. When glucose or sodium was removed from the bathing solutions, the pH_m values began to rise. Metabolic inhibitors (1 mM iodoacetate or 2,4-dinitrophenol) also caused the pH_m values to rise. Said et al. [78] hypothesized that a Na^+/H^+ antiporter mechanism, dependent on cellular metabolism, was responsible for the acid pH microclimate.

The tips of villi have the lowest pH_m values, whereas the crypt regions have $pH_m > 8$ values [70]. Most remarkable was that an alkaline microclimate (pH_m 8) was observed in the human stomach, whose bulk pH_b is generally about 1.7. In the stomach and duodenum, the near-neutral microclimate pH was attributed to the secretion of HCO_3^- from the epithelium [70].

2.4 INTRACELLULAR pH ENVIRONMENT

Asokan and Cho [83] reviewed the distribution of pH environments in the cell. Much of what is known in the physiological literature was determined using pH-sensitive fluorescent molecules and specific functional inhibitors. The physiological pH in the cytosol is maintained by plasma membrane-bound H^+-ATPases, ion exchangers, as well as the Na^+/K^+-APTase pumps. Inside the organelles, pH microenvironments are maintained by a balance between ion pumps, leaks, and internal ionic equilibria. Table 2.1 lists the approximate pH values of the various cellular compartments.

2.5 TIGHT-JUNCTION COMPLEX

Many structural components of the tight junctions (TJs) have been defined since 1992 [85–97]. Lutz and Siahaan [95] reviewed the protein structural components of the TJ. Figure 2.7 depicts the occludin protein complex that makes the water pores so restrictive. Freeze-fracture electronmicrographs of the constrictive region of the TJ show net-like arrays of strands (made partly of the cytoskeleton) circumscribing the cell, forming a division between the apical and the basolateral

TABLE 2.1 Intracellular pH Environment

Intracellular Compartment	pH
Mitocondria	8.0
Cytosol	7.2–7.4
Endoplasmic reticulum	7.1–7.2
Golgi	6.2–7.0
Endosomes	5.5–6.0
Secretory granules	5.0–6.0
Lysosomes	4.5–5.0

Source: Ref. 83.

sides. A region 10 strands wide forms junctions that have very small pore openings; fewer strands produce leakier junctions. The actual cell-cell adhesions occur in the cadheren junctions, located further away from the apical side. Apparently three calciums contiguously link 10-residue portions of cadheren proteins spanning from two adjoining cell walls, as depicted in Fig. 2.7 [95]. Calcium-binding agents can open the junctions by interactions with the cadheren complex.

2.6 STRUCTURE OF OCTANOL

Given the complexities of the phospholipid bilayer barriers separating the luminal contents from the serosal side, it is remarkable that a simple 'isotropic' solvent system like octanol has served so robustly as a model system for predicting transport properties [98]. However, most recent investigations of the structure of water-saturated octanol suggest considerable complexity, as depicted in Fig. 2.8 [99,100]. The 25 mol% water dissolved in octanol is not uniformly dispersed. Water clusters form, surrounded by about 16 octanols, with the polar hydroxyl groups pointing to the clusters and intertwined in a hydrogen-bonded network.

The aliphatic tails form a hydrocarbon region with properties not too different from the hydrocarbon core of bilayers. The clusters have an interfacial zone

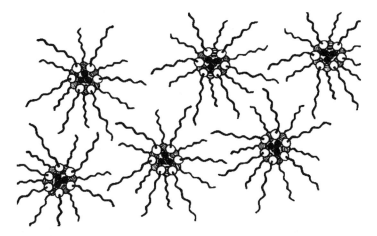

Figure 2.8 Modern structure of wet octanol, based on a low-angle X-ray diffraction study [100]. The four black circles at the center of each cluster represent water molecules. The four hydrogen-bonded water molecules are in turn surrounded by about 16 octanol molecules (only 12 are shown), H-bonded mutually and to the water molecules. The aliphatic tails of the octanol molecules form a hydrocarbon region largely free of water molecules. It is thought that ion-paired drug molecules are located in the water–octanol clusters, and thus can readily diffuse through the "isotropic" medium. For example, filters impregnated with octanol show substantial permeability of charged drug species. However, permeabilities of charged drugs in filters impregnated with phospholipid–alkane solutions are extremely low. [Avdeef, A., *Curr. Topics Med. Chem.*, **1**, 277–351 (2001). Reproduced with permission from Bentham Science Publishers, Ltd.]

between the water interior and the octanol hydroxyl groups. Since water can enter octanol, charged drug molecules need not shed their solvation shells upon entry into the octanol phase. Charged drugs, paired up with counterions (to maintain charge neutrality in the low dielectric medium of octanol, $\epsilon = 8$), can readily diffuse in octanol. Phospholipid bilayers may not have a comparable mechanism accorded to charged lipophilic species, and free diffusion may not be realizable.

2.7 BIOPHARMACEUTICS CLASSIFICATION SYSTEM

The transport model considered in this book, based on permeability and solubility, is also found in the biopharmaceutics classification system (BCS) proposed by the U.S. Food and Drug Administration (FDA) as a bioavailability–bioequivalence (BA/BE) regulatory guideline [101–110]. The BCS allows estimation of the likely contributions of three major factors—dissolution, solubility, and intestinal permeability—which affect oral drug absorption from immediate-release solid oral products. Figure 2.9 shows the four BCS classes, based on high and low designations of solubility and permeability. The draft document posted on the FDA website details the methods for determining the classifications [106]. If a molecule is classed as highly soluble, highly permeable (class 1), and does not have a narrow therapeutic index, it may qualify for a waiver of the very expensive BA/BE clinical testing.

The solubility scale is defined in terms of the volume (mL) of water required to dissolve the highest dose strength at the lowest solubility in the pH 1–8 range, with 250 mL as the dividing line between high and low. So, high solubility refers to complete dissolution of the highest dose in 250 mL in the pH range 1–8. Permeability is the major rate-controlling step when absorption kinetics from the GIT is controlled

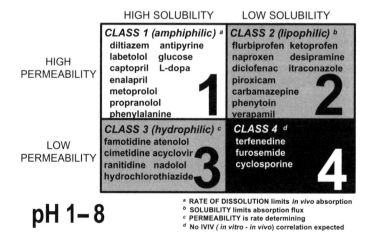

Figure 2.9 Biopharmaceutics classification system [101–110]. Examples are from Refs. 102 and 104. [Avdeef, A., *Curr. Topics Med. Chem.*, **1**, 277–351 (2001). Reproduced with permission from Bentham Science Publishers, Ltd.]

by drug biopharmaceutical factors and not by formulation factors. Extending the BCS to low-permeability drugs would require that permeability and intestinal residence time not be affected by excipients [110].

Permeability in the BCS refers to human jejunal values, where "high" is $\geq 10^{-4}$ cm/s and "low" is below that value. Values of well-known drugs have been determined in vivo at pH 6.5 [56]. The high permeability class boundary is intended to identify drugs that exhibit nearly complete absorption (>90% of an administered oral dose) from the small intestine. The class boundary is based on mass balance determination or in comparison to an intravenous reference dose, without evidence suggesting instability in the gastrointestinal tract. Intestinal membrane permeability may be measured by in vitro or in vivo methods that can predict extent of drug absorption in humans. It is curious that so little emphasis is placed on the pH dependence of permeability assessment, given that the small intestine is a pH gradient spanning about 5–8.

The rapid dissolution class boundary is defined in terms of the in vitro dissolution being greater than 85% in 30 min in 900 mL aqueous media at pH 1, 4.5, and 6.8, using USP Apparatus I (100 rpm) or Apparatus II [50 rpm (revolutions/min)] [104]. A similar guideline has been introduced in the European Union [105]. Examples of molecules from the various four classes are presented in Fig. 2.9 [102,104].

CHAPTER 3

CHARGE STATE

Weak acids and bases ionize in solutions to varying extent, depending on pH. This, in turn, affects the distribution of the chemicals in solution and affects their availability to enter biological reactions. The characteristic thermodynamic parameter relating the pH to the charge state of a molecule is the ionization constant, pK_a [34,35]. Knowledge of the pK_a of a substance is widely useful. It can predict the absorption, distribution, and excretion of medicinal substances. For example, urine pH (normally 5.7–5.8) can be altered (with oral doses of NH_4Cl or $NaHCO_3$) to satisfy reabsorption of uncharged species for therapeutic reasons, or to ease excretion of ionized species in toxicological emergencies [111]. Weak acids may be excreted in alkaline urine and weak bases may be eliminated in acidic urine, a principle that may be lifesaving with overdoses of barbiturates, amphetamines, and narcotics, for example. Knowledge of the pK_a of a substance can be used in maximizing chemical reaction or synthesis yields. For example, solvent extraction can be best applied in a pH region where the synthesized molecule is uncharged. Interpretations of kinetic measurements can depend on the pK_a of a reactant.

The method of choice for the measurement of ionization constants is potentiometry [35,112–119]. Special circumstances warrant the determination of the pK_a by UV spectrophotometry [120–143], capillary electrophoresis (CE) [144–147], and a chromatographic technique [148]. In principle, UV and CE methods are more sensitive and less sample-demanding than is the pH-metric method. That not withstanding, the latter method is preferred because it is so much better developed,

Absorption and Drug Development: Solubility, Permeability, and Charge State. By Alex Avdeef
ISBN 0-471-423653. Copyright © 2003 John Wiley & Sons, Inc.

and is very strongly supported commercially [Sirius]. Currently, the UV method is under vigorous development, and is also supported commercially [131–143]. The CE method is in the orphan stage, with apparently little interest shown by the manufacturers of CE equipment, although that may soon change. A small and enthusiastic user base exists, however. Many other techniques have been used, but the methods described above are best suited for pharmaceutical applications.

3.1 CONSTANT IONIC MEDIUM REFERENCE STATE

The ionization reactions for acids, bases, and ampholytes (diprotic) may be represented by the generic forms

$$HA \rightleftarrows A^- + H^+ \qquad K_a = \frac{[A^-][H^+]}{[HA]} \qquad (3.1)$$

$$BH^+ \rightleftarrows B + H^+ \qquad K_a = \frac{[B][H^+]}{[BH^+]} \qquad (3.2)$$

$$XH_2^+ \rightleftarrows XH + H^+ \qquad K_{a1} = \frac{[XH][H^+]}{[XH_2^+]} \qquad (3.3)$$

$$XH \rightleftarrows X^- + H^+ \qquad K_{a2} = \frac{[X^-][H^+]}{[XH]} \qquad (3.4)$$

Listed after the reactions are the corresponding equilibrium quotients. The law of mass action sets the concentration relations of the reactants and products in a reversible chemical reaction. The negative log (logarithm, base 10) of the quotients in Eqs. (3.1)–(3.4) yields the familiar Henderson–Hasselbalch equations, where "p" represents the operator "–log:"

$$pK_a = pH + \log \frac{[HA]}{[A^-]} \qquad (3.5)$$

$$pK_a = pH + \log \frac{[BH^+]}{[B]} \qquad (3.6)$$

$$pK_{a1} = pH + \log \frac{[XH_2^+]}{[XH]} \qquad (3.7)$$

$$pK_{a2} = pH + \log \frac{[XH]}{[X^-]} \qquad (3.8)$$

Equations (3.5)–(3.8) indicate that when the concentration of the free acid, HA (or conjugate acid, BH^+), equals that of the conjugate base, A^- (or free base, B), the pH has the special designation, pK_a. If the pH is two units lower than the pK_a for an acid, Eq. (3.5), $[HA]/[A^-] = 100$, and the uncharged species accounts for 100/101

(99%) of the total substance in solution. If the pH is two units higher than the pK_a, then it is the anion that accounts for 99% of the total.

Ibuprofen (HA) has a pK_a 4.45 ± 0.04 [149] determined at 25°C and ionic strength I 0.15 M (fixed by KCl). Chlorpromazine (B) has a pK_a 9.24 ± 0.01 at 25°C, I 0.15 M (NaCl) [150]. Morphine (XH) has pK_{a1} 8.17 ± 0.01 and pK_{a2} 9.26 ± 0.01 at 25°C, I 0.15 M (NaCl) [151].

All equilibrium constants in the present discussion are based on the *concentration* (not activity) scale. This is a perfectly fine thermodynamic scale, provided the ionic strength of the solvent medium is kept fixed at a "reference" level (and therefore sufficiently higher than the concentration of the species assayed). This is known as the "constant ionic medium" thermodynamic state. Most of the results reported these days are determined in 0.15 M KCl or NaCl, the physiological level, because of standardization in the available commercial instruments. If the ionic strength is changed, the ionization constant may be affected. For example, at ionic strength of 0.001 M, morphine pK_a values were determined to be 8.13 ± 0.01 and 9.46 ± 0.01 [151]. The change in the second constant illustrates the need to report the ionic strength (and the temperature, since constants are also temperature-dependent) [34,35].

The ionic strength dependence of ionization constants can be predicted by the Debye–Hückel theory [34,35]. In the older literature, values were reported most often at "zero sample and ionic strength" and were called the "thermodynamic" constants. The constants reported at 0.15 M ionic medium are no less thermodynamic. Nevertheless, a result determined at 0.15 M KCl background, can be corrected to another background salt concentration, provided the ionic strength is within the limitations of the theory (<0.3 M for the Davies [152] variant of the Debye–Hückel expression). It is sometimes convenient to convert constants to "zero ionic strength" to compare values to those reported in older literature. A general ionic strength correction equation is described in the literature [112,118,153].

3.2 pK_a DATABASES

The "blue book" compilations [154–158] are probably the most comprehensive sources of ionization constants collected from the literature (up to the end of 1970s). These are recommended for experts in the field. On the other hand, the "red books" contain critically selected values [159]. The six-volume set has been put into electronic form in cooperation with NIST (National Institute of Standards and Technology), and is available at a very reasonable price [160]. A two-volume set of critically determined constants is available from Sirius Analytical Instruments Ltd., and covers molecules of particular interest to the pharmaceutical community [161,162]. In Section 3.8 at the end of this chapter, a list of "gold standard" pK_a values of mostly drug-like molecules is presented (see Table 3.1), with many of the values determined by the author since the early 1970s.

3.3 POTENTIOMETRIC MEASUREMENTS

In pH-metric titration, precisely known volumes of a standardized strong acid (e.g., HCl) or base (e.g., KOH or NaOH) are added to a vigorously-stirred solution of a protogenic substance, during which pH is continuously measured with a precision combination glass electrode, in a procedure confined to the interval pH 1.5–12.5. The substance (50–500 μM or higher) being assayed is dissolved in 2–20 mL of water or in a mixed solvent consisting of water plus an organic water-miscible cosolvent [e.g., methanol, dimethylsulfoxide (DMSO), acetonitrile, or 1,4-dioxane]. An inert water-soluble salt (0.15 M KCl or NaCl) is added to the solution to improve the measurement precision, and to mimic the physiological state. Usually, the reaction vessel is thermostated at 25°C and a blanket of a heavy inert gas (argon, but *not* helium) bathes the solution surface.

The plot of pH against titrant volume added is called a *potentiometric titration curve*. Figure 3.1a shows two examples. The shape of such a curve can suggest the amount of substance present and its characteristic acid–base ionization properties. The left curve in Fig. 3.1a represents a strong acid–base titration, containing no sample species. The curve on the right side of Fig. 3.1a is that of morphine-6-glucuronide (M6G), which has three pK_a values (XH$_3^+$ $\rightleftarrows$ XH$_2^\pm$ $\rightleftarrows$ XH$^-$ $\rightleftarrows$ X^{2-}) [151]. The inflection points corresponding to where the slope in such plots is max-imum in size are called endpoints (pH 7 in the left curve, pH 5.5 and 10 in the right curve). At the endpoint the sample is almost completely in one state of ionization (e.g., XH$_2^\pm$ zwitterion at pH 5.5). The inflection points where the slope is at a mini-mum size designate regions of maximum buffering (pH 8.8 in the morphine meta-bolite curve). At such a point the molecule is present in two states of protonation of equal concentration (pH = pK_a), *unless two or more overlapping* pK_a *values are in the buffer region*. So by inspection of Fig. 3.1a, one can say that a pK_a of M6G may be ~8.8. (We will see in the next section that such a simple interpretation of the titration curve can lead to the wrong conclusion, because M6G has two overlapping pK_a values centered about pH 8.8.) Where are the other pK_a values of M6G? Unfor-tunately, a titration curve does not always reveal all the pK_a values that a molecule may have. To reveal the other two pK_a values of M6G and to test for overlapping pK_a values, it is necessary to transform the titration curves into Bjerrum plots [112,116,118,153,163–165].

3.3.1 Bjerrum Plots

The Bjerrum plots are probably the most important graphical tools in the initial stages of titration data analysis. Since one knows how much strong acid and strong base have been added to the solution at any point and also how many dissociable protons the sample substance brings to the solution, one knows the *total* hydrogen ion concentration in solution, despite what equilibrium reactions are taking place. By measuring the pH (and after converting it into p$_c$H = −log[H$^+$]), one knows the *free* hydrogen ion concentration [H$^+$]. The difference between the total and the free concentrations is equal to the concentration of the *bound* hydrogen ions. The latter

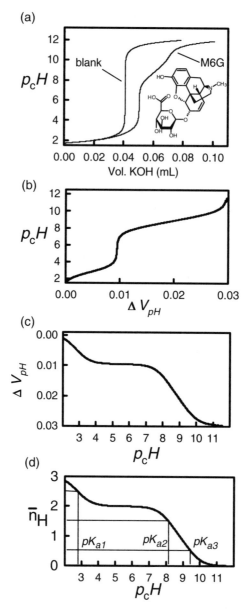

Figure 3.1 Four-step construction of the Bjerrum difference plot for a three-pK_a molecule, whose constants are obscured in the simple titration curve (see text): (a) titration curves; (b) isohydric volume differences; (c) rotated difference plot; (d) Bjerrum plot. [Avdeef, A., *Curr. Topics Med. Chem.*, **1**, 277–351 (2001). Reproduced with permission from Bentham Science Publishers, Ltd.]

concentration divided by that of the sample gives the average number of bound hydrogen atoms per molecule of substance $\bar{n}_H$. The Bjerrum curve is a plot of $\bar{n}_H$ versus p_cH.

Operationally, such a plot can be obtained by subtracting a titration curve containing no sample ("blank" titration; left curve in Fig. 3.1a) from a titration curve with sample (right curve in Fig. 3.1a) at fixed values of pH. The resultant difference plot is shown in Fig. 3.1b. The plot is then rotated (Fig. 3.1d), to emphasize that $\bar{n}_H$ is the dependent variable and pH is the independent variable [163]. The volume differences can be converted to proton counts as described in the preceding paragraph, to obtain the final form, shown in Fig. 3.1d.

The Bjerrum plot in Fig. 3.1d reveals all the pK_a terms as p_cH values at half-integral $\bar{n}_H$ positions. The three pK_a values of M6G are evident: 2.8, 8.2, and 9.4. In contrast to this, deducing the constants by simple inspection of the titration curves is not possible (Fig. 3.1a): (1) the low pK_a is obscured in Fig. 3.1a by the buffering action of water and (2) the apparent pK_a at pH 8.8 is misleading. M6G has two overlapping pK_a terms, whose *average* value is 8.8. M6G nicely illustrates the value of Bjerrum analysis. With Bjerrum analysis, overlapping pK_as pose no difficulty. Figure 3.2a shows an example of a 6-pK_a molecule, vancomycin [162,166]. Figure 3.2b shows an example of a 30-pK_a molecule, metallothionein, a small heavy-metal-binding protein, rich in sulfhydryl groups [167]. (The reader is challenged to identify the six ionization sites of vancomycin.)

3.3.2 pH Definitions and Electrode Standardization

To establish the *operational* pH scale [168–170], the pH electrode can be calibrated with a single aqueous pH 7 phosphate buffer, with the ideal Nernst slope assumed. Because the $\bar{n}_H$ calculation requires the "free" hydrogen ion concentration (as described in the preceding section) and because the concentration scale is employed for the ionization constants, an additional electrode standardization step is necessary. That is where the operational scale is converted to the *concentration* scale p_cH ($= -\log [H^+]$) using the four-parameter equation [116,119,171,172]

$$pH = \alpha + k_s\, p_cH + j_H[H^+] + \frac{j_{OH}K_w}{[H^+]} \qquad (3.9)$$

where K_w is the ionization constant of water [173]. The four parameters are empirically estimated by a weighted nonlinear least-squares procedure using data from alkalimetric titrations of known concentrations of HCl (from pH 1.7 to 12.3) or standard buffers [116,174–180]. Typical aqueous values of the adjustable parameters at 25°C and 0.15 M ionic strength are $\alpha = 0.08 \pm 0.01$, $k_s = 1.001 \pm 0.001$, $j_H = 1.0 \pm 0.2$, and $j_{OH} = -0.6 \pm 0.2$. Such a scheme extends the range of accurate pH measurements and allows pK_a values to be assessed as low as 0.6 (caffeine [161]) and as high as 13.0 (debrisoquine [162]).

(a)

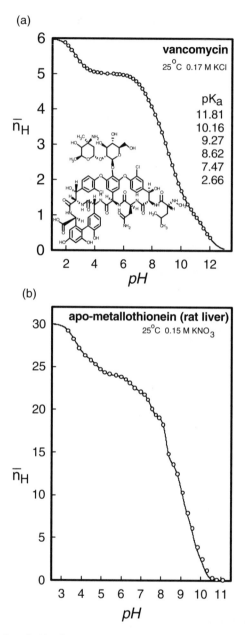

(b)

Figure 3.2 Example of (a) 6-pK_a molecule Bjerrum plot (vancomycin [166]) and (b) 30-pK_a molecule plot (apometallothionein [167]). [Avdeef, A., *Curr. Topics Med. Chem.*, **1**, 277–351 (2001). Reproduced with permission from Bentham Science Publishers, Ltd.]

3.3.3 The "Solubility Problem" and Cosolvent Methods

Since many new substances of interest are very poorly soluble in water, the assessment of the pK_a in aqueous solution can be difficult and problematic. Potentiometry can be a quick technique for such assessment, provided the solubility of the substance is at least 100 μM. (Solutions as dilute as 10 μM can still be analyzed, but special attention must be given to electrode calibration, and ambient carbon dioxide must be excluded.) If the substance is soluble to only 1–10 μM and possesses a pH-sensitive UV chromophore, then spectrophotometry can be applied. CE methods may also be useful since very small sample quantities are required, and detection methods are generally quite sensitive.

If the compound is virtually insoluble (<1 μM), then a pH-metric mixed-solvent approach can be tried [112]. For example, the pK_a of the antiarrhythmic amiodarone, 9.06 ± 0.14, was estimated from water–methanol mixtures, though the intrinsic solubility of the molecule is ~0.008 μM (6 ng/mL) [pION].

The most frequently explored solvent systems are based on water–alcohol mixtures [119,164,166,181–210]. DMSO–water [211–215], dioxane–water [216–220], and other systems [221,222] have been explored. Where possible, methanol is the solvent of choice, because its general effect on pK_a values has been studied so extensively. It is thought to be the least error-prone of the common solvents.

Mixed-solvent solutions of various cosolvent–water proportions are titrated and p_sK_a (the apparent pK_a) is measured in each mixture. The aqueous pK_a is deduced by extrapolation of the p_sK_a values to zero cosolvent. This technique was first used by Mizutani in 1925 [181–183]. Many examples may be cited of pK_a estimated by extrapolation in mixtures of methanol [119,161,162,191,192,196,200], ethanol [184,188–190,193], propanol [209], DMSO [212,215], dimethylformamide [222], acetone [221], and dioxane [216]. Plots of p_sK_a versus weight percent organic solvent, $R_w = 0 - 60$ wt%, at times show either a "hockey-stick" or a "bow" shape [119]. For $R_w > 60$ wt%, S-shaped curves are sometimes observed. (Generally, p_sK_a values from titrations with $R_w > 60$ wt% are not suitable for extrapolation to zero cosolvent because KCl and other ion pairing interferes significantly in the reduced dielectric medium [223].)

For values of $R_w < 60$ wt%, the nonlinearity in p_sK_a plots can be ascribed partly to electrostatic long-range ion–ion interactions. Extensions of the Born electrostatic model, drawing on Bjerrum's theory of ion association [223], were introduced by Yasuda [194] and Shedlovsky [201]. It was recognized that equilibrium quotients in mixed solvents of varying proportions ought explicitly to incorporate the concentration of water, since constancy in water activity cannot be expected in cosolvent mixtures. It was thus proposed that the plot of $p_sK_a + \log [H_2O]$ versus $1/\epsilon$ should produce a straight line for solutions with dielectric constant ϵ, > 50, which for methanol at 25°C means $R_w < 60$ wt%. The slope in such a plot is expected to be inversely proportional to the average ionic diameter of the solvated molecule [201]. The Yasuda–Shedlovsky procedure is now widely used to assess pK_a values of very sparingly soluble pharmaceutical compounds [119,150,166,172, 224,225].

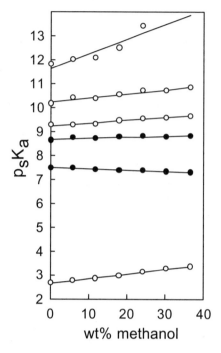

Figure 3.3 The six apparent ionization constants of vancomycin plotted as a function of weight % methanol. Unfilled circles denote acid groups, and filled circles denote basic groups. Acids usually are indicated by positive slopes and bases, by negative slopes. [Avdeef, A., *Curr. Topics Med. Chem.*, **1**, 277–351 (2001). Reproduced with permission from Bentham Science Publishers, Ltd.]

3.3.4 Use of Cosolvents for Water-Soluble Molecules

As the dielectric constant of the solvent mixture decreases, the pK_a of an acid increases and the pK_a of a base decreases. In a multiprotic molecule, this can be a useful property in identifying the ionization groups. Figure 3.3 shows how the pK_a values of vancomycin are affected by changing dielectric constant [62,166]. The p_sK_a/R_w curves with positive slopes were assigned to the carboxylic group and the phenolic residues (structure in Fig. 3.2a), and the two remaining curves, one with a distinct negative slope, were assigned to bases (primary amine on the vancosamine moiety and the secondary amine on the right side of the molecule pictured in Fig. 3.2a). The nonlinear appearance of the highest pK_a in Fig. 3.3 is notably improved in a Yasuda–Shedlovsky plot [162].

It is conceivable that the lowest descending pK_a and the lowest ascending pK_a may cross as R_w approaches 100% [162]. It is interesting that the dielectric constant for pure methanol is about 32, the same value associated with the surface of phospholipid bilayers (in the region of the phosphate groups). This point is further explored later.

3.4 SPECTROPHOTOMETRIC MEASUREMENTS

The most effective spectrophotometric procedures for pK_a determination are based on the processing of whole absorption curves over a broad range of wavelengths, with data collected over a suitable range of pH. Most of the approaches are based on mass balance equations incorporating absorbance data (of solutions adjusted to various pH values) as dependent variables and equilibrium constants as parameters, refined by nonlinear least-squares refinement, using Gauss–Newton, Marquardt, or Simplex procedures [120–126,226].

For an ionizable molecule, the refinement model can be posed as

$$A_{ik} = \sum_{j}^{\text{species}} c_{ij}\epsilon_{jk} \tag{3.10}$$

where A_{ik} is the calculated absorbance at the k wavelength in the i spectrum. Different values of i denote spectra collected at different pH levels. The molar absorptivity of the j species at the k wavelength is denoted by ϵ_{jk}, and the concentration of the j species at the i pH is c_{ij}. "Species" here refers to the different charge-state forms of a molecule. The values of c_{ij} are functions of the total sample concentration and the ionization constants; these are calculated as in procedures for the pH-metric refinement of constants [118]. One can estimate pK_a values, intelligently guess the values of ϵ_{jk}, and use these to calculate values of A_{ik}. In the calculation, the objective is to minimize the sum of the residuals between the calculated and observed absorbances,

$$S = \sum_{k}^{\text{species}} \sum_{i}^{\text{spectra(pH)}} \frac{(A_{ik}^{\text{obs}} - A_{ik}^{\text{calc}})^2}{\sigma_{ik}^2} \tag{3.11}$$

where σ_{ik} are the estimated uncertainties in the measured values of absorbances. Mathematically imposed constraints prevent the calculation of negative values of absorbances [227]. The "best" set of refined pK_a constants are those that minimize S.

In complicated equilibria, uninformed guessing of pK_a values and ϵ_{jk} can be unsettling. Elegant mathematical methods have evolved to help this process of supervised calculation. Since not all species in a multiprotic compound possess detectable UV chromophores or sometimes more than one species have nearly identical molar absorptivity curves, methods had to be devised to assess the number of spectrally active components [121]. With ill-conditioned equations, damping procedures are required [122]. Gampp et al. [127] considered principal-component analysis (PCA) and evolving factor analysis (EFA) methods in deciding the presence and stoichiometries of the absorbing species.

Tam and others [131–135,137,138,140–143,228,229] developed a very effective generalized method for the determination of ionization constants and molar absorptivity curves of individual species, using diode-array UV spectrophotometry, coupled to an automated pH titrator. Species selection was effected by target factor analysis (TFA), and EFA methods were used. Multiprotic compounds with overlapping pK_a values were investigated. Binary mixtures of ionizable compounds were considered [141]. Assessment of microconstants has been reported [138,140]. The use of cosolvents allowed the deconvolutions of 12 microconstants of cetirizine, a 3-pK_a molecule [142]. Validation studies, comparing the TFA method to the first derivative technique, were reported [132,137].

A 96-well microtiter plate high-throughput method, called *spectral gradient analysis* (SGA), based on a pH gradient flow technique with diode-array UV detection has been reported [135,136,139]. A universal buffer, consisting of citric acid, phosphate, tris(hydroxymethyl)-aminomethane, and *n*-butylamine, was developed in an acidified and an alkaline form [139]. Mixture of the two forms in a flowing stream produced a pH gradient very linear in time. The SGA method was successfully validated using 110 structurally unrelated compounds [135]. Poorly soluble molecules still pose a challenge to the SGA method, although this problem is being vigorously addressed by the manufacturer.

Apparently similar flowstream universal buffers have been developed by Alibrandi and others [128,129] for assessing kinetic parameters, such as the pH-dependent hydrolysis of acetylsalicylic acid. The pH–time curves are not as linear as in the SGA system. Other reports of continuous flow pH gradient spectrophotometric data have been described, with application to rank-deficient resolution of solution species, where the number of components detected by rank analysis is lower than the real number of components of the system [130]. The linear pH–time gradient was established in the flowstream containing 25 mM H_3PO_4 by the continuous addition of 100 mM Na_3PO_4.

At *p*ION's analytical services laboratory, the pK_a of a molecule (whose structure may not be known beforehand) is first measured by the TFA method, because very little sample is consumed. (Sometimes there is not much more than 1 mg of sample with which to work.) Only when the analysis of the data proves problematic do we repeat the measurement, the second time using potentiometry, where more sample is required. If any indication of precipitation is evident, either DMSO or methanol is added to the titrated solution and the titration is repeated 3 times (using the same sample), with additional water added between the repeats, to obtain different R_w values of the mixed solvent solutions. It has been our experience that if the TFA method fails and more sample is available, the follow-up pH-metric method always works.

3.5 CAPILLARY ELECTROPHORESIS MEASUREMENTS

CE determination of pK_a is new, compared to the other techniques [144–147]. It has the advantage of being a rather universal method since different detection systems can be coupled to CE. Because it is a separation technique, sample impurities seldom are a problem. A fused-silica capillary, with an inner diameter of 50–75 μm and 27–70 cm in length is filled with a dilute aqueous buffer solution (ionic strength

0.01–0.05 M) [144]. About 10 nL of a sample solution, whose concentration is ~50 μM, is gathered at one end of the capillary, and a 20–30-kV potential is applied between the ends of the capillary dipped into each of two beakers. Sample consumption is roughly 0.2 ng per injection. Sample species migrate according to their charge and fluid drag. Apparent electrophoretic mobility is determined, which is related to the migration time, the length of the capillary, and the applied voltage. The mobility of ionizable compounds is dependent on the fraction of the compound in the charged form. This, in turn, depends on the pK_a. The plot of the apparent mobility versus pH has a sigmoidal shape, with the midpoint pH equal to the pK_a. The practical range for buffer pH in CE is 2–3 at the low end and 11–12 at the high end. When UV detection is used, the limit of detection for a molecule having the molar absorptivity of benzoic acid at 220 nm is ~2 μM [144]. Ishihama et al. [145] were able to determine the pK_a of multiprotic molecules by CE, one molecule having seven ionization groups. They reported a 10 μM limit of detection for verapamil. Its reported pK_a, 8.89, compares well to that determined by potentiometry, 9.07 [pION].

Ishihama et al. [147] have describe a rapid screening method for determining pK_a values of pharmaceutical samples by pressure-assisted CE, coupled with a photodiode array detector. Each CE run was completed in less than 1 min, so a 96-well microtiter plate could be measured in one day. Determinations of the pK_a values of 82 drugs illustrated this interesting new method.

Since most drug discovery projects deal with very sparingly soluble compounds, the usual CE sample concentration would lead to precipitation. The handling of "real" drug candidate molecules is poorly developed in CE applications, in comparison to the most robust potentiometric method.

3.6 CHROMATOGRAPHIC pK_a MEASUREMENT

Oumada et al. [148] described a new chromatographic method for determining the aqueous pK_a of drug compounds that are sparingly soluble in water. The method uses a rigorous intersolvent pH scale in a mobile phase consisting of a mixture of aqueous buffer and methanol. A glass electrode, previously standardized with common aqueous buffers, was used to measure pH online. The apparent ionization constants were corrected to a zero-cosolvent pH scale. Six sparingly soluble nonsteroidal antiinflammatory weak acids (diclofenac, flurbiprofen, naproxen, ibuprofen, butibufen, fenbufen) were used successfully to illustrate the new technique.

3.7 pK_a MICROCONSTANTS

In certain types of multiprotic molecules it is possible that chemically different species of the same stoichiometric composition are formed [142,230–244]. The pH-metric titration technique cannot distinguish between such tautomeric species. In such cases the determined pK_a is a composite constant, a macroconstant. The

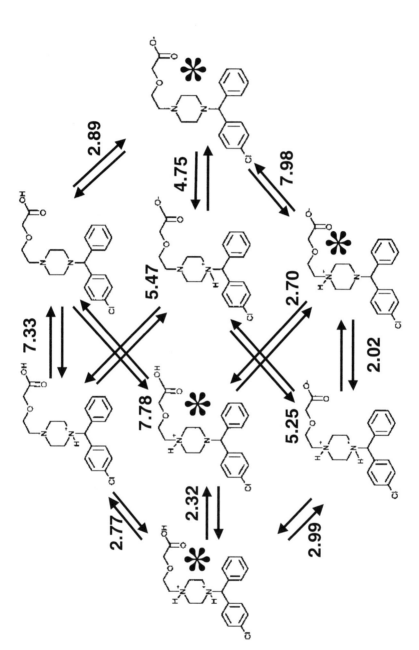

Figure 3.4 Microspeciation of cetirizine, a three-pK$_a$ molecule [142]. The numeric quantities refer to micro-pK$_a$ values. The asterisks denote the principal species at various pH states. [Avdeef, A., *Curr. Topics Med. Chem.*, **1**, 277–351 (2001). Reproduced with permission from Bentham Science Publishers, Ltd.]

thermodynamic experiment is a proton-counting technique. It cannot identify the site in the molecule from which the proton comes. It can only be said that a proton emerges from somewhere in the molecule. On the other hand, microconstants are characteristic of individual species, of which there may be more than one with the same composition.

Various relationships between macro- and microconstants have been derived in the cited literature. The microspecies and microconstants of cetirizine (triprotic molecule with macroconstant pK_a values 2.12, 2.90, and 7.98) are shown in Fig. 3.4, based on the impressive work of Tam and Quéré [142]. The microspecies denoted by an astrisk in Fig. 3.4 are the principal species present in solution. As pH increases, the protonated nitrogen nearest the phenyl groups is the first center to shed charge. The corresponding dication ⇌ monocation reaction has the micro-pK_a 2.32. The next principal center to shed a proton is the carboxylic group, leading to the formation of a zwitterion (micro-pK_a 2.70). The highest-pH principal deprotonation consists of the protonated nitrogen nearest the carboxylate group losing its proton (micro-pK_a 7.98) to form the anionic species on the right side of Fig. 3.4.

In cetirizine, the carboxylic group has four different micro-pK_a values in the range, 2.70–5.47, depending on the neighboring-group charge state. The nitrogen nearest the phenyl groups has the micro-pK_a values in the range 2.02–7.33. The other nitrogen has values in the range 2.77–7.98.

3.8 pK_a "GOLD STANDARD" FOR DRUG MOLECULES

About 250 experimentally determined pK_a values of drugs and some agrochemicals are listed in Table 3.1. These have been critically selected to represent high-quality results. Most of these constants have been determined either at Sirius or *p*ION since 1990, with many personally determined by the author.

TABLE 3.1 Critically Selected Experimental pK_a Values of Drug Molecules

Compound	pK_a	t^a(°C)	I(M)	Ref.
1-Benzylimidazole	6.70	—	0.11	119,153
2-Aminobenzoic acid	4.75, 2.15	—	0.15	161, p. 8
2-Naphthoic acid	4.18	—	0.15	26
2,4-Dichlorophenoxyacetic acid	2.64	—	0.15	161, p. 63
3-Bromoquinoline	2.74	—	0.15	150
3-Chlorophenol	9.11	—	0	150
3-Aminobenzoic acid	4.53, 3.15	—	0.15	161, p. 25
3,4-Dichlorophenol	8.65	—	0	150
3,5-Dichlorophenol	8.22	—	0	150
4-Butoxyphenol	10.26	—	0	119,150
4-Phenylbutylamine	10.50	—	0.15	149

TABLE 3.1 (*Continued*)

Compound	pK_a	$t^a(°C)$	I(M)	Ref.
4-Aminobenzoic acid	4.62, 2.46	—	0.15	161, p. 105
4-Chlorophenol	9.46	—	0	150
4-Me-umbilleferyl-β-ᴅ-glucuronide	2.82	—	0.15	151
4-Methoxyphenol	10.27	—	0	150
4-Iodophenol	9.45	—	0	150
4-Ethoxyphenol	10.25	—	0	119,150
4-Propoxyphenol	10.27	—	0	150
4-Pentoxyphenol	10.13	—	0	150
5-Phenylvaleric acid	4.56	—	0.15	149
6-Acetylmorphine	9.55, 8.19	—	0.15	151
α-Methyl-DOPA	12.66, 10.11, 8.94, 2.21	—	0.15	56
Acebutolol	9.52	—	0.15	362
Acetaminophen	9.63	—	0.15	166,357
Acetic acid	4.55	—	0.15	119
Acetylsalicylic acid	3.50	—	0.15	161, p. 167
Acyclovir	9.23, 2.34	—	0.15	—[b]
Albendazole sulfoxide	9.93, 3.28	—	0.15	166
Allopurinol	9.00	37	0.15	385
Alprenolol	9.51	—	0.15	362
Amiloride	8.65	—	0.15	26,—[b]
Aminophenazone (aminopyrine)	5.06	—	0.15	357
Amiodarone	9.06	—	0.15	—[b]
Amitriptyline	9.49	—	0.15	—[b]
Amitrole	10.72, 4.19	—	0	265
Amlodipine	9.26	—	0.15	—[c]
Amoxicillin	9.53, 7.31, 2.60	—	0.15	—[b]
Ampicillin	7.14, 2.55	—	0.15	162,p. 133
Amylobarbitone	8.07	—	0	150
Antipyrine (phenazone)	1.44	—	0.15	56
Ascorbic acid	11.62, 4.05	—	0.15	357
Aspartic acid	9.67, 3.66, 1.94	—	0.17	161, p. 120
Atenolol	9.54	—	0.15	362
Atropine	9.84	—	0.15	—[b]
Azithromycin	9.69, 8.65	—	0.17	358,—[b]
Bentazone	2.91	—	0	265
Benzocaine	2.39	—	0.15	162, p. 25
Benzoic acid	3.98	—	0.15	153,474
Benzydamine	9.27	—	0.15	472
Bisoprolol	9.57	—	0.15	362
Bromocriptine	5.40	—	0.15	509
Buprenorphine	9.62, 8.31	—	0.15	151
Buspirone	7.60	—	0.15	357
Butobarbitone	8.00	—	0	150
Caffeine	0.60	—	0.15	161, p. 26
Carazolol	9.52	—	0.15	362

TABLE 3.1 (*Continued*)

Compound	pK_a	$t^a(°C)$	$I(M)$	Ref.
Carbenicillin	3.25, 2.22	—	0.11	162, p. 109
Carbomycin B	7.55	—	0.17	358
Carbomycin A	7.61	—	0.17	358
Carvedilol	7.97	—	0.15	362
Cefalexin	7.14, 2.53	—	0.15	166
Celiprolol	9.66	—	0.15	150
Chlorpromazine	9.24	—	0.15	26
Chlorsulfuron	3.63	—	0	265
Cimetidine	6.93	—	0.15	474
Ciprofloxacin	8.62, 6.16	—	0.15	—[b]
Citric acid	5.59, 4.28, 2.88	—	0.17	347
Clarithromycin	8.99	—	0.17	358
Clopyralid	2.32	—	0	265
Clozapine	7.90, 4.40	—	0.15	509
Codeine phosphate	8.22	—	0.15	151
Debrisoquine	13.01	—	0.18	161, p. 119
Deprenyl	7.48	—	0.15	162, p. 26
Deramciclane	9.61	—	0.15	166
Desipramine	10.16	—	0.15	—[b]
Desmycarosyl carbomycin A	8.44	—	0.17	358
Desmycosin	8.36	—	0.17	358
Diacetylmorphine	7.96	—	0.15	151,312
Diazepam	3.40	—	—	—
Diclofenac	3.99	—	0.15	26,149
Diltiazem	8.02	—	0.15	474
Diphenhydramine	9.10	—	0.15	—[b]
Disopyramide	10.32	—	0.15	—[b]
DOPA	12.73, 9.81, 8.77, 2.21	—	0.15	56
Doxycycline	11.54, 8.85, 7.56, 3.21	—	0.15	—[b]
Enalapril	5.42, 2.92	—	0.15	474
Enalaprilat	7.84, 3.17, 1.25	—	0.15	56
Ephedrine	9.65	—	0.15	166
Ergonovine	6.91	—	0.15	—[b]
Erythromycin	8.80	—	0.15	—[b]
Erythromycylamine	9.95, 8.96	—	0.17	358
Erythromycylamine-11,12-carbonate	9.21, 8.31	—	0.17	358
Ethirimol	11.06, 5.04	—	0	265
Famotidine	11.19, 6.74	—	0.15	473
Fenpropimorph	7.34	—	0	265
Flamprop	3.73	—	0	265
Fluazifop	3.22	—	0	265
Flufenamic acid	4.09	—	0.15	—[b]
Flumequine	6.27	—	0.15	161, p. 19
Fluoxetine	9.62	37	0.15	385
Flurbiprofen	4.03	—	0.15	472,473

TABLE 3.1 (*Continued*)

Compound	pK_a	$t^a(°C)$	I(M)	Ref.
Fluvastatin	4.31	—	0.15	56
Folinic acid	10.15, 4.56, 3.10	—	0.15	—[b]
Fomesafen	3.09	—	0	265
Furosemide	10.63, 3.52	—	0.15	26,473
Gly-Gly-Gly	7.94, 3.23	—	0.16	161, p. 126
Gly-Gly-Gly-Gly	7.88, 3.38	—	0.16	161, p. 127
Glyphosate	10.15, 5.38, 2.22, 0.88	—	0.17	162, p. 46
Haloperidol	8.65	—	0.15	—[b]
Hexachlorophene	11.40, 3.90	—	0.15	161, p. 32
Hydrochlorothiazide	9.95, 8.76	—	0.15	473
Hydroxyzine	7.52, 2.66	—	0.16	161, p. 146
Ibuprofen	4.45	—	0.15	149,172,473
Imazapyr	11.34, 3.64, 1.81	—	0	265
Imazaquin	11.14, 3.74, 2.04	—	0	265
Imazethapyr	3.91, 2.03	—	0	265
Imidacloprid	11.12, 1.56	—	0	265
Imipramine	9.51	—	0.15	—[b]
Indomethacin	4.42	—	0.15	26,—[b]
Ioxynil	4.08	—	0	265
Ketoprofen	3.98	—	0.15	473
Labetalol	9.42, 7.48	—	0.15	473
Leucine	9.61, 2.38	—	0.15	56
Lidocaine	7.95	—	0.15	149
Maleic hydrazide	5.79	—	0	265
Mannitol	13.50	—		312
Mecoprop	3.21	—	0	265
Mefluidide	4.79	—	0	265
Mellitic acid	6.04, 5.05, 4.00, 2.75, 1.69, 1.10	—	0.2	153
Meloxicam	3.43	—	0.15	162, p. 112
Metformin	2.93	—	0.15	—[b]
Methotrexate	5.39, 4.00, 3.31	—	0.15	—[b]
Metipranolol	9.54	—	0.15	362
Metolazone	9.70	—	0.15	509
Metoprolol	9.56	—	0.15	362
Metsulfuron, methyl	3.64	—	0	265
Mexiletine	9.14	—	0.15	166
Miconazole	6.07	—	0.15	26
Morphine-3β-D-glucuronide	8.21, 2.86	—	0.16	151
Morphine-6β-D-glucuronide	9.42, 8.22, 2.77	—	0.16	151
Morphine	9.26, 8.18	—	0.15	151,166
Moxonidine	7.36	—	0.15	385
N-Methylaniline	4.86	—	0.15	150
N-Methyl-D-glucamine	9.60	—	0.15	225
Nadolol	9.69	—	0.15	474

TABLE 3.1   (*Continued*)

Compound	pK$_a$	$t^a(°C)$	I(M)	Ref.
Nalidixic acid	6.01	—	0.15	—[b]
Naloxone	9.44, 7.94	—	0	334
Naproxen	4.18	—	0.15	473
Neomycin B	9.33, 8.78, 8.18,	—	0.15	—[b]
	7.64, 7.05, 5.69	23		
Nicotine	8.11, 3.17	—	0.15	161, p. 36
Niflumic acid	4.44, 2.26	—	0.15	161, p. 18
Nitrazepam	10.37, 3.02	—	0.15	161, p. 169
Nizatidine	6.75, 2.44	37	0.15	385
Norcodeine	9.23	—	0.15	151
Norfloxacin	8.51, 6.23	—	0.15	—[b]
Normorphine	9.80	—	0.15	151
Nortriptyline	10.13	—	0.15	26
Ofloxacin	8.31, 6.09, 0.77	—	0.15	161, p. 9
Olanzapine	7.80, 5.44	37	0.15	385
Oleandomycin	8.84	—	0.17	358
Ontazolast	4.20	—	0.15	—[b]
Oxprenolol	9.57	—	0.15	362
p-F-Deprenyl	7.42	—	0.15	162, p. 28
Papaverine	6.39	—	0.15	166
Paromomycin	8.90, 8.23, 7.57,	37	0.15	385
	7.05, 5.99			
Penbutolol	9.92	—	0.15	362
Pentachlorophenol	4.69	—	0	265
Pentobarbitone	8.18	—	0	150
Pericyazine	8.76	—	0.15	150
Phe-Phe	7.18, 3.20	—	0.15	162, p. 6
Phe-Phe-Phe	7.04, 3.37	—	0.15	162, p. 12
Phenazopyridine	5.15	—	0.15	26
Phenobarbital	7.49	—	0	166,150
Phenol	10.01	—	0	150
Phenylalanine	9.08, 2.20	—	0.15	161, p. 116
Phenytoin	8.21	—	0.15	473
Potassium phosphate	11.80, 6.81, 2.01	—	0.17	162, p. 162
Phosphoserine	9.75, 5.64, 2.13	—	0.15	162, p. 79
Phthalic acid	4.92, 2.70	—	0.15	347
Pilocarpine	7.08	—	0.15	357
Pindolol	9.54	—	0.15	362
Pirimicarb	4.54	—	0	265
Pirimiphos, methyl	3.71	—	0	265
Piroxicam	5.07, 2.33	—	0.15	26
Prazosin	7.11	—	0.15	—[b]
Primaquine	10.03, 3.55	—	0.15	—[b]
Probenecid	3.01	—	0.15	26
Procaine	9.04, 2.29	—	0.16	149

TABLE 3.1   (*Continued*)

Compound	pK_a	$t^a(°C)$	I(M)	Ref.
Promethazine	9.00	—	0.15	—[b]
Propamocarb	9.48	—	0	265
Propoxyphene	9.06	—	0.15	474
Propranolol	9.53	—	0.15	26,149,362
Prostaglandin E$_1$	4.87	—	0.15	161, p. 40
Prostaglandin E$_2$	4.77	—	0.15	161, p. 46
Pyridoxine	8.87, 4.84	—	0.16	161, p. 19
Quinalbarbitone	8.09	—	0	150
Quinine	8.55, 4.24	—	0.15	172,474
Quinmerac	3.96	—	0	265
Quinoline	4.97	—	0.15	150
Ranitidine	8.31, 2.11	—	0.15	—[b]
Repromicin	8.83	—	0.17	358
Rifabutine	9.37, 6.90	37	0.15	385
Rivastigmine	8.80	—	0.15	509
Rosaramicin	8.79	—	0.17	358
Roxithromycin	9.27	—	0.15	162, p. 107
Salicylic acid	13.35, 2.88	—	0.15	166
Serotonin	10.91, 9.97	—	0.15	—[c]
Sethoxydim	4.58	—	0	265
Sotalol	9.72, 8.28	—	0.15	162, p. 167
Sucrose	12.60	—		312
Sulfamethazine	7.80, 2.45	—	0	150
Sulfanilamide	10.43, 2.00	—	0.17	161, p. 64
Sodium sulfate	1.33	—	0.17	162, p. 136
Sulfasalazine	10.51, 7.95, 2.65	—	0.15	—[b]
Tamoxifen	8.48	—	0.15	—[b]
Terbinafine	7.05	37	0.15	385
Terbutaline	11.02, 9.97, 8.67	—	0.15	162, p. 36
Terfenadine	9.86	—	0.15	26
Tetracaine	8.49, 2.39	—	0.15	149
Theophylline	8.55	—	0.15	162, p. 128
Thiabendazole	4.64, 1.87	—	0	265
Ticarcillin	3.28, 2.89	—	0.11	162, p. 109
Tilmicosin	9.56, 8.18	—	0.17	358
Timolol	9.53	—	0.15	362
Tralkoxydim	4.98	—	0	265
Triazamate acid	3.49	—	0	265
Trimethoprim	7.07	—	0.15	—[b]
Trovafloxacin	8.11, 5.90	—	0.15	474
TrpPphe	7.30, 3.18	—	0.15	162, p. 2
Trp-Trp	7.27, 3.38	—	0.15	162, p. 8
Tryptophan	9.30, 2.30	—	0.16	162, p. 10
Tylosin	7.73	—	0.17	358
Tyrosine	10.12, 9.06, 2.20	—	0.16	161, p. 112

TABLE 3.1   (*Continued*)

Compound	pK_a	t^a(°C)	I(M)	Ref.
Uracil	13.28, 9.21	—	0.16	162, p. 121
Valsartan	4.70, 3.60	—	0.15	509
Vancomycin	11.86, 10.16, 9.26,	—	0.17	162, p. 32
	8.63, 7.49, 2.66			
Verapamil	9.07	—	0.15	—[b]
Warfarin	4.82	—	0.15	149
Xipamide	10.47, 4.58	37	0.15	385
Zidovudine	9.53	—	0.15	—[b]
Zopiclone	6.76	37	0.15	385

[a]Temperature 25°C, unless otherwise noted.
[b]*p*ION.
[c]Sirius Analytical Instruments.

CHAPTER 4

PARTITIONING INTO OCTANOL

In all other sections of this book, we use the term K_p to represent the partition coefficient and K_d, the apparent partition coefficient. These terms were chosen to avoid symbol conflict when discussing permeability and diffusivity. Since this chapter and Chapter 5 are devoted primarily to partition coefficients, we will use the most common terminology: P for partition coefficients and D for apparent (pH-dependent) partition coefficients. [Other symbols for these parameters have been used in the literature, including P_{OW} (oil-water partition), K_{OW}, PC, and APC.]

Central to the Hansch analysis [17,98] is the use of $\log P$ or $\log D$ to predict biological activity. Much literature has been published about the measurement and applications of these parameters [17,23,24,57,98–100,224,225,243,245–265]. Two conferences have been dedicated to the topic [266,267]. Several studies [245,246,268] describe how to measure $\log P / \log D$: which techniques to use, what pitfalls to look out for, what lipid : water volumes to consider, the value of GLP—in other words, how to do it right. The structure of octanol became better understood [99,100]. Issues of water drag were investigated [247,248]. Partition solvents other than octanol (CHCl₃, various alkanes, PGDP, and 1,2–dichloroethane) were explored for the effect of their hydrogen bonding donor/acceptor properties [17,151,249,261,269]. Seiler's [250] concept of $\Delta \log P$ was further tested [251,252,257]. Methods to predict H-bond factors from two-dimensional structures were expanded [254–260]. Hydrogen bonding was prodded as "the last mystery in drug design" [253]. The concept of "molecular chameleons," proposed by Testa and others, was applied to the study of intramolecular effects

Absorption and Drug Development: Solubility, Permeability, and Charge State. By Alex Avdeef
ISBN 0-471-423653.   Copyright © 2003 John Wiley & Sons, Inc.

in morphine glucuronide conformational-sensitive partitioning [151,262,263]. A case was made for the return of olive oil, as a model solvent in the prediction of partitioning into adipose tissue [264].

Today almost every practicing pharmaceutical scientist knows the difference between $\log P$ and $\log D$ [229,270–276]. Better understanding of the partitioning behavior of ampholytes and charged species emerged [277–291]. The concept of the micro-$\log P$ was formalized [224,243,273,275]. Rapid high-performance liquid chromatography (HPLC) methods for determining $\log P$ were fine-tuned [292–298]. Immobilized artificial membrane (IAM) chromatography [47,299–311], liposome chromatography [312–319], and capillary electrophoresis [320–322] evoked considerable interest. An accurate (compared to shake-flask) and fast (2 h) method using dialysis tubing to separate the aqueous phase from the octanol phase was reported [323]. Potentiometric methods of $\log P$ determination matured and achieved recognition [25,112,149–151,153,161,162,166,172,224,225,250,268,269, 275,324–363]. Some remarkable new insights were gained about the membrane interactions of charged amphiphilic species from the study of drug partitioning into liposomes (Chapter 5). The need for high-throughput measurements led to the scaling down of several techniques to the 96-well microtiter plate format [294].

4.1 TETRAD OF EQUILIBRIA

The topic of drug partitioning between water and lipids concerns chemical equilibria. For a monoprotic weak acid (and base), the partitioning equilibria may be represented as

$$HA \rightleftarrows HA_{(ORG)} \qquad (B \rightleftarrows B_{(ORG)}) \tag{4.1}$$

As mentioned in Chapter 3, the law of mass action sets the concentration relations of the reactants and products. So, the equilibrium constants, termed the partition coefficients, are the quotients

$$P_{HA} = \frac{[HA_{(ORG)}]}{[HA]} \qquad \left(P_B = \frac{[B_{(ORG)}]}{[B]}\right) \tag{4.2}$$

where [HA] ([B]) is the free-acid (free-base) aqueous concentration, moles/liter aqueous solution, and the ORG-subscripted term is the concentration in the oil phase, moles/liter *of organic solvent* [347]. When the partition coefficient is determined directly, usually the aqueous concentration is determined analytically (UV or HPLC), and the oil-phase counterpart is inferred through mass balance [245]. Not only the neutral species, but the charged species can partition into the organic phase (such as octanol), although usually to a much lesser extent:

$$A^- \rightleftarrows A^-_{(ORG)} \qquad (BH^+ \rightleftarrows BH^+_{(ORG)}) \tag{4.3}$$

$$P_A = \frac{[A^-_{(ORG)}]}{[A^-]} \qquad \left(P_{BH} = \frac{[BH^+_{(ORG)}]}{[BH^+]}\right) \tag{4.4}$$

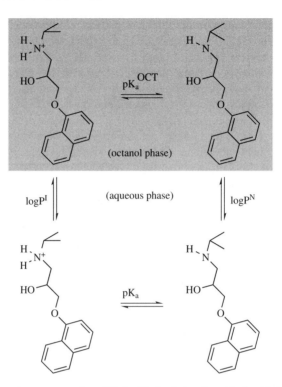

Figure 4.1 Octanol–water tetrad equilibria. [Avdeef, A., *Curr. Topics Med. Chem.*, **1**, 277–351 (2001). Reproduced with permission from Bentham Science Publishers, Ltd.]

To distinguish partition coefficients of neutral species from ionized species, the notation $\log P^N$ and $\log P^I$ may be used, respectively, or the symbol C or A may be used as a substitute for superscript I, denoting a cation or anion, respectively. [362].

It is convenient to summarize the various reactions in a box diagram, such as Fig. 4.1 [17,275,280], illustrated with the equilibria of the weak base, propranolol. In Fig. 4.1 is an equation labeled pK_a^{oct}. This constant refers to the octanol pK_a, a term first used by Scherrer [280]. When the concentrations of the uncharged and the charged species *in octanol* are equal, the *aqueous* pH at that point defines pK_a^{oct}, which is indicated for a weak acid as

$$HA_{(ORG)} \rightleftharpoons A^-_{(ORG)} + H^+ \qquad K_a^{oct} = \frac{\left[A^-_{(ORG)}\right][H^+]}{\left[HA_{(ORG)}\right]} \qquad (4.5)$$

Characteristic of a box diagram, the difference between the partition coefficients is equal to the difference between the two pK_a values [229,275,280,362]:

$$diff(\log P^{N-I}) = \log P^N - \log P^I = \left| pK_a^{oct} - pK_a \right| \qquad (4.6)$$

In a box diagram, if any three of the equilibrium constants are known, the fourth may be readily calculated from Eq. (4.6), taking into account that octanol causes the pK_a of weak acids to increase, and that of weak bases to decrease.

In mixtures containing high lipid : water ratios, HCl will appreciably partition into solutions with pH <2.5, as will KOH when pH >11.5 [162,284]. General box diagrams reflecting these caveats have been discussed [275].

4.2 CONDITIONAL CONSTANTS

The constants in Eqs. (4.4) and (4.5) are *conditional* constants. Their value depends on the background salt used in the *constant ionic medium reference state* (Section 3.1). In the partition reactions considered, the ionized species migrating into the oil phase is accompanied by a counterion, forming a charge-neutral ion pair. The lipophilic nature and concentration of the counterion (as well as that of the charged drug) influences the values of the the the ion pair constants. This was clearly illustrated [277] in the study of the partitioning of the charged form of chlor-promazine into octanol at pH 3.9 (pK_a 9.24 [150]) in the 0.125 M background salt concentrations: $P^I = 56$ (KBr), 55 (NaPrSO$_3$), 50 (KNO$_3$), 32 (KCl, NaCl), 31 (NH$_4$Cl), 26 (Me$_4$NCl), 25 (NaEtSO$_3$), 19 (Et$_4$NCl), 16 (Pr$_4$NCl), 15 (Na$_2$SO$_4$, NaMeSO$_3$), 13 (KCl + 2M urea), and 5 (no extra salt used), suggesting the counterion lipophilicity scale: Br$^-$ > PrSO$_3^-$ > NO$_3^-$ > Cl$^-$ > EtSO$_3^-$ > SO$_4^{2-}$, MeSO$_3^-$. An additional example along this line was described by van der Giesen and Janssen [279], who observed the relationship $\log P^I = 1.00 \log[\mathrm{Na}^+] + 0.63$ for warfarin at pH 11, as a function of sodium concentration. In all the following discussions addressing ion pairs, it is be assumed that 0.15 M KCl or NaCl is the background salt, unless otherwise indicated.

4.3 log *P* DATABASES

A large list of log *P* values has been tabulated by Leo et al. in a 1971 review [364]. Commercial databases are available [365–369]. The best known is the Pomona College MedChem Database [367], containing 53,000 log *P* values, with 11,000 confirmed to be of high quality, the "log *P*-star" list. (No comparably extensive listing of log *D* values has been reported.) Table 4.1 lists a set of "gold standard" octanol–water $\log P^N$, $\log P^I$ and $\log D_{7.4}$ values of mostly drug-like molecules, determined by the pH-metric method.

4.4 log *D*

The distribution ratio *D* is used only in the context of ionizable molecules [229,270–276]. Otherwise, *D* and *P* are the same. The partition coefficient *P*, defined in Eq. (4.2), refers to the concentration ratio of a *single* species. In contrast, the distribution coefficient D can refer to a collection of species and can depend on pH. In the most general sense, D is defined as the sum of the concentrations of all charge-state forms of a substance dissolved in the lipid phase divided by the sum of those dissolved in water. For a simple multiprotic molecule X, the distribution ratio is defined as

$$D = \frac{([X_{(ORG)}]' + [XH_{(ORG)}]' + [XH_{2(ORG)}]' + \cdots)/([X] + [XH] + [XH_2] + \cdots)}{r}$$

(4.7)

where r is the lipid–water volume ratio, $v(ORG)/v(H_2O)$. The primed quantity is defined in concentration units of moles of species dissolved in the organic phase per liter of *aqueous* phase. Assuming a diprotic molecule and substituting Eqs. (3.7), (3.8), (4.2), and (4.4) into Eq. (4.7) yields

$$D = \frac{P_A + P_{HA}10^{+(pK_{a2}-pH)} + P_{H_2A}10^{+(pK_{a2}+pK_{a1}-2pH)}}{1 + 10^{+(pK_{a2}-pH)} + 10^{+(pK_{a2}+pK_{a1}-2pH)}}$$

(4.8)

where P_A refers to the ion pair partition coefficient of the dianion; P_{HA}, to that of the anion, and P_{H_2A}, to the partition coefficient of the neutral species. If no ion pair partitioning takes place, then Eq. (4.8) further simplifies to

$$\log D = \log P^N - \log\{1 + 10^{-(pK_{a2}+pK_{a1}-2pH)} + 10^{-(pK_{a1}-pH)}\}$$

(4.9)

Note that the distribution coefficient depends only on pH, pK_a values, and P (not on concentration of sample species). Equation (4.7) is applicable to all lipophilicity calculations. Special cases, such as eq. 4.9, have been tabulated [275].

Figures 4.2a, 4.3a, and 4.4a show examples of lipophilicity profiles, $\log D$ versus pH, of an acid (ibuprofen), a base (chlorpromazine), and an ampholyte (morphine). The flat regions in Figs. 4.2a and 4.3a indicate that the $\log D$ values have reached the asymptotic limit where they are equal to $\log P$: at one end, $\log P^N$ and at the other end, $\log P^I$. (The morphine example in Fig. 4.4a is shown free of substantial ion pair partitioning.) The other regions in the curves have the slope of either -1 (Fig. 4.2a) or $+1$ (Fig. 4.3a) or ± 1 (Fig. 4.4a). Ibuprofen has the octanol–water $\log P_{HA}$ 3.97 (indicated by the flat region, pH < 4, Fig. 4.2a) and the ion pair $\log P_A$ -0.05 in 0.15 M KCl (flat region, pH > 7) [161]. Chlorpromazine has $\log P_B$ 5.40 and an ion-pair $\log P_{BH}$ 1.67, also in 0.15 M KCl (Fig. 4.3a) [161]. Ion pairing becomes significant for pH < 6 with the base. The equation that describes the sigmoidal curve, valid for monoprotic acids and bases for the entire pH range, is

$$\log D = \log(P_X + P_{XH}10^{+pK_a-pH}) - \log(1 + 10^{+pK_a-pH})$$

(4.10)

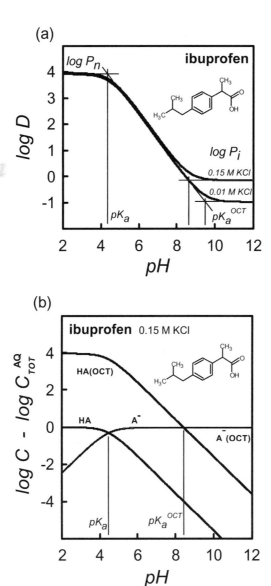

Figure 4.2 (a) Lipophilicity profile of a weak acid at two values of background salt and (b) log–log speciation plot at 0.15 M KCl. [Avdeef, A., *Curr. Topics Med. Chem.*, **1**, 277–351 (2001). Reproduced with permission from Bentham Science Publishers, Ltd.]

For a weak acid, $P_{XH} > P_X$ and the log D curve decreases with pH; for a weak base, $P_X > P_{XH}$, and the log D curve increases with pH, according to this equation.

An additional and useful property of lipophilicity profiles is that the pK_a values are indicated at points where the horizontal asymptote lines intersect the diagonal lines (where $d \log D / d$ pH $= 0.5$ [275]). In Fig. 4.2a, the pK_a and pK_a^{oct} (see Fig. 4.1)

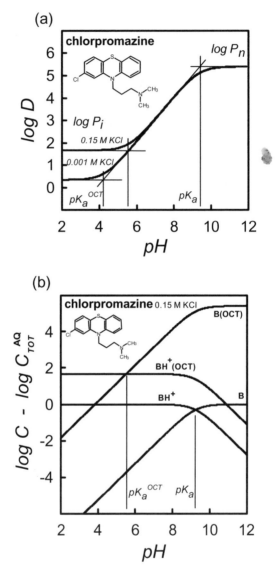

Figure 4.3 (a) Lipophilicity profile of a weak base at two values of background salt and (b) log–log speciation plot at 0.15 M KCl. [Avdeef, A., *Curr. Topics Med. Chem.*, **1**, 277–351 (2001). Reproduced with permission from Bentham Science Publishers, Ltd.]

values are 4.45 and 8.47, respectively; in Fig. 4.3a, the two values are 9.24 and 5.51, respectively. Since pK_a^{oct} is associated with ion pairing, its value depends on the ionic strength, as discussed above. This is clearly evident in Figs. 4.2a and 4.3a.

It may surprise some that for a diprotic molecule with overlapping pK_a values the region of maximum $\log D$ (0.76 in Fig. 4.4a) does *not* equal $\log P$; a displaced horizontal line in Fig. 4.4a indicates the $\log P$ to be 0.89 for morphine [161,162].

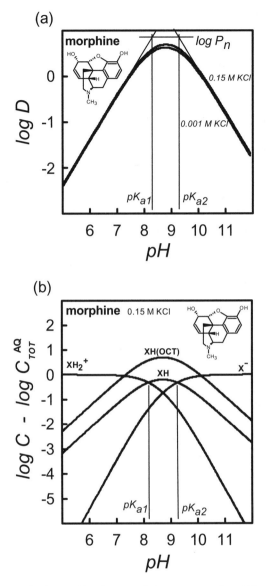

Figure 4.4   (a) Lipophilicity profile of an ampholyte at two values of background salt and (b) log–log speciation plot at 0.15 M KCl. [Avdeef, A., *Curr. Topics Med. Chem.*, **1**, 277–351 (2001). Reproduced with permission from Bentham Science Publishers, Ltd.]

Figures 4.2b, 4.3b, and 4.4b are log–log speciation plots, indicating the concentrations of species in units of the total aqueous sample concentration. (Similar plots were described by Scherrer [280].) The uppermost curve in Fig. 4.2b shows the concentration of the uncharged species in octanol, as a function of pH. If only uncharged species permeate across lipid membranes, as the pH-partition hypothesis suggests, then this curve deserves attention, perhaps more so than the $\log D$ curve (unless the active site is in the apical membrane outer leaflet of the epithelial cell surface, where permeation of the membrane by the charged species is not necessary). That curve is like that of the $\log D$ curve, but with the ion-pair component removed.

4.5 PARTITIONING OF QUATERNARY AMMONIUM DRUGS

The octanol–water partitioning behavior of orally active quaternary ammonium drugs (which are always charged in the physiological pH range), such as propantheline, trantheline, homidium, and neostigmine, was reported by Takács-Novák and Szász [291]. Propantheline has 10% oral absorption, whereas neostigmine is very poorly absorbed from the GIT [370]. Consistent with this, the octanol–water $\log P$ of the bromide salts range from -1.1 to < -3 [291]. However, in the presence of a 50-fold excess of the bile salt deoxycholate, the homidium apparent partition coefficient, $\log P$, elevates to $+2.18$. Similarly heightened numbers were seen when the quaternary drugs were combined with prostaglandin anions, suggesting a possible role of endogenous lipophilic counterions in the GI absorption of the quaternary ammonium drugs.

4.6 log D OF MULTIPROTIC DRUGS AND THE COMMON-ION EFFECT

Ion pair partitioning effects with simple salts should no longer be surprising, given the examples presented above. Partitioning of multiprotic molecules, however, warrants additional consideration. The partitioning behavior of charged molecules, including zwitterions (peptide and other kinds) and ordinary ampholytes, has been intriguing [229,276,278,282,283,285–289,371]. These molecules are sometimes charged over the physiological pH range. Scherrer proposed a classification system for ampholytes based on their pK_a–pK_a^{oct} relationships [276]. It is an important topic to understand, since the oral absorption of such molecules can be poor, and methods to overcome it are the focus of many efforts.

When the $\log D$/pH measurement of a peptide is performed by the shake-flask or the partition chromatography method (using hydrophilic buffers to control pH), usually the shape of the curve is that of a *parabola* (see Ref. 371 and Fig. 1 in Ref. 282), where the maximum $\log D$ value corresponds to the pH at the isoelectric point (near pH 5–6). Surprisingly, when the potentiometric method is used to characterize the same peptide [275], the curve produced is a *step function*, as indicated by the thick line in Fig. 4.5 for dipeptide Trp-Phe.

Both results (parabola vs. step) are correct, even though there is a big difference in the profiles. The explanation for the difference lies in charged-species partitioning: the counterion (from background salt *or buffer*) plays an ineluctable role. In the potentiometric method, pH is controlled by adding HCl or KOH, to a solution that has a 0.15 M physiological level of salt (KCl or NaCl). Thus, the partitioning

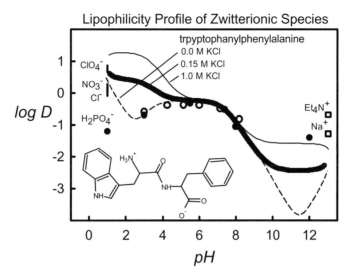

Figure 4.5 Potentiometrically determined [162] lipophilicity profiles of a dipeptide, showing the effect of background salt concentrations. The unfilled symbols [282] and the filled symbols [371] are based on shake-flask measurements. [Avdeef, A., *Curr. Topics Med. Chem.*, **1**, 277–351 (2001). Reproduced with permission from Bentham Science Publishers, Ltd.]

medium always has at least 0.15 M K^+ and Cl^- with which to associate into ion pairs. The effect of buffers in shake-flask or HPLC assays is not always taken into account in discussions of results. We can see in Figs. 4.2a and 4.3a, that the log D profiles take on different values when the background salt is reduced from 0.15 to 0.001–0.01 M. In Fig. 4.5, we indicate what happens to the log D curve when three different levels of salt are used. Very good match to the "anomalous" values, indicated by open and closed symbols, is found [282,371]. The upward turns in the dashed curve in Fig. 4.5 for pH >11.5 and <2.5 are due to the common-ion effect of the salt introduced by the titrant: K^+ (from KOH) and Cl^- (from HCl), respectively.

In studies of the salt dependence of peptides, an attempt was made to look for evidence of ion triplet formation [162], as suggested by the work of Tomlinson and Davis [278]. Phe-Phe-Phe was used as a test tripeptide, and it was reasoned that by performing the octanol–water partitioning in an aqueous solution containing different levels of salt (0.02–0.50 M KCl), one might see the zwitterion log P show the salt dependence that is to be expected of an ion triplet formation. None was evident (other than for the cation at low pH and the anion at high pH, as expected of simple ion-extraction reactions) [162]. An interesting explanation was suggested Dr. Miloň Tichý [1995, unpublished], based on conformational analysis of the structure of the tripeptide in water, that Phe-Phe-Phe can form a cyclic structure, with an intramolecular (internally-compensated) electrostatic bond, $(-CO_2^- \cdots {}^+NH_3-)$, formed between the two ends of the molecule. A highly stabilized ring structure may be more stable than a $K^+ \cdots {}^-O_2C)$—$(NH_3^+ \cdots Cl^-$ ion triplet.

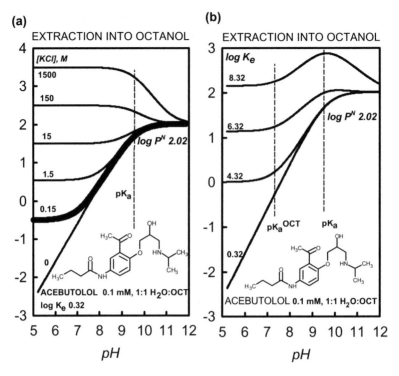

Figure 4.6 Hypothetical lipophilicity profiles: (a) fixed extraction constant with differing salt concentrations; (b) fixed salt concentration with differing extraction constants. [Avdeef, A., *Curr. Topics Med. Chem.*, **1**, 277–351 (2001). Reproduced with permission from Bentham Science Publishers, Ltd.]

The next example, shown in Fig. 4.6a, is the amusing consequence of continually increasing the concentration of background salt (beyond its aqueous solubility—just to make the point) to the shape of log D/pH profile for acebutolol (whose normal 0.15 M salt curve [362] is indicated by the thick line in Fig. 4.6a). The base-like (cf. Fig. 4.3a) lipophilicity curve shape at low levels of salt can become an acid-like shape (cf. Fig. 4.2a) at high levels of salt! An actual example of a dramatic reversal of character is the ionophore monensin, which has a log P^I (in a background of Na^+) 0.5 *greater* than log P^N [276,281].

To cap off the topic of salt dependence, is the following example (also using acebutolol), which will indeed surprise most readers, at first. It is possible to have a peak in a log D/pH profile of a monoprotic molecule! In Fig. 4.6b, we simulated the case by assuming that the level of salt was kept constant and *equal* to the concentration of the sample, and proceeded to explore what should happen if the log of the extraction constant K_e [162,225,275,277]

$$BH^+ + Cl^- \rightleftarrows BH^+Cl^-_{(ORG)} \qquad K_e = \frac{[BH^+Cl^-_{(ORG)}]}{[BH^+][Cl^-]} \qquad (4.11)$$

were raised from the value 0.32 [362] to higher values. The $\log D$ profile eventually develops a peak at $pH = pK_a$ and the series of curves in Fig. 4.6b all have the same pK_a^{oct}, whose value is equal to $pK_a - \log P^N$, namely, 7.5 [Eq. (4.10) is inadequate to explain the phenomenon]. Similarly shaped curves were reported by Krämer et al. [368], who considered the partitioning of propranolol into liposomes (containing free fatty acids) that had *surface charge* that was pH-dependent. In the present case of salt-induced extraction, the maximum point in Fig. 4.6b is not sustainable as pH increases past the pK_a, because the concentration of the charged sample component diminishes, in accordance with the pK_a.

4.7 SUMMARY OF CHARGED-SPECIES PARTITIONING IN OCTANOL–WATER

Excluding effects not in the scope of this book, such as interfacial transport of charged species driven by electrical potentials, the main lesson of the partitioning studies of charged drugs is that the charged molecule needs to be accompanied by a counterion in order for the ion pair to enter a lipid phase such as octanol. Later, it will become apparent that it must not be taken for granted that charged species enter other lipid phases as they do octanol. The peculiar structure of octanol (Fig. 2.8) may facilitate the entry of ion pairs in a way that may be impossible in a phospholipid bilayer, for example (covered below).

Scherrer observed [280,281], as have others [161,162,275], that for a large number of ordinary charged species partitioning into octanol in the presence of aqueous solutions containing 0.15 M KCl or NaCl, weak-acid salts have values of $diff(\log P^{N-I})$ equal to ~4, and that weak-base salts have *diff* values equal to ~3. These are helpful numbers to keep in mind when predicting the values of $\log P^I$ when $\log P^N$ is known.

Scherrer identified the conditions where *diff* 3–4 may be transgressed: (1) if the drug has several polar groups or a large polar surface over which charge can be delocalized, then smaller values of *diff* are observed; (2) hydroxyl groups adjacent to amines or carboxylic groups stabilize ion pairs, leading to lower *diff* values; and (3) steric hindrance to solvation leads to higher values of *diff*, as seen with tertiary amines, compared to primary ones [280,281].

4.8 ION PAIR ABSORPTION OF IONIZED DRUGS— FACT OR FICTION?

A review article with this title appeared in 1983 [369]. It's an old question, one not fully resolved: What does the charged-species partitioning seen in octanol–water systems have to do with biological systems? If getting to the receptor site involves passing through many lipid membranes, and if the pH partition hypothesis is to hold, the answer to the question is a resounding "Nothing." If the active site is in the outer leaflet of the apical membrane and the drug is orally introduced, or if ocular or skin absorption is considered [372,373], the answer is "Maybe something." We will return to this question in several instances in the next sections, for its answer warrants serious consideration.

4.9   MICRO-log P

We considered micro-pK_a values in Section 3.6. A parallel concept applies to partition coefficients (of multiprotic molecules); namely, if an ionizable substance of a particular stoichiometric composition can exist in different structural forms, then it is possible for each form to have a different micro-log P [224,243,273,275]. When log P is determined by the potentiometric method (below), the constant determined is the macro-log P. Other log P methods may also determine only the macroscopic constant.

Niflumic acid, which has two pK_a values, was studied both pH-metrically and spectroscopically using the shake-flask method [224]. The monoprotonated species can exist in two forms: (1) zwitterion, $XH^{\pm}$ and (2) ordinary (uncharged) ampholyte, XH^0. The ratio between the two forms (tautomeric ratio) was measured spectroscopically to be 17.4. On assuming that a negligible amount of zwitterion $XH^{\pm}$ partitions into octanol, the calculated micro-log P for XH^0 was 5.1, quite a bit higher than the macro-log P 3.9 determined pH-metrically in 0.15 M NaCl. It is noteworthy that the distribution coefficient D is the same regardless of whether the species are described with microconstants or macroconstants [275].

4.10   HPLC METHODS

HPLC log P techniques, first described by Mirrlees et al. [374] and Unger et al., [375], are probably the most frequently used methods for determining log P. The directly measured retention parameters are hydrophobicity indices, and need to be converted to a log P scale through the use of standards. The newest variants, breadths of scope, and limitations have been described in the literature [292–298]. A commercial automated HPLC system based on an extension of the approach described by Slater et al. [150] has just introduced by Sirius (www.sirius-analytical.com).

4.11   IAM CHROMATOGRAPHY

A very promising method, immobilized artificial membrane (IAM) chromatography, was developed by Pidgeon and co-workers [299–304,307], where silica resin was modified by covalent attachment of phospholipid-like groups to the surface. The retention parameters mimic the partitioning of drugs into phospholipid bilayers. The topic has been widely reviewed [47,298,307,309–311].

4.12   LIPOSOME CHROMATOGRAPHY

A method where phospholipids are entrapped in the pores of resin beads, in the forms of multilamellar vesicles, has been described [313–319,376]. In some ways, the idea is similar to that of IAM chromatography, even though the resin is modified differently. The retention indices correlate very well with the partition coefficients measured in liposome–water systems (described below).

4.13   OTHER CHROMATOGRAPHIC METHODS

Capillary electrophoresis (CE) (see Section 3.5) has been used to determine partition coefficients [320–322]. Lipid vesicles or micelles are added to the buffer whose pH is adjusted to different values. Since drug molecules partition to a different extent as a function of pH, the analysis of mobility vs pH data yields log *P* values.

Centrifugal partition chromatography (CPC) has been used to characterize the partitioning behavior of hydrophilic molecules, where log *D* values as low as −3 can be obtained [371,377–379]. It is not as popular a method as it used to be, apparently due to instrumental challenges. Cyclic voltammetry (CV) has become the new method used to get access to very low log *D* values, with partition coefficients reported as low as −9.8 [261,269,362].

4.14   pH–METRIC log *P* METHOD

In 1952, Dyrssen (using a radiometer titrator) performed the first dual-phase titrations to determine oil–water partition coefficients [324]. In a series of papers on solvent extraction of metal complexes, he and co-workers [324–331] measured neutral and ion pair log *P* of compounds, studied dimerization reactions of dialkylphosphates in aqueous as well as chloroform solutions, used log *D*/pH plots, and derived a method for deducing the pK_a of water-insoluble molecules from knowledge of their log *P*, later called the *PDP* method [112]. In 1963, Brändström [332], using a pH-stat titrator, applied the log *P* methods to pharmaceutical problems. In the mid-1970s, the technique was "reborn." Seiler described a method where the pK_a and log *P* were determined simultaneously from a single titration [250]. At about the same time, working independently, Koreman and Gur'ev [333], Kaufman et al. [334], and Johansson and Gustavii [335,336] published in this area. Gur'ev and co-workers continued to apply the method, but their work was not well known outside of Russian literature [337–343]. Clarke and others [344,345,350,351] presented a comprehensive treatment of the technique, and applied it to mono-, di- and triprotic substances. Numerical differentiation and matrix algebra were used to solve a number of simultaneous equations. Both graphical and refinement procedures for dealing with ion pair formation were devised. A dual-phase microtitration system has been described [361]. The rigorous development of the pH-metric method continued in a commercial setting by Avdeef and colleagues [25,112,149–151,153,161,162,224,225,275,346–349,352,357,362].

The pH-metric technique consists of two linked titrations. Typically, a pre-acidified 100–500 μM solution of a weak acid is titrated with standardized 0.5 M KOH to some appropriately high pH; octanol (or any other useful organic partition solvent that is immiscible with water) is then added (in low relative amounts for lipophilic molecules and high amounts for hydrophilic molecules), and the dual-solvent mixture is titrated with standardized 0.5 M HCl back to the starting pH. After each titrant addition, pH is measured. If the weak acid partitions into the octanol phase, the two assays show nonoverlapping titration curves. The greatest divergence between the two curves occurs in the buffer region. Since the pK_a is approximately equal to the pH at the midbuffer inflection point, the two-part assay yields two constants: pK_a and p_oK_a, where p_oK_a is the apparent constant derived from the octanol-containing segment of data. A large difference between pK_a and p_oK_a indicates a large value of $\log P$.

Bjerrum analysis (Section 3.3.1) is used for initial processing of the titration data. Figure 4.7a shows the Bjerrum plots of the two segments of the titration of a weak acid, phenobarbital [150]. The solid curve corresponds to the octanol-free segment, and the dotted curve corresponds to the curve obtained from the octanol-containing data, where r, the octanol–water volume ratio, is 1 in the example. As said before (Sec. 3.3.1), the pK_a and p_oK_a may be read off the curve at half-integral values of $\bar{n}_H$. From the difference between pK_a and p_oK_a, one obtains [347]

$$P_{HA} = \frac{10^{+(p_oK_a - pK_a)} - 1}{r} \tag{4.12}$$

Figure 4.7b shows an example of a weak base, diacetylmorphine (heroin) [151]. The partition coefficient for the weak base is derived from

$$P_B = \frac{10^{-(p_oK_a - pK_a)} - 1}{r} \tag{4.13}$$

If the two phases are equal in volume (1 : 1) and the substance is lipophilic, a very simple relationship can be applied to determine $\log P$;

$$\log P_{HA} \approx (p_oK_a^{1:1} - pK_a) \qquad (\log P_B \approx -(p_oK_a^{1:1} - pK_a)) \tag{4.14}$$

Note that for a weak acid, the octanol causes the Bjerrum curve to shift in the direction of higher pH, whereas for a weak base, octanol causes the shift to lower values of pH. Equation (4.14) may be applied to the molecules in Fig. 4.7, and $\log P$ deduced from the shifts in the curves.

For diprotic molecules, 12 different characteristic shift patterns have been identified for cases where two species may partition simultaneously into the lipid phase [347]. Three of these cases are shown in Fig. 4.8, picking familiar drug substances as examples. Once the approximate constants are obtained from Bjerrum analysis,

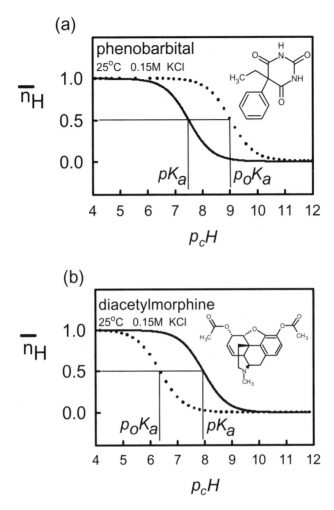

Figure 4.7 Octanol–water Bjerrum plots for a monoprotic (a) acid and (b) base. The volumes of octanol and water are equal, so that the difference between the apparent pK_a and the true pK_a is about equal to the partition coefficient. [Avdeef, A., *Curr. Topics Med. Chem.*, **1**, 277–351 (2001). Reproduced with permission from Bentham Science Publishers, Ltd.]

they may be further refined by a weighted [117] nonlinear least-squares procedure [153].

The pH-metric procedure has been validated against the standard shake-flask method [150,357], and many studies using it have been reported [56,149–151,153,161,162,224,225,229,246,250,268,269,275,276,280,281,324–363]. Determinations of values of log *P* as low as −2 and as high as +8 have been documented [161,162,352]. The published literature clearly indicates that the Dyrssen technique is a reliable, versatile, dynamic, and accurate method for measuring log *P*. It may lack the speed of HPLC methods, and it cannot go as low in log *P* as the CV

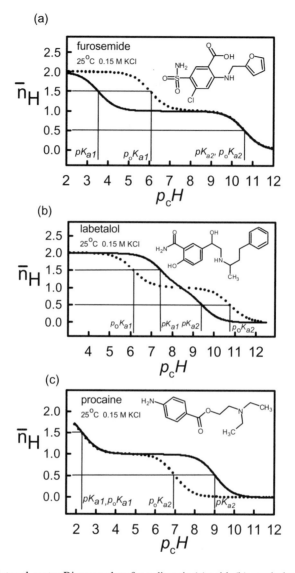

Figure 4.8 Octanol–water Bjerrum plots for a diprotic (a) acid, (b) ampholyte, and (c) base. The volumes of octanol and water are equal, so that the difference between the apparent pK_a and the true pK_a is about equal to the partition coefficient. [Avdeef, A., *Curr. Topics Med. Chem.*, **1**, 277–351 (2001). Reproduced with permission from Bentham Science Publishers, Ltd.]

method, but all in all, it is well positioned to replace the shake-flask procedure as the primary validation method for ionizable molecules. What keeps it from being the "gold standard," its Achilles' heel, is that the sample molecules must be ionizable and have a pK_a in the measurable pH range.

4.15 HIGH-THROUGHPUT log P METHODS

Several efforts have been made to increase the throughput of the traditional log P methods by scaling down to a 96–well microtiter plate format [294]. The generic fast gradient HPLC methods look promising (see Section 4.10). The commercial HPLC system (see Section 4.10) shows promise of industrywide standardization. Immobilized liposome and IAM chromatography methods can also be fast (see Sections 4.11 and 4.12) All the chromatography methods suffer from being essentially series-based assays.

Parallel methods using scanning 96/384-well plate UV spectrophotometers are inherently faster [292]. They will become 50-fold faster with the imminent introduction of diode-array plate readers.

4.16 OCTANOL–WATER log P^N, log P^I, AND log $D_{7.4}$ "GOLD STANDARD" FOR DRUG MOLECULES

About 300 values of octanol–water log P^N, log P^I, and log $D_{7.4}$ of drugs and some agrochemicals are listed in Table 4.1. These have been critically selected to represent high-quality results. Most of these constants have been determined at Sirius or pION since 1991, with many personally determined by the author.

TABLE 4.1 Critically Selected Experimental log P^N, log P^I, and log $D_{7.4}$ of Drug Molecules[a]

Compound	log P^N	log $P^I(+)$	log $P^I(-)$	log $D_{7.4}$	Ref.
1-Benzylimidazole	1.60	—	—	—	112, p. 70
2,4-Dichlorophenoxyacetic acid	2.78	—	−0.87	−0.82	161, p. 63
2-Aminobenzoic acid	1.26	—	—	−1.31	161, p. 8
3,4-Dichlorophenol	3.39	—	—	—	150
3,5-Dichlorophenol	3.63	—	—	3.56	150
3-Aminobenzoic acid	0.34	−0.93	—	−2.38	161, p. 25
3-Bromoquinoline	2.91	—	—	2.91	150
3-Chlorophenol	2.57	—	—	2.56	150
4-Aminobenzoic acid	0.86	−0.40	—	−1.77	161, p. 105
4-Butoxyphenol	2.87	—	—	—	150
4-Chlorophenol	2.45	—	—	—	150
4-Ethoxyphenol	1.81	—	—	—	150
4-Iodophenol	2.90	—	—	—	150
4-Methoxyphenol	1.41	—	—	—	150
4-Methylumbilleferyl-β-D-glucuronide	−0.39	—	—	—	151

TABLE 4.1   (*Continued*)

Compound	$\log P^N$	$\log P^I(+)$	$\log P^I(-)$	$\log D_{7.4}$	Ref.
4-Pentoxyphenol	3.26	—	—	—	150
4-Phenylbutylamine	2.39	−0.45	—	−0.62	149
4-Propoxyphenol	2.31	—	—	—	150
5-Phenylvaleric acid	2.92	—	−0.95	1.69	149
6-Acetylmorphine	1.55	−0.42	—	0.61	151
Acebutolol	2.02	−0.50	—	−0.09	362
Acetaminophen	0.34	—	—	0.34	357
Acetic acid	−0.30	—	—	−2.88	—[c]
Acetophenone	1.58	—	—	1.58	296
Acetylsalicylic acid	0.90	—	—	−2.25	161, p. 167
Alprazolam	2.61	—	—	2.08	550
Alprenolol	2.99	0.21	—	0.86	362
Aminopyrine (aminophenazone)	0.85	—	—	0.63	357
Amiodarone	7.80	4.02	—	6.10	—[b]
Amitriptyline	4.62	0.16	—	2.80	—[b]
Amitrole	−0.97	—	—	—	265
Amlodipine	3.74	1.09	—	2.25	—[c]
Amoxicillin	−1.71	−1.22	−1.56	−2.56	56
Ampicillin	−2.17	−1.15	−1.31	−1.85	162, p. 133
Amylobarbitone	2.01	—	—	—	150
Antipyrine (phenazone)	0.56	—	—	0.56	56
Ascorbic acid	−1.85	—	—	−4.82	357
Atenolol	0.22	—	—	−2.01	362
Atropine	1.89	−1.99	—	−0.66	—[b]
Azithromycin	3.87	0.23	—	0.33	—[b]
Bentazone	2.83	—	—	—	265
Benzocaine	1.89	—	—	1.90	162, p. 25
Benzoic acid	1.96	—	—	−1.25	150
Betamethasone	2.06	—	—	2.10	550
Bifonazole	4.77	—	—	4.77	296
Bisoprolol	2.15	−1.22	—	—	362
Bromazepam	1.65	—	—	1.65	296
Bromocriptine	4.20	—	—	4.20	509
Bumetanide	4.06	—	—	−0.11	561
Buprenorphine	4.82	0.09	—	3.75	151
Bupropion	3.21	—	—	2.61	561
Buspirone	2.78	—	—	—	357
Butobarbitone	1.58	—	—	—	150
Caffeine	−0.07	—	—	−0.07	296
Captopril	1.02	—	—	−2.00	561
Carazolol	3.73	0.77	—	1.58	362
Carbamazepine	2.45	—	—	2.45	56
Carbomycin A	3.04	—	—	—	358
Carbomycin B	3.52	—	—	—	358

TABLE 4.1 (*Continued*)

Compound	log P^N	log $P^I(+)$	log $P^I(-)$	log $D_{7.4}$	Ref.
Carvedilol	4.14	1.95	—	3.53	362
Cefadroxil	−0.09	—	—	−1.77	561
Cefalexin	0.65	—	—	−1.00	561
Cefixime	0.11	—	—	−0.79	550
Cefoxitin	1.55	—	—	−0.60	550
Celiprolol	1.92	—	—	−0.16	150
Chlorambucil	3.70	—	—	0.61	550
Chloramphenicol	1.14	—	—	1.14	296
Chloroquine	4.69	—	—	0.89	550
Chlorothiazide	−0.24	—	—	−0.05	561
Chlorpheniramine	3.39	—	—	1.41	296
Chlorpromazine	5.40	1.67	—	3.45	161, p. 163
Chlorprothixene	6.03	—	—	3.71	550
Chlorsulfuron	1.79	—	—	—	265
Chlortalidone	−0.74	—	—	0.78	550
Cimetidine	0.48	—	—	0.34	—[b]
Ciprofloxacin	−1.08	−1.69	—	−1.12	—[b]
Citric acid	−1.64	—	—	—	161, p. 168
Clarithromycin	3.16	—	—	—	358
Clofibrate	3.65	—	—	3.39	561
Clonazepam	3.02	—	—	2.45	550
Clonidine	1.57	—	—	0.62	296
Clopyralid	1.07	—	—	−2.95	265
Clotrimazole	5.20	—	—	5.20	296
Clozapine	4.10	—	—	3.13	509
Cocaine	3.01	—	—	1.07	550
Codeine	1.19	—	—	0.22	151
Coumarin	1.39	—	—	1.44	550
Cromolyn	1.95	—	—	-1.15	561
Dapsone	0.94	—	—	0.68	550
Debrisoquine	0.85	−0.87	—	−0.87	161, p. 119
Deprenyl	2.90	−0.95	—	2.49	162, p. 26
Desipramine	3.79	0.34	—	1.38	—[b]
Desmycarosyl carbomycin A	0.30	—	—	—	358
Desmycosin	1.00	—	—	—	358
Diacetylmorphine	1.59	—	—	—	151
Diclofenac	4.51	—	0.68	1.30	162, p. 146
Diethylstilbestrol	5.07	—	—	5.07	296
Diflunisal	4.32	—	—	0.37	550
Diltiazem	2.89	—	—	2.16	—[b]
Diphenhydramine	3.18	−0.52	—	1.39	—[b]
Disopyramide	2.37	—	—	−0.66	—[b]
Doxorubicin	0.65	—	—	−0.33	550
Doxycycline	0.42	0.09	−0.34	0.23	—[b]
Enalaprilmaleate	0.16	−0.10	—	−1.75	—[b]

TABLE 4.1 (*Continued*)

Compound	$\log P^N$	$\log P^I(+)$	$\log P^I(-)$	$\log D_{7.4}$	Ref.
Enalaprilat	−0.13	−0.99	−1.07	−2.74	56
Ephedrine	1.13	−0.96	—	−0.77	162, p. 131
Ergonovine	1.67	−0.51	—	1.54	—[b]
Erythromycin	2.54	−0.43	—	1.14	—[b]
Erythromycylamine	3.00	—	—	—	358
Erythromycylamine-11,12-carbonate	2.92	—	—	—	358
Ethinylestradiol,17-α	3.42	—	1.29	3.42	—[b]
Ethirimol	2.22	—	—	—	265
Etofylline	−0.49	—	—	−0.27	550
Etoposide	1.97	—	—	1.82	561
Famotidine	−0.81	−0.54	—	−0.62	—[b]
Fenpropimorph	4.93	—	—	—	265
Flamprop	3.09	—	—	—	265
Fluazifop	3.18	—	—	—	265
Fluconazole	0.50	—	—	0.50	296
Flufenamic acid	5.56	—	1.77	2.45	—[b]
Flumazenil	1.64	—	—	1.21	561
Flumequine	1.72	—	—	0.65	161, p. 19
Fluocortolone	2.06	—	—	2.10	550
Flurbiprofen	3.99	—	—	0.91	—[b]
Fluvastatin	4.17	—	1.12	1.14	56
Fomesafen	3.00	—	—	—	265
Furosemide	2.56	—	—	−0.24	—[b]
Gabapentin	−1.25	—	—	−2.00	561
Griseofulvin	2.18	—	—	2.18	296
Guanabenz	3.02	—	—	1.40	561
Haloperidol	3.67	1.32	—	3.18	—[b]
Heptastigmine	4.82	—	—	0.17	550
Homidium bromide	−1.10	—	—	−1.10	291
Hydrochlorothiazide	−0.03	—	−1.59	−0.18	—[b]
Hydrocortisone-21-acetate	2.19	—	—	2.19	296
Hydroflumethiazide	0.54	—	—	0.31	550
Hydroxyzine	3.55	0.99	—	3.13	161, p. 146
Ibuprofen	4.13	—	−0.15	1.44	149
Imazapyr	0.22	—	—	—	265
Imazaquin	1.86	—	—	—	265
Imidacloprid	0.33	—	—	0.33	265
Imipramine	4.39	0.47	—	2.17	—[b]
Indomethacin	3.51	—	−2.00	0.68	—[b]
Ioxynil	3.43	—	—	—	265
Ketoconazole	4.34	—	—	3.83	561
Ketoprofen	3.16	—	-0.95	−0.11	—[b]
Ketorolac	1.265	—	—	−0.27	561
Labetalol	1.33	—	—	1.08	—[b]

TABLE 4.1   (*Continued*)

Compound	log P^N	log $P^I(+)$	log $P^I(-)$	log $D_{7.4}$	Ref.
Lasinavir	3.30	—	—	—	509
Leucine	−1.55	−1.58	−2.07	−1.77	56
Lidocaine	2.44	−0.52	—	1.72	149
Lorazepam	2.48	—	—	2.39	550
Lormetazepam	2.72	—	—	2.72	296
Maleic hydrazide	−0.56	—	—	—	265
Mebendazole	2.42	—	—	3.28	550
Mecoprop	3.21	—	—	—	265
Mefluidide	2.02	—	—	—	265
Meloxicam	3.43	−0.03	—	0.12	162, p. 112
Melphalan	−0.52	—	—	−2.00	561
Metergoline	4.75	—	—	3.50	550
Methotrexate	0.54	—	−0.92	−2.93	—[b]
Methylprednisolone	2.10	—	—	2.10	561
Methylthioinosine	0.09	—	—	0.09	296
Methysergide	1.95	—	—	2.13	550
Metipranolol	2.81	−0.26	—	0.55	362
Metoclopramide	2.34	—	—	0.41	550
Metolazone	4.10	—	—	4.10	509
Metoprolol	1.95	−1.10	—	−0.24	362
Metronidazole	−0.02	—	—	−0.02	296
Metsulfuron, methyl-	1.58	—	—	—	265
Morphine sulfate	0.89	−2.05	—	−0.06	151
Morphine-3β-D-glucuronide	−1.10	—	—	−1.12	151
Morphine-6β-D-glucuronide	−0.76	—	—	−0.79	151
Moxonidine	0.90	−0.20	—	—	385
N-Me-deramcylane iodide	−1.12	—	—	−1.12	291
N-Me-quinidine iodide	−1.31	—	—	−1.31	291
Nadolol	0.85	—	—	−1.43	362
Naloxone	2.23	—	—	1.09	550
Naphthalene	3.37	—	—	3.37	296
Naproxen	3.24	—	−0.22	0.09	—[b]
Nicotine	1.32	—	—	0.45	161, p. 36
Nifedipine	3.17	—	—	3.17	296
Niflumic acid	3.88	2.48	0.44	1.43	224
Nifuroxime	1.28	—	—	1.28	296
Nitrazepam	2.38	1.21	0.64	2.38	161, p. 169
Nitrendipine	3.59	—	—	3.50	550
Nitrofurantoin	−0.54	—	—	−0.26	550
Nitrofurazone	0.23	—	—	0.23	296
N-Methylaniline	1.65	—	—	—	150
N-Methyl-D-glucamine	−1.31	—	—	−3.62	225
Norcodeine	0.69	—	—	−1.26	151
Nordiazepam	3.15	—	—	3.01	550
Norfloxacin	1.49	—	—	−0.46	550

TABLE 4.1 (*Continued*)

Compound	$\log P^N$	$\log P^I(+)$	$\log P^I(-)$	$\log D_{7.4}$	Ref.
Normorphine	−0.17	—	—	−1.56	151
Nortriptyline	4.39	1.17	—	1.79	—[b]
Ofloxacin	−0.41	—	−0.84	−0.34	161, p. 9
Oleandomycin	1.69	—	—	—	358
Omeprazole	1.80	—	—	2.15	550
Oxprenolol	2.51	−0.13	—	0.18	362
Papaverine	2.95	−0.22	—	2.89	162, p. 30
Penbutolol	4.62	1.32	—	2.06	362
Penicillin V	2.09	—	—	−0.62	561
Pentachlorophenol	5.12	—	—	—	265
Pentamidine	2.08	—	—	−0.19	550
Pentobarbitone	2.08	—	—	—	150
Pentoxifylline	0.38	—	—	0.33	550
Pericyazine	3.65	—	—	—	150
p-F-Deprenyl	3.06	−0.58	—	2.70	162, p. 28
Phenazopyridine	3.31	1.41	—	3.31	—[b]
Phenobarbital	1.53	—	—	1.51	150
Phenol	1.48	—	—	—	150
Phenylalanine	−1.38	−1.41	—	−1.37	161, p. 116
Phenylbutazone	3.47	—	—	0.47	550
Phenytoin	2.24	—	—	2.17	—[b]
Phe-Phe	−0.63	−0.05	—	−0.98	162, p. 6
Phe-Phe-Phe	0.02	0.82	−0.55	−0.29	162, p. 12
Pilocarpine	0.20	—	—	—	357
Pindolol	1.83	−1.32	—	−0.36	362
Pirimicarb	1.71	—	—	—	265
Pirimiphos, methyl-	3.27	—	—	—	265
Piroxicam	1.98	0.96	−0.38	0.00	162, p. 110
Prazosin	2.16	—	—	1.88	561
Prednisolone	1.69	—	—	1.83	550
Prednisone	1.56	—	—	1.44	550
Primaquine	3.00	1.14	—	1.17	—[b]
Probenecid	3.70	—	−0.52	−0.23	—[b]
Procainamide	1.23	—	—	−0.36	550
Procaine	2.14	−0.81	—	0.43	149
Progesterone	3.48	—	—	3.48	561
Promethazine	4.05	—	—	2.44	—[b]
Propamocarb	1.12	—	—	—	265
Propantheline bromide	−1.07	—	—	−1.07	291
Propoxyphene	4.37	—	—	2.60	—[b]
Propranolol	3.48	0.78	—	1.41	362
Proquazone	3.13	—	—	3.21	550
Prostaglandin E_1	3.20	—	−0.33	0.78	225
Prostaglandin E_2	2.90	—	−0.54	0.41	225
Proxyphylline	−0.14	—	—	−0.07	550

TABLE 4.1   (*Continued*)

Compound	log P^N	log $P^I(+)$	log $P^I(-)$	log $D_{7.4}$	Ref.
Pyridoxine	−0.50	−1.33	—	−0.51	161, p. 19
Pyrimethamine	2.87	—	—	2.44	550
Quinalbarbitone	2.39	—	—	—	150
Quinidine	3.44	—	—	2.41	550
Quinine	3.50	0.88	—	2.19	162, p. 128
Quinmerac	0.78	—	—	—	265
Quinoline	2.15	—	—	2.15	150
Ranitidine	1.28	—	—	−0.53	550
Repromicin	2.49	—	—	—	358
Rifabutine	4.55	2.80	—	—	385
Rifampin	0.49	—	—	0.98	550
Rivastigmine	2.10	—	—	—	509
Rosaramicin	2.19	—	—	—	358
Roxithromycin	3.79	1.02	—	1.92	162, p. 107
Rufinamide	0.90	—	—	—	509
Saccharin	0.91	—	—	−1.00	550
Salicylic acid	2.19	—	—	−1.68	—[c]
Serotonin	0.53	−1.66	—	−2.17	—[c]
Sethoxydim	4.38	—	—	—	265
Sotalol	−0.47	−1.43	—	−1.19	162, p. 167
Sulfadiazine	−0.12	—	—	−0.60	550
Sulfamethazine	0.89	—	—	—	150
Sulfasalazine	3.61	—	0.14	0.08	—[b]
Sulfinpyrazone	2.32	—	—	−0.07	550
Sulfisoxazole	1.01	—	—	−0.56	550
Sulindac	3.60	—	—	0.12	550
Suprofen	2.42	—	—	−0.30	550
Tacrine	3.32	—	—	0.34	550
Tamoxifen	5.26	−2.96	—	4.15	—[b]
Terazosin	2.29	—	—	1.14	561
Terbutaline	−0.08	−1.97	−2.05	−1.35	162, p. 36
Terfenadine	5.52	1.77	—	3.61	—[b]
Tetracaine	3.51	0.22	—	2.29	149
Theophylline	0.00	—	—	0.00	162, p. 128
Thiabendazole	1.94	—	—	1.94	265
Thiamphenicol	−0.27	—	—	−0.27	296
Tilmicosin	3.80	—	—	—	358
Timolol	2.12	−0.94	—	0.03	362
Tolnaftate	5.40	—	—	5.40	296
Tralkoxydim	4.46	—	—	—	265
Tranexamic acid	−1.87	—	—	−3.00	561
Trazodone	1.66	—	—	2.54	296
Triazamate acid	1.62	—	—	—	265
Trimethoprim	0.83	−0.88	—	0.63	—[b]
Trovafloxacin	0.15	−0.65	—	0.07	—[b]

TABLE 4.1 *(Continued)*

Compound	$\log P^N$	$\log P^I(+)$	$\log P^I(-)$	$\log D_{7.4}$	Ref.
Trp-Phe	−0.28	0.33	−2.44	−0.50	162, p. 2
Trp-Trp	−0.10	0.49	−0.99	−0.40	162, p. 8
Tryptophan	−0.77	−0.55	−1.57	−0.77	162, p. 10
Tylosin	1.63	—	—	—	358
Valsartan	3.90	—	—	—	509
Verapamil	4.33	0.71	—	2.51	—[b]
Warfarin	3.54	—	0.04	1.12	149

[a] Measurements at 25°C, 0.15 M ionic strength.
[b] pION.
[c] Sirius Analytical Instruments.

CHAPTER 5

PARTITIONING INTO LIPOSOMES

The octanol–water partition model has several limitations; notably, it is not very "biological." The alternative use of liposomes (which are vesicles with walls made of a phospholipid bilayer) has become more widespread [149,162,275, 380–444]. Also, liposomes contain the main ingredients found in all biological membranes.

5.1 TETRAD OF EQUILIBRIA AND SURFACE ION PAIRING (SIP)

Figure 5.1 shows a tetrad of equilibrium reactions related to the partitioning of a drug between an aqueous environment and that of the bilayer formed from phospholipids. (Only half of the bilayer is shown in Fig. 5.1.) By now, these reaction types might be quite familiar to the reader. The subscript "mem" designates the partitioning medium to be that of a vesicle formed from a phospholipid bilayer. Equations (4.1)–(4.4) apply. The pK_a^{mem} in Fig. 5.1 refers to the "membrane" pK_a. Its meaning is similar to that of pK_a^{oct}; when the concentrations of the uncharged and the charged species *in the membrane phase* are equal, the *aqueous* pH at that point defines pK_a^{mem}, which is described for a weak base as

$$BH^+_{(mem)} \rightleftarrows B_{(mem)} + H^+ \qquad K_a^{mem} = \frac{[B_{(mem)}][H^+]}{[BH^+_{(mem)}]} \qquad (5.1)$$

Absorption and Drug Development: Solubility, Permeability, and Charge State. By Alex Avdeef
ISBN 0-471-423653. Copyright © 2003 John Wiley & Sons, Inc.

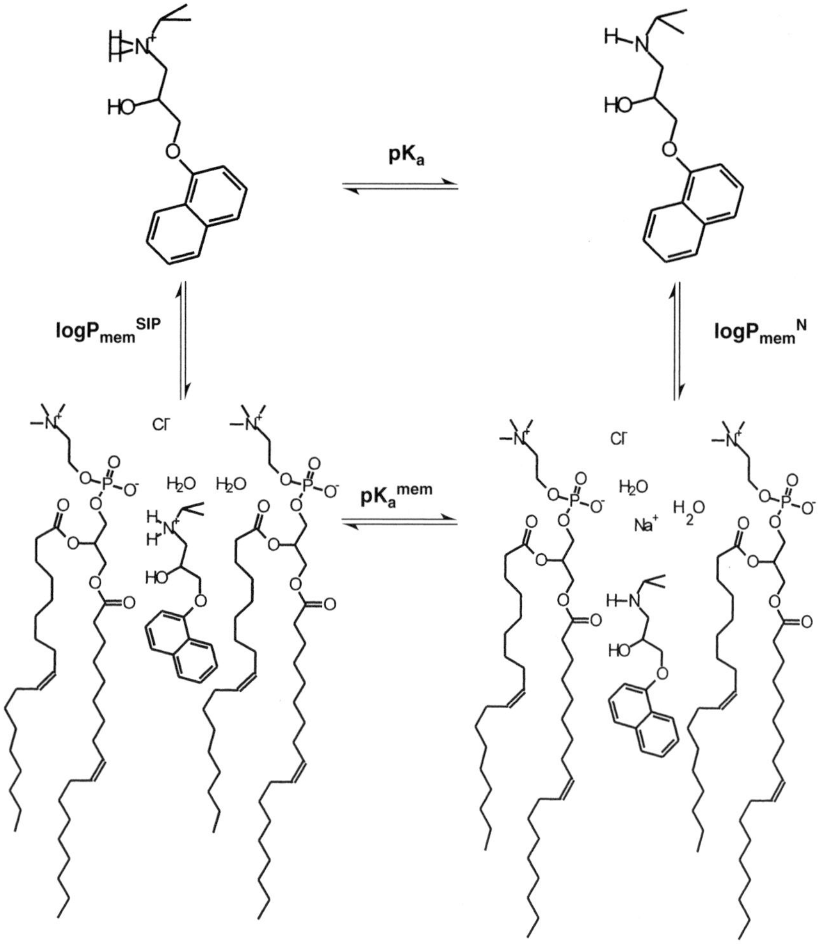

Figure 5.1 Phospholipid membrane–water tetrad equilibria. Only half of a bilayer is shown. [Avdeef, A., *Curr. Topics Med. Chem.*, **1**, 277–351 (2001). Reproduced with permission from Bentham Science Publishers, Ltd.]

The salt dependence of constants discussed in Section 4.2 also applies to the pK_a^{mem} and $\log P_{mem}^{SIP}$ constants. Although they are *conditional*, the dependence on ionic strength is subtle [433,442]. It is thought that when a charged drug migrates into the lipid environment of a liposome, the counterion that at first accompanies it may be exchanged with the zwitterionic phosphatidylcholine head groups, as suggested in Fig. 5.1. As the nature of the ion pair may be different with liposome partitioning, the term *surface ion pair* (SIP) is used to denote it. We use the term $diff_{mem}$ to designate the difference between the neutral species partitioning and the surface ion pair partitioning [see Eq. (4.6)].

5.2 DATABASES

There are no convenient databases for liposome log P values. Most measured quantities need to be ferreted from original publications [149,162,376,381–387,443]. The handbook edited by Cevc [380] is a comprehensive collection of properties of phospholipids, including extensive compilations of structural data from X-ray crystallographic studies. Lipid-type distributions in various biological membranes have been reported [380,388,433].

5.3 LOCATION OF DRUGS PARTITIONED INTO BILAYERS

Based on the observed nuclear Overhauser effect in a $^{31}P\{^{1}H\}$ nuclear magnetic resonance (NMR) study of egg phosphatidylcholine (eggPC) bilayers, Yeagle et al. [399] concluded that the N-methyl hydrogens were in close proximity to phosphate oxygens in neighboring phospholipids, suggesting that the surface of the bilayer was a "shell" of interlocking (intermolecular) electrostatic associations. Added cholesterol bound below the polar head groups, and did not interact with them directly. However, its presence indirectly broke up some of the surface structure, making the surface more polar and open to hydration.

Boulanger et al. [420,421] studied the interactions of the local anesthetics procaine and tetracaine with eggPC multilamellar vesicles (MLV, 52–650 mM), as a function of pH, using deuterium nmr as a structural probe. They proposed a three-site model, similar to that in Fig. 5.1, except that the membrane-bound species (both charged and uncharged) had two different locations, one a weakly bound surface site (occupied at pH 5.5), and the other a strongly bound deeper site (occupied at pH 9.5). Membrane partition coefficients were estimated for both sites. Westman et al. [422] further elaborated the model by applying the Gouy–Chapman theory. When a charged drug partitions into the bilayer, a Cl^- is likely bound to the surface, to maintain charge neutrality. They found unexpected low values of $diff_{mem}$ of 0.77 for tetracaine and 1.64 for procaine (see Section 4.7). Kelusky and Smith [423], also using deuterium NMR, proposed that at pH 5.5, there was an electrostatic bond formed between the protonated drug and the phosphate groups, ($\equiv P-O^- \cdots {}^+H_3N-$), and a hydrogen bond formed between the aminobenzene proton and the acyl carbonyl oxygen. At pH 9.5, the ionic bond breaks as the secondary amine moves deeper into the interior of the bilayer; however, the aminobenzene H bond, ($=CO \cdots H_2N-$), continues to be an anchoring point.

Bäuerle and Seelig [395] studied the structural aspects of amlodipine (weak base, primary amine pK_a 9.26 [162]) and nimodipine (nonionizable) binding to phospholipid bilayers, using NMR, microcalorimetry, and zeta-potential measurements. They were able to see evidence of interactions of amlodipine with the cis double bond in the acyl chains. They saw no clear evidence for ($\equiv P-O^- \cdots {}^+H_3N-$) electrostatic interactions.

Herbette and co-workers [425–428,445] studied the structures of drugs bound to liposomes using a low-angle X-ray diffraction technique. Although the structural details were coarse, it was apparent that different drugs position in different locations of the bilayer. For example, amlodipine is charged when it partitions into a bilayer at physiological pH; the aromatic dihydropyridine ring is buried in the vicinity of the carbonyl groups of the acyl chains, while the $-NH_3^+$ end points toward the aqueous phase, with the positive charge located near the phosphate negative-charge oxygen atoms [426–428]. A much more lipophilic molecule, amiodarone (weak base with pK_a 9.1 [pION]), positioned itself closer to the center of the hydrocarbon interior [425].

5.4 THERMODYNAMICS OF PARTITIONING: ENTROPY- OR ENTHALPY-DRIVEN?

Davis et al. [394] studied the thermodynamics of the partitioning process of substituted phenols and anisoles in octanol, cyclohexane, and dimyristoylphosphatidylcholine (DMPC) at 22°C (which is *below* the gel–liquid transition temperature of DMPC). Table 5.1 shows the results for 4-methylphenol. The phenol partitioned into the lipid phases in the order DMPC > octanol > cyclohexane, as indicated by ΔG_{tr}. Thus, the free energy of transfer into DMPC was greater than into octanol or cyclohexane. Partitioning was generally-entropy driven, but the components of the free energy of transfer were greatly different in the three lipid systems (Table 5.1). Octanol was the only lipid to have an exothermic heat of transfer (negative enthalpy), due to H-bond stabilization of the transferred solute, not found in cyclohexane. Although ΔH_{tr} in the DMPC system is a high positive number (endothermic), not favoring partitioning into the lipid phase, the entropy increase (+114.1 eu) was even greater, more than enough to offset the enthalpy destabilization, to end up an entropy-driven process. The large ΔH_{tr} and ΔS_{tr} terms in the DMPC system are due to the disruption of the ordered gel structure, found below the transition temperature.

The partition of lipophilic drugs into lipid phases is often believed to be entropy-driven, a hydrophobic effect. Bäuerle and Seelig [395] studied the thermodynamics of amlodipine and nimodipine binding to phospholipid bilayers (*above* the transition temperature) using highly sensitive microcalorimetry. The partitioning of the drugs into the lipid bilayer was enthalpy-driven, with ΔH_{tr} -38.5 kJ mol^{-1} bound amlodipine. The entropy of transfer is *negative*, contrary to the usual interpretation

TABLE 5.1 Energy of Transfer (kJ/mol) into Lipid Phase for 4-Methylphenol

Component	DMPC	Octanol	Cyclohexane
ΔH_{tr}	+92.0	−7.3	+18.6
$T\Delta S_{tr}$	+114.1	+9.2	+22.2
ΔG_{tr}	−22.1	−16.5	−3.6

of the hydrophobic effect. Thomas and Seelig [397] found the partitioning of the Ca^{2+} antagonist, flunarizine (a weak base), also to be predominantly enthalpy-driven, with $\Delta H_{tr} -22.1 \text{ kJ mol}^{-1}$, again at odds with the established ideas of entropy-driven partitioning of drugs. The same surprise was found for the partitioning of paclitaxil [398]. These observations thus appear to suggest that drugs partition into membrane phases because they are lipophilic, and not because they are hydrophobic! This needs to be investigated more extensively, using microcalorimetry.

5.5   ELECTROSTATIC AND HYDROGEN BONDING IN A LOW-DIELECTRIC MEDIUM

Section 3.3.4 pointed out that cosolvents alter aqueous ionization constants; as the dielectric constant of the mixture decreases, acids appear to have higher pK_a values and bases appear (to a lesser extent than acids) to have lower values. A lower dielectric constant implies that the force between charged species increases, according to Coulomb's law. The equilibrium reaction in Eq. (3.1) is shifted to the left in a decreased dielectric medium, which is the same as saying that pK_a increases. Numerous studies indicate that the dielectric constant in the region of the polar head groups of phospholipids is ~32, the same as the value of methanol. [381,446–453] Table 5.2 summarizes many of the results.

These and other values [381,406] allow us to depict the dielectric spectrum of a bilayer, shown in Fig. 5.2. Given this view, one can think of the phospholipid bilayer as a dielectric *microlamellar* structure; as a solute molecule positions itself closer to the center of the hydrocarbon region, it experiences lower dielectric field (Fig. 5.2). At the very core, the value is near that of vacuum. A diatomic molecule of Na^+Cl^- in vacuum would require more energy to separate into two distinct ions than that required to break a single carbon–carbon bond!

This means that ions will not easily enter the interior of bilayers without first forming contact ion pairs. It is reasonable to imagine that simple drug–counterion pairs, such as $(BH^+ \cdots Cl^-)$ will undergo exchange of charge pairs (BH^+ for Na^+ originally in the vicinity of $\equiv PO^-$) on entering the head-group region, to form, for example, $(\equiv PO^- \cdots {}^+HB)$, with the release of Na^+ and Cl^-, as depicted in Fig. 5.1. We called such an imagined pairing SIP in Section 5.1 [149].

An interesting hypothesis may be put forward. The interfacial pK_a^{mem} (Fig. 5.1) that a solute exhibits depends on the dielectric environment of its location in the bilayer. Simple isotropic water-miscible solvents may be used to approximate pK_a^{mem}. Pure methanol (ϵ 32), may do well for the bilayer zone containing the phosphate groups; pure 1,4-dioxane (ϵ 2) may mimic some of the dielectric properties of the hydrocarbon region. It appears that p_sK_a values of several weak bases, when extrapolated to 100% cosolvent, do approximate pK_a^{mem} values [119,162,172]. Fernández and Fromherz made favorable comparisons using dioxane [448]. This idea is of considerable practical use, and has been largely neglected in the literature.

**TABLE 5.2 Dielectric Constants of Water–Lipid Interfaces
(Expanded from Ref. 453)**[a]

Type	Site	Method	ϵ	Ref.
Unilamellar vesicles (PC, αT)[a]	Polar head/acyl core	Chemical reaction, αT-DPPH	26	446
Unilamellar vesicles PC	Polar head/acyl core	Fluorescence polarization (DSHA)	33	381
Unilamellar vesicles PC+10% cholesterol	Polar head/acyl core	Fluorescence polarization (DSHA)	40	381
Unilamellar vesicles PC+20% stearylamine	Polar head/acyl core	Fluorescence polarization (DSHA)	43	381
Unilamellar vesicles PC+20% cardiolipin	Polar head/acyl core	Fluorescence polarization (DSHA)	52	381
Unilamellar vesicles, PC	Hydrocarbon core	Fluorescence polarization (AS)	2	381
Multilamellar PC	Polar head/bulk water	Fluoresecence polarization (ANS)	32	447
Multilamellar PC	Polar head/acyl core	Fluorescence polarization (NnN'-DOC)	25	447
Unilamellar vesicles (PC, DPPC)	Polar head/acyl core	Fluorescence depolarization (DSHA)	32	450
Unilamellar vesicles (PC,αT)	Polar head/acyl core	Chemical reaction, αT-DPPH	29–36	453
GMO bilayers	Polar head/acyl core	Electrical time constant	30–37	451
Micelles (CTAB, SDS, Triton-X100)	Aqueous surface	Fluorescence (HC, AC)	32	448
Micelles (various types)	Aqueous surface	Fluorescence (p-CHO)	35–45	449
Micelles (SDES, SDS, STS)	Aqueous surface	Absorption wavelength maximum	29–33	452

[a]*Abbreviations*: αT = α-tocopherol, AC = aminocoumarin, ANS = 1-anilino-8-naphthalenesulfonic acid, CTAB = cetyltrimethylammonium bromide, DPPC = dipalmitoylphosphatidylcholine, DPPH = 1,1-diphenyl-2-picrylhydrazyl, DSHA = N-dansylhexadecylamine, GMO = glycerol monooleate, HC = hydrocoumarin, N,N'-DOC = N,N'-di(octadecyl)oxacarbocyanine, PC = phosphatidylcholine, p-CHO = pyrene caroboxaldehyde, SDES = sodium decyl sulfate, SDS = sodium dodecyl sulfate, STS = sodium tetradecyl sulfate.

The molecular view of the interactions of drug molecules with phospholipid bilayers, suggested graphically in Fig. 5.1, has (1) an electrostatic component of binding with the head groups, which depends on the dielectric constant; (2) a hydrogen bonding component, since the phospholipids are loaded with strong H-bond acceptors ready to interact with solutes having strong H-bond donor groups; and (3) a hydrophobic/lipophilic component. Interactions between drugs and bilayers are like that of a solute and a 'fuzzy, delocalized'' receptor with the microlamellar zones (Fig. 5.2) *electrostatic···H bond···hydrophobic*. It is useful to explore this idea, and we will do so below.

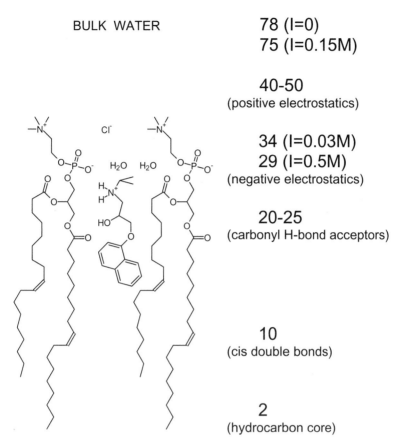

BULK WATER

78 (I=0)
75 (I=0.15M)

40-50
(positive electrostatics)

34 (I=0.03M)
29 (I=0.5M)
(negative electrostatics)

20-25
(carbonyl H-bond acceptors)

10
(cis double bonds)

2
(hydrocarbon core)

Figure 5.2 Approximate dielectric properties of a phospholipid bilayer, compiled from a number of sources, summarized in Table 5.2. [Avdeef, A., *Curr. Topics Med. Chem.*, **1**, 277–351 (2001). Reproduced with permission from Bentham Science Publishers, Ltd.]

5.6 WATER WIRES, H⁺/OH⁻ CURRENTS, AND THE PERMEABILITY OF AMINO ACIDS AND PEPTIDES

The stability of vesicular pH gradients (between the inner and outer aqueous solutions) depends on processes that can allow protons to permeate across phospholipid barriers. Phospholipid bilayers are thought not to be permeable to charged species (per the pH partition hypothesis). However, some studies suggest H⁺/OH⁻ permeability to be surprisingly high, as high as 10^{-4} cm/s, greatly exceeding that of about 10^{-12} cm/s for Na⁺ [409–419]. Biegel and Gould [409] rapidly changed the pH (acid pulse measurements) of a suspension of small unilamellar vesicles (SUVs; soybean PC) from the equilibrated pH 8.2 to the external pH 6.7, and monitored the rate of influx of H⁺ into the vesicles. (The pH inside of vesicles can be measured by fluorescent probes [409,419].) It took several minutes for the internal pH to drop from pH 8.2 to 7.4. This time was long because charge transfer led to buildup of a potential difference across the membrane (Donnan potential), which was slow to dissipate. The time was dropped to about 300 ms in the presence of a K^+ ionophore, valinomycin, an antiporter type of effect. The proton ionophore, bis(hexafluoroacetonyl)acetone, dropped the reequilibration time down to <1 ms.

Discussions of the possible mechanisms of H^+ transport ensued. It was pointed out that the solubility of water in n-alkanes was high enough to suggest the participation of membrane-dissolved water in the transport mechanism. Biegel and Gould [409] predicted that the SUVs used in their study could have 30–40 H_2O molecules dissolved in the bilayer hydrocarbon (HC) core. Meier et al. [411] measured the concentration of water in the HC interior of bilayes to be about 100 mM. Two short reviews discussed proton conductance: Nagle [412] defended the position that "water wires" inside the HC core can explain H^+ conductance; Gutknecht [413] questioned that view, proposing that fatty acid impurities can also explain the phenomenon, in a flip-flop movement of the neutralized weak acid. Proton carriers such as CO_2 or H_2CO_3 could also be involved [415]. The last word has not been said on this topic.

Using liposomes made from phospholipids as models of membrane barriers, Chakrabarti and Deamer [417] characterized the permeabilities of several amino acids and simple ions. Phosphate, sodium and potassium ions displayed effective permeabilities 0.1–1.0×10^{-12} cm/s. Hydrophilic amino acids permeated membranes with coefficients 5.1–5.7×10^{-12} cm/s. More lipophilic amino acids indicated values of 250–410×10^{-12} cm/s. The investigators proposed that the extremely low permeability rates observed for the polar molecules must be controlled by bilayer fluctuations and transient defects, rather than normal partitioning behavior and Born energy barriers. More recently, similar magnitude values of permeabilities were measured for a series of enkephalin peptides [418].

5.7 PREPARATION METHODS: MLV, SUV, FAT, LUV, ET

Working with liposomes requires considerable care, compared to octanol. Handling of liposomes is ideally done under an inert atmosphere at reduced temperatures. Prepared suspensions ought to be stored frozen when not used. Air oxidation of cis double bonds is facile; hydrolysis of esters to form free fatty acids (FFAs) is usually a concern. The best commercial sources of phospholipids have <0.1% FFA. Procedurally, a dry chloroform solution of a phospholipid is placed in a round-bottomed glass flask. Argon is allowed to blow off the chloroform while the flask is vortexed; a thin multilamellar layer forms on the glass surface. After evacuation of the residual chloroform, a buffer is added to the flask, and the lipid is allowed to hydrate under vortexing agitation, with argon gas protecting the lipid from air oxidation. A suspension of multilamellar vesicles (MLVs; diameter >1000 nm) forms in this way. [162] Small unilamellar vesicles (SUV, 50 nm diameters) can be made by vigorous sonication of MLVs. [385,386] Hope and coworkers developed procedures for preparing large unilamellar vesicles (LUV, 100–200 nm diameter) by an extrusion technique (ET), starting from the MLV suspension [389–391]. Freeze-and-thaw (FAT) steps are needed to distribute buffer salts uniformly between the exterior aqueous solution and the aqueous solution trapped inside vesicles [390]. Methods for determining volumes of liquid trapped inside the vesicles have been discussed [392]. When liposome surfaces are modified by covalent attachment of polyethylene glycol (PEG) polymer, the so-called stealth liposomes can evade the body's immune system, and stay in circulation for a long time, acting like a Trojan horse bearing drugs [393]. Such systems have been used in drug delivery [391,393]. Ordinary liposomes carrying drugs are quickly dismembered by the immune system.

For partition studies, only SUV [385,386] or LUV [149] should be used; MLVs have many layers of trapped solution, which usually cause hysteresis effects [162].

5.8   EXPERIMENTAL METHODS

The determination of partition coefficients using liposomes as a lipid phase require that the sample be equilibrated with a suspension of liposomes, followed by a separation procedure, before the sample is quantitated in the fraction free of the lipid component.

Miller and Yu [444] used an ultrafiltration method to separate the drug-equilibrated liposomes from the aqueous solution, in a study of the effect of cholesterol and phosphatidic acid on log P^N_{mem} and log $P^{\text{SIP}}_{\text{mem}}$ values of pentobarbitone, as a function of pH. Herbette and colleagues [425–428] and Austin et al., [441,442] and others [433] used ultrafiltration/centrifugation to separate the drug-laden liposomes from the aqueous solution. Wunderli-Allenspach's group [435–438] and others [381,383,384] used equilibrium dialysis for the separation step, the "gold standard" method [311]. It is the gentlest (and slowest) procedure. One reported high-throughput method may speed things up [454]. An interesting new method is based on the use of phospholipid-impregnated porous resin [317,318,376]. Trapped MLVs form in the rehydrated resin. Drug samples are allowed to equilibrate with the suspended particles, and then the solution is simply filtered. The filtrate is assayed for the unbound sample. No separation of phases is required when the NMR method is used [439,440]. Line broadening as a function of pH was used to determine partitioning into liposomes.

The pH-metric method, which also requires no phase separation, has been used to determine drug–liposome partitioning [149,162,385–387]. The method is the same as that described in Section 4.14, except that FAT-LUV-ET liposomes are used in place of octanol. SUV liposomes have also been used [385,386]. To allow for pH gradients to dissipate (Section 5.6) in the course of the titration, at least 5–10 min equilibration times are required between successive pH readings.

5.9   PREDICTION OF log P_{mem} FROM log P

In a very comprehensive study, Miyoshi et al. [381] measured log P_{mem}^N of 34 substitued phenols using four eggPC liposome systems: (1) lecithin, (2) lecithin + 10 mol% cholesterol, (3) lecithin + 20 mol% cardiolipin (negative charge), and (4) lecithin + 20 mol% stearylamine (positive charge). They probed the dielectric properties of the interfacial and the hydrocarbon core regions of the four systems using N-dansylhexadecylamine (DSHA) and anthroylstearic acid (AS) fluorescent probes. Phenol concentrations ranged from 10 to 100 µM; the unilamellar liposome suspensions, 5 mg/mL, were prepared in a 40 mM aspartate buffer at pH 6. Equilibrium dialysis (12 h) was used for the partition coefficients determination. Fujita's group [381] found that surface polarity increases with charged lipids; interfacial dielectric constants, ϵ (see Table 5.2), were estimated as 33 (unmodified), 40 (cholesterol), 43 (stearylamine), and 52 (cardiolipin). (There was minimal effect in the hydrocarbon core: ϵ 2.1, 1.9, 2.0, 2.0, respectively.) As ϵ increased, the membrane surface becomes more hydrated, with weakened inter-head-group interactions. Cholesterol appears to lead to tigher chain packing, weaker inter-head-group interactions, producing a more hydrated surface (see Section 5.3). The membrane log P_{mem}^N values were compared to those of the octanol–water system, log P_{oct}^N, with the following (quantitative structure–property relation (QSPR)) derived

$$\delta = \log P_{mem}^N - \log P_{oct}^N = 0.82 - 0.18 \log P_{oct}^N + 0.08\,\mathrm{HB} - 0.12\,\mathrm{VOL} \qquad (5.2)$$

where HB refers to H-bond donor strength (HB = $pK_a^H - pK_a^R$, where pK_a^H is the reference phenol value), and VOL is related to a steric effect. For a substituted phenol with a log P_{oct}^N near zero, the log P_{mem}^N value is ~0.82. This "membrane advantage" factor is sensitive to ionic strength effects, and may be indicative of an electrostatic interaction. As the octanol log P value increases, the δ factor decreases from the 0.82 base level, as the negative coefficient −0.18 suggests, which can be interpreted to mean that the membrane is less lipophilic than octanol (more alkane like). The H-bonding coefficient, +0.08, indicates that the H-bond acceptor property in membranes is greater than that of octanol, and strong H-bond donor phenols will show higher membrane partitioning, compared to octanol. The last term in Eq. (5.2) indicates that membranes do not tolerate steric hindrance as well as octanol; bulky di-ortho substituents produces higher VOL values.

Figure 5.3 illustrates the key features of the Fujita study. In relation to the reference phenol in frame (a), frames (b), and (c) illustrate the effect of H-bonding, and frames (d) and (e) illustrate steric hindrance. Given that the H-bond donor strength of (b) is greater than that of (c), since pK_a (b) $<pK_a$ (c), the relative membrane partitioning, δ, increases in (b) and decreases in (c), relative to (a). Similarly, steric hindrance in (d) produces negative δ, compared to (e).

A plot of δ versus log P_{oct}^N of 55 substituted phenols, combining the data from Fujita's group [381] with those of Escher et al. [382,383] is shown in Fig. 5.4. The slope-intercept parameters listed in the figure are close the the values in Eq. (5.2).

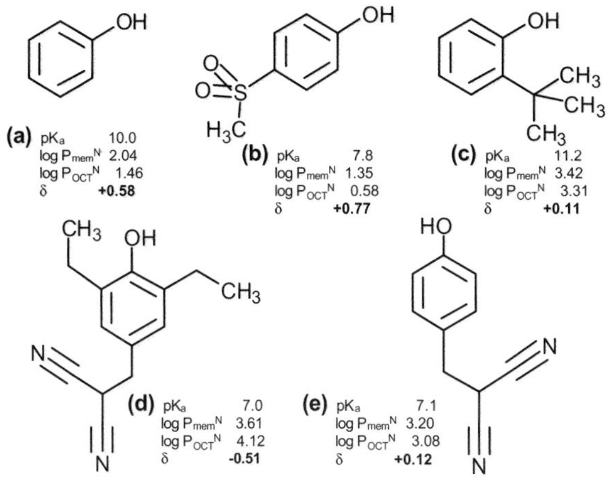

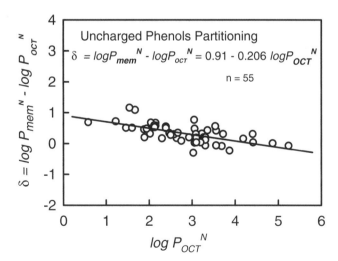

(a) pKₐ   10.0
log P_{mem}^N 2.04
log P_{OCT}^N 1.46
δ   +0.58

(b) pKₐ   7.8
log P_{mem}^N 1.35
log P_{OCT}^N 0.58
δ   +0.77

(c) pKₐ   11.2
log P_{mem}^N 3.42
log P_{OCT}^N 3.31
δ   +0.11

(d) pKₐ   7.0
log P_{mem}^N 3.61
log P_{OCT}^N 4.12
δ   -0.51

(e) pKₐ   7.1
log P_{mem}^N 3.20
log P_{OCT}^N 3.08
δ   +0.12

Figure 5.3   The effect of hydrogen bonding and steric hindrance on the difference between liposome–water and octanol–water partition coefficients δ; increased H-bond donor strength and decreased steric hindrance favor membrane partitioning in the substituted phenols [381]. [Avdeef, A., *Curr. Topics Med. Chem.*, **1**, 277–351 (2001). Reproduced with permission from Bentham Science Publishers, Ltd.]

The homologous series of (*p*-methylbenzyl)alkylamines [387] indicates an interesting δ/log P_{oct}^N plot, shown in Fig. 5.5. The slope factor of the smaller members of the series, −1.02, is larger than that of the phenol series. The value being near 1 indicates that log P_{mem}^N is invariant with the octanol partition constant—the

Figure 5.4   Comparing liposome–water to octanol–water partition coefficients of a series of uncharged substituted phenols [381–383]. [Avdeef, A., *Curr. Topics Med. Chem.*, **1**, 277–351 (2001). Reproduced with permission from Bentham Science Publishers, Ltd.]

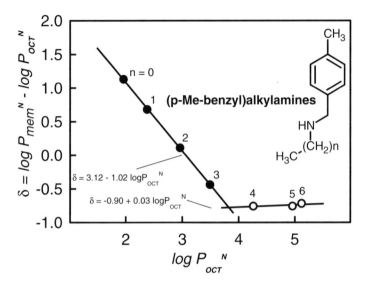

Figure 5.5   Comparing liposome–water to octanol–water partition coefficients of a series of uncharged substituted benzylalkylamines [387]. The membrane partitioning of the smaller members of the series ($n = 0 \cdots 3$) is thought to be dominated by electrostatic and H-bonding effects (enthalpy-driven), whereas the partitioning of the larger members is thought to be directed by hydrophobic forces (entropy-driven) [387]. [Avdeef, A., *Curr. Topics Med. Chem.*, **1**, 277–351 (2001). Reproduced with permission from Bentham Science Publishers, Ltd.]

membrane partitioning does not change for $n = 0$–3 in the series. For $n = 4$–6 the octanol and membrane partition coefficients change at about the same rate. For longer-chain members of the series, the partitioning in both solvent systems expresses hydrophobicity (entropy-driven). However, for the short-chain members, various electrostatic and polar interactions play a role, and partitioning in the membrane system is not sensitive to the length of the chain (enthalpy-driven). It would be illuminating to subject this series to a precision microcalorimetric investigation.

When unrelated compounds are examined [149,162,385,386,429], exclusive of the phenols and the amines just considered, the variance of the relationship is considerably higher, but the general trend is evident, as seen in Fig. 5.6; the higher the octanol–water partition coefficient, the smaller is the δ difference between membrane and octanol partitioning. The slope of the relationship is Fig. 5.6 is about twice that found for phenols. For molecules with log P_{oct}^N between 2 and 4, δ values are close to zero, indicating that the partition coefficients for many drug molecules are about the same in octanol as in phospholipid bilayers [149]. However, outside this interval, the differences can be substantial, as the next examples show. For *hydrophilic* molecules, the membrane partition coefficient is surprisingly high, in comparison to that of octanol. For example, acyclovir has log $P^N = -1.8$ in octanol–water but $+1.7$ in liposome–water, indicating a δ of $+3.5$ log units. Similar trends are found for other hydrophilic molecules, such as famotidine or zidovudine (Fig. 5.6). Atenolol and xamoterol also have notably high log P_{mem}^N values [433].

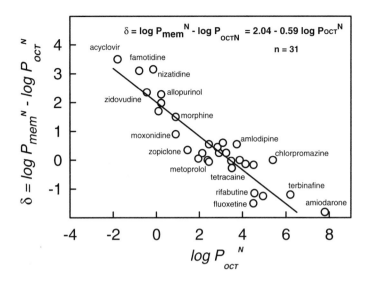

Figure 5.6 The difference between liposome–water and octanol–water partitioning as a function of the octanol–water partition coefficient for a series of unrelated structures [149,385,386,429]. For example, acyclovir partitions into liposomes over 3000 times more strongly than into octanol, and amiodarone partitions into liposomes 100 times more weakly than into octanol. [Avdeef, A., *Curr. Topics Med. Chem.*, **1**, 277–351 (2001). Reproduced with permission from Bentham Science Publishers, Ltd.]

At the opposite extreme is the example of amiodarone. The log octanol partition coefficient is 7.8 [162], whereas the membrane constant is reported as 6.0 [429], surprisingly, almost two orders of magnitude smaller ($\delta = -1.8$).

Although the relationship in Fig. 5.6 is somewhat coarse, it is still useful in predictions. Since octanol–water log P prediction programs are omnipresent and adequately reliable, it can now be said that they can predict membrane–water partitioning, by using the equation in Fig. 5.6. Better yet, if one measures the value of log P_{oct}^N, one can estimate the membrane partition coefficient with the confidence of the variance expressed in Fig. 5.6.

5.10 log D_{mem}, $diff_{mem}$, AND THE PREDICTION OF log P_{mem}^{SIP} FROM log P^I

In the preceding section, we explored the relationship between log P_{oct}^N and log P_{mem}^N. We now focus on the partitioning of the charged species into phospholipid bilayer phases. More surprises are in store.

Figure 5.7 shows lipophilicity profiles (log D vs. pH) for an acid (warfarin), a base (tetracaine), and an ampholyte (morphine). The dashed curves correspond to the values determined in octanol–water and the solid curves, to values in liposome–water. As is readily apparent, the major differences between octanol and liposomes

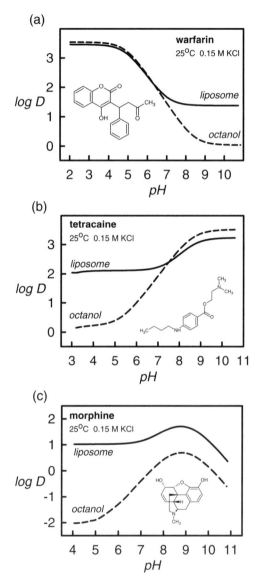

Figure 5.7 Comparison of liposome–water (solid lines) to octanol–water (dashed lines) lipophilicity profiles for a weak acid, a weak base, and an ampholyte. [Avdeef, A., *Curr. Topics Med. Chem.*, **1**, 277–351 (2001). Reproduced with permission from Bentham Science Publishers, Ltd.]

occur in the pH regions where charged-species partitioning takes place. In Section 4.7 we noted that octanol–water $diff(\log\ P^{N-I})$ values for simple acids were ~4 and for simple bases ~3. When it comes to liposome–water partitioning, the "*diff* 3–4" rule appears to slip to the "*diff* 1–2" rule. This is evident in Figs. 5.7a,b. The smaller $diff_{mem}$ values in membrane systems have been noted for some time, for example, with reported $diff_{mem} = 0$ for tetracaine, 1 for procaine and lidocaine [455], and $diff_{mem} = 1.45$ for tetracaine [424]. Miyazaki et al. [396] considered $diff_{mem}$ values of 2.2 for acids and 0.9 for bases in their study of dimyristoylphosphatidylcholine (DMPC) bilayer dispersions. Other studies indicated similar $diff$ values [149,383–386,433–438,441,442]. It seems that charged species partition into membranes about 100 times more strongly than suggested by octanol.

Alcorn et al. [433] studied the partitioning of proxicromil (acid: pK_a 1.93, log P_{oct}^N ~5, log P_{oct}^I 1.8 [456]) in MLV liposomes prepared from reconstituted brush-border lipids (slightly negatively charged at pH 7.4). Membrane partition coefficients were determined by the centrifugation (15 min, 150 kg) method. It was observed that in 0.15 M NaCl background, proxicromil showed a nearly constant log D_{mem} (3.0–3.5)at pH 4–9, which was unexpected, given the pK_a. However, when the background salt was lowered to 0.015 M, the expected curve shape (log D_{mem} 3.5 at pH 3 and 1.5 at pH 9) was observed, similar to that in Fig. 5.7a. Interestingly, the researchers took the solutions at pH 8 and titrated them with NaCl and LiCl. The log D_{mem} seen in the 0.15 M NaCl medium was reestablished by titration (more easily with NaCl than LiCl). The ionic strength dependence can be explained by the Gouy–Chapman theory [406,407]. The sample concentration (1.67 mM) was high enough to cause a buildup of negative charge on the surface. Without the high 0.15 M NaCl to shield the surface charge, sample anion-anion electrostatic repulsion on the membrane surface prevented the complete partitioning of the drug, making it appear that log D_{mem} was lowered. The Na^+ titration reduced the surface charge, allowing more anionic drug to partition. The fact that the Na^+ titrant is more effective than the Li^+ titrant can be explained by the higher hydration energy of Li^+, making it less effective at interacting with the membrane surface [400]. Incidently, we predict the log P_{mem}^N of proxicromil using the relationship in Fig. 5.6 to be ~4, in very acceptable agreement with the observed value.

In well-designed experiments, Pauletti and Wunderli-Allensbach [435] studied the partitioning behavior of propranolol in eggPC at 37°C, and reported log D_{mem} for pH 2–12. SUVs were prepared by the controlled detergent method. The equilibrium dialysis method was used to determine the partition coefficients, with propranolol concentration (10^{-6} to 10^{-9} M) determined by liquid scintillation counting. The lipid concentration was 5.2 mM. Internal pH of liposomes was checked by the fluorescein isothiocyanate method. Gradients in pH were dissipated within 5 min after small pH changes in the bulk solution. The lipophilicity curve they obtained is very similar in shape to that of tetracaine, shown in Fig. 5.7b. The log $P_{mem}^N = 3.28$ and log $P_{mem}^{SIP} = 2.76$ values indicate $diff_{mem} = 0.52$.

Austin et al. [441] reported the partitioning behavior of amlodipine, 5-phenylvaleric acid, 4-phenylbutylamine, and 5-hydroxyquinoline at 37°C in 1–100 mg mL^{-1} DMPC SUVs. The ultrafiltration (10 kDa cutoff) with mild (1.5 kg) centrifugation method was used to determine partition coefficients. Sample concentrations were $3–8 \times 10^{-5}$ M. Most remarkably, $diff_{mem} = 0.0$ was observed for amlodipine. A similarly low value of 0.29 was reported for 4-phenylbutylamine. Furthermore, the partitioning behavior was unchanged by ionic strength changes in the interval 0.0–0.15 M, seemingly in contradiction to the effect observed by Alcorn and co-workers. They proposed that charged molecules associated with the charged head

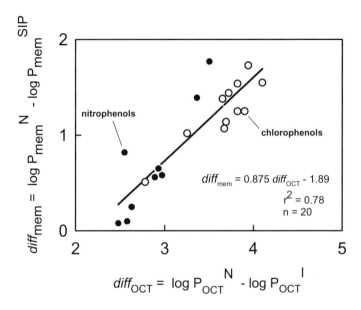

Figure 5.8 Comparison of liposome *diff* to octanol *diff* functions of substituted phenols [382,383]. [Avdeef, A., *Curr. Topics Med. Chem.*, **1**, 277–351 (2001). Reproduced with permission from Bentham Science Publishers, Ltd.]

groups of the phospholipids, an effect they preferred not to call "ion pairing." Undeniably, the nature of the charged-species partitioning into phospholipid bilayers is different from that found in octanol.

In a later study, Austin et al. [442] effectively were able to reconcile the ionic strength differences between their study and that of Alcorn et al. [433], using a Gouy–Chapman model. When the drug concentration in the membrane is plotted against the drug concentration in water, the resultant hyperbolic curve shows a lessening slope ($\log D$) with increasing drug concentration (10^{-6} to 10^{-4} M) when there is no background salt. This is consistent with the interpretation that surface-bound charged drug repulsion attenuates additional charged drug partitioning. Bäuerle and Seelig [395] and Thomas and Seelig [397] observed hyperbolic curves with drug concentrations exceeding 1 μM. The addition of 0.15 M NaCl mitigates the effect substantially, allowing for higher drug concentrations to be used.

Avdeef et al. [149] and Balon et al. [385,386] reported $\log P_{\mathrm{mem}}^{N}$ and $\log P_{\mathrm{mem}}^{\mathrm{SIP}}$ values of a number of drugs, determined by the pH-metric method, using both LUVs and SUVs, in a background of 0.15 M KCl.

Escher and colleagues [383,384] reported SIP values for a large series of substituted phenols, using DOPC SUVs and the equilibrium dialysis/centrifugation method. Figure 5.8 is a plot of $diff_{\mathrm{mem}}$ versus $diff_{\mathrm{oct}}$ for the series of phenols studied by Escher. It appears that knowing the octanol *diff* values can be useful in predicting the membrane values, and for phenols the relationship is described by

$$diff_{\mathrm{mem}} = 0.88\,diff_{\mathrm{oct}} - 1.89 \qquad (5.3)$$

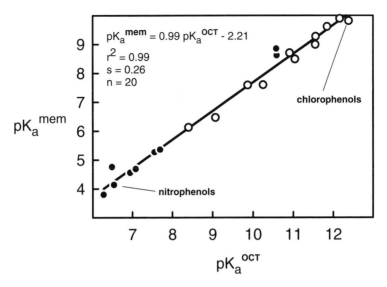

Figure 5.9   The remarkable relationship between the octanol and the membrane pK_a values of a series of substituted phenols [382,383]. [Avdeef, A., *Curr. Topics Med. Chem.*, **1**, 277–351 (2001). Reproduced with permission from Bentham Science Publishers, Ltd.]

The offset of 1.89 indicates that surface ion pairing in membranes is about 100 times greater than that of octanol. Scherrer suggested that comparisons of pK_a^{oct} to pK_a^{mem} may be more predictive [276]. Indeed, this is true for the phenols, as indicated in Fig. 5.9. It is remarkable that the relation for the phenols

$$pK_a^{mem} = 0.99\,pK_a^{oct} - 2.21 \qquad (5.4)$$

has an r^2 of 0.99, for 20 compounds. Again, we see the 2.21 offset, indicating that the 100-fold slippage from the *diff* 3-4 rule to the *diff* 1-2 rule. This harbors good prediction relations.

The well-behaved prediction of charged phenol partitioning is less certain when carried over to unrelated structures, as shown in Fig. 5.10, for the molecules reported by Avdeef et al. [149] and Balon et al. [385,386].

5.11   THREE INDICES OF LIPOPHILICITY: LIPOSOMES, IAM, AND OCTANOL

Taillardat-Bertschinger et al. [311] explored the molecular factors influencing retention of drugs on IAM columns, compared to partitioning in liposomes and octanol. Twenty-five molecules from two congeneric series (β-blockers and (*p*-methylbenzyl)alkylamines; see Fig. 5.5) and a set of structurally unrelated drugs were used. Liposome-buffer partitioning was determined by the equilibrium dialysis and the pH-metric methods. The intermethod agreement was good, with r^2 0.87.

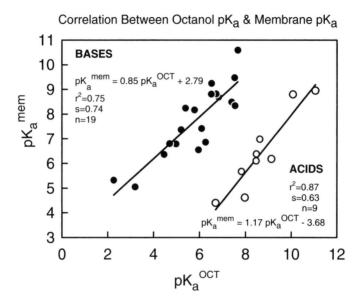

Figure 5.10 Comparison of membrane to octanol pK_a values of compounds with unrelated structures [149,385,386]. [Avdeef, A., *Curr. Topics Med. Chem.*, **1**, 277–351 (2001). Reproduced with permission from Bentham Science Publishers, Ltd.]

However, when the IAM log $k_{IAMw}^{7.0}$ were compared to liposome log $D_{mem,7.0}$, there was no direct correlation when all of the compounds were used. It was clear that multi–mechanisms were operative, in the Lipinski sense [1].

For the series of large molecules, such as, β blockers or the long-chain (*p*-methylbenzyl)alkylamines, IAM retention correlated with liposome partitioning. Hydrophobic recognition forces was thought to be responsible for the partitioning process. In addition, the formation of an H bond between the hydroxy group of the β blocker and the ester bond of phospholipids (Fig. 5.1) may explain why the β blockers partitioned into the liposomes more strongly than the alkylamines. For the more hydrophilic short-chain (*p*-methylbenzyl)alkylamines ($n = 0$–3 in Fig. 5.5), the balance between electrostatic and hydrophilic interactions was different in the IAM and liposome systems. Electrostatic interactions are thought to play only a minor role for the IAM retention of the model solutes, presumably due to the smaller density of phospholipids in IAM resin surfaces, compared to liposomes. The solute's capacity to form H-bonds, which is important for partitioning in liposomes, plays only a minor role in the IAM system.

5.12 GETTING IT WRONG FROM ONE-POINT log D_{mem} MEASUREMENT

In the early literature, it was a common practice to make a single measurement of log D, usually at pH 7.4, and use a simplified version of Eq. (4.10) (with log P^l

neglected) along with the known pK_a to calculate log P^N. (The practice may still persist today. We have intentionally omitted these simplified equations in this book.) Most of the time this produced the correct log P^N, often because ion pairing was not extensive at the pH of measurement. This is true for the β blockers whose pK_a is about 9.5; the *diff* 3–4 rule would suggest that ion pair partitioning should be extensive only below pH 6.5.

With liposome partitioning, however, the rule slips to *diff* 1–2. This means SIP partitioning starts at about pH 8.5 for weak bases whose pK_a values are near 9.5 (e.g., Figs. 5.7b, 5.11). So, all who published "anomalous" values of log P^N_{mem} may need to get out their slide rules [429–432]! (What we know now was not known then.)

5.13   PARTITIONING INTO CHARGED LIPOSOMES

Wunderli-Allenspach's group reported several partition studies where drugs interacted with liposomes that were charged [368,436–438]. Although not entirely surprising, it was quite remarkable that propranolol partitions into negatively charged liposomes with log P^N_{mem} 3.49 and log P^{SIP}_{mem} 4.24 [438], compared to values determined with neutral liposome values log P^N_{mem} 3.27 and log P^{SIP}_{mem} 2.76 [435]. Negatively charged liposomes can enhance the surface ion pair (SIP) partitioning of positively charged propranolol by a factor of 30. The unusually-shaped lipophilicity profile is shown in Fig. 5.11, for the system where negative charge is imparted by 24 mol% oleic acid in the eggPC.

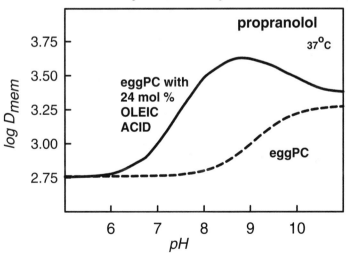

Figure 5.11 Lipophilicity profiles of propranolol in liposome–water (dashed curve) and liposome–water, where the liposome phase had 24 mol% FFA, imparting a negative charge to the surface above pH 6 [436]. [Avdeef, A., *Curr. Topics Med. Chem.*, **1**, 277–351 (2001). Reproduced with permission from Bentham Science Publishers, Ltd.]

Since the FFA is an anion >pH 7 and propranolol a cation <pH 9, there is a window of opportunity between pH 7 and 9 for electrostatic attraction of propranolol into the membrane phase, as indicated in Fig. 5.11. Note how similar the curve shapes in Fig. 5.11 are to some of the curves in Fig. 4.6b.

5.14 pK_a^{mem} SHIFTS IN CHARGED LIPOSOMES AND MICELLES

Ionizable molecules embedded in the surfaces of lipids, such as octanol (see Fig. 2.8), liposomes (see Fig. 5.2), or micelles, will have their apparent pK_a values shifted. With *neutral* lipids, the pK_a of an acid increases and the pK_a of a base decreases. This is due to the effect of the decreased dielectric constant in the interfacial zone, as we have already discussed in various sections.

An additional (electrostatic) shift occurs if the lipid vesicles or micelles have a charged surface, according to the expression suitable for monoprotic acids and bases

$$pK_a^{mem} = pK_a \pm diff_{mem} - \frac{F\phi}{2.3RT} \qquad (5.5)$$

where the terms have their usual meanings, with $\pm$ being $+$ sign for acids, $-$ sign for bases [396,404,406, 407,448,457,458]. At 25°C and using mV (millivolt) units to express the surface potential, ϕ, the rightmost term in eq. 5.5 becomes $\phi/59.16$. The rationale for the electrostatic term goes like this: if the surface is negatively charged, then it will attract protons into the interfacial zone, such that the interfacial pH will be lower than the bulk pH, by the amount of $|\phi/59.16|$. A proton fog envelops the negatively charged vesicle. Since the proton concentration is in the pK_a expression [Eq. (3.1)], the apparent pK_a changes accordingly.

Consider negatively charged liposomes made from a mixture of phosphatidylcholine (PC) and phosphatidylserine (PS). Unlike the zwitterionic head group of PC (invariant charge state, pH > 3), the head group of PS has two ionizable functions for pH > 3: the amine and the carboxylic acid. In physiologically neutral solution, the PS group imparts a negative charge to the liposome (from the phosphate). Titrations of PS-containing liposomes reveal the pK_a values of 5.5 for the carboxylic acid group and 11.5 for the amine group [403]. When the head-group molecule itself (free of the acyl HC chains), phosphoserine (Fig. 5.12), is titrated, the observed pK_a values for the two sites are 2.13 and 9.75, respectively [162].

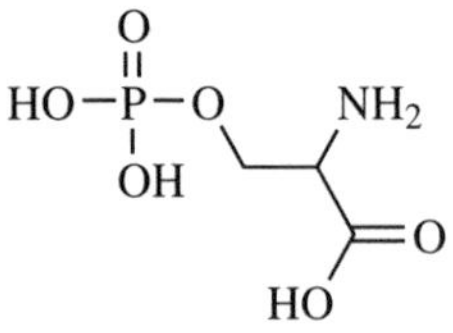

Figure 5.12 Phosphoserine.

According to the *diff* 1–2 rule, we should have expected to see the pK_a 4.13 (carboxylate) and 8.75 (amine), but the liposome titration shows something else. Instead, we have an "anomalous" additional shift of +1.37 for the carboxylic group and a +2.75 for the amine group. These extra shifts are due to the negatively charged surface of the liposomes! We can estimate, using Eq. (5.5), that when the carboxylic group is titrated in the PS liposome, the surface charge is −81 mV (pH 5.5), and when the amine group is titrated, the surface charge drops to −163 mV (pH 11.5). Conversely, if we had a way of estimating surface charge,

TABLE 5.3 Critically Selected Experimental Liposome–Water Partition Coefficients

Compound	log P_{mem}^N	log P_{mem}^{SIP}	t^a (°C)	Ref.
4-Phenylbutylamine	2.39	2.48	—	149
5-Phenylvaleric acid	3.17	1.66	—	149
Acetylsalicylic acid	2.40	1.60	37	385
Acyclovir	1.70	2.00	37	385
Allopurinol	2.50	2.70	37	385
Amiloride	1.80	1.60	37	385
Amlodipine	4.29	4.29	—	—[b]
Atenolol	2.20	1.00	37	385
Carvedilol	4.00	4.20	—	—[b]
Chlorpromazine	5.40	4.45	—	—[b]
Diclofenac	4.34	2.66	—	149
Famotidine	2.30	1.70	37	385
Fluoxetine	3.00	2.20	37	385
Furosemide	3.00	1.90	37	385
Ibuprofen	3.87	1.94	—	149
Lidocaine	2.39	1.22	—	149
Metoprolol	2.00	1.25	37	385
Miconazole	3.70	2.90	37	385
Morphine	1.89	1.02	—	—[b]
Moxonidine	1.80	1.30	—	385
Nizatidine	3.00	2.80	37	385
Olanzapine	3.70	2.70	37	385
Paromomycin	1.70	1.20	37	385
Procaine	2.38	0.76	—	149
Propranolol	3.45	2.61	—	149
Rifabutine	3.40	3.50	37	385
Terbinafine	5.00	3.00	37	385
Tetracaine	3.23	2.11	—	149
Warfarin	3.46	1.38	—	149
Xipamide	3.30	1.70	37	385
Zidovudine	1.90	2.40	37	385
Zopiclone	1.80	1.40	37	385

[a]Temperature 25°C, unless otherwise stated.
[b]Sirius Analytical Instrument Ltd.

TABLE 5.4 Liposome–Water Partition Coefficients of Substituted Phenols and Other Compounds

Compound	$\log P_{\text{mem}}^{N}$	$\log P_{\text{mem}}^{\text{SIP}}$	Note/Ref.
Phenol	1.97	—	—[a]
2-Cl-phenol	2.78	1.57	—[a,b]
3-Cl-phenol	2.78	—	—[a]
4-Cl-phenol	2.92	2.43	—[a,b]
2,4-Di-Cl-phenol	3.54	2.41	—[a,b]
2,6-Di-Cl-phenol	2.83	1.09	—[a,b]
3,4-Di-Cl-phenol	3.82	2.82	—[b]
2,4,5-Tri-Cl-phenol	4.35	2.80	—[b]
2,4,6-Tri-Cl-phenol	3.82	2.59	—[a,b]
3,4,5-Tri-Cl-phenol	4.72	3.18	—[b]
2,3,4,5-Tetra-Cl-phenol	4.88	3.63	—[b]
2,3,4,6-Tetra-Cl-phenol	4.46	3.39	—[b]
penta-Cl-phenol	5.17	3.79	—[b]
2-NO$_2$-phenol	2.09	0.70	—[b]
4-NO$_2$-phenol	2.72	0.95	—[b]
2,4-Di-NO$_2$-phenol	2.73	1.94	—[a,b]
2,6-Di-NO$_2$-phenol	1.94	1.84	—[b]
2-Me-4,6-di-NO$_2$-phenol	2.69	2.46	—[a,b]
4-Me-2,6-di-NO$_2$-phenol	2.34	2.26	—[b]
2-s-Bu-4,6-di-NO$_2$-phenol	3.74	3.33	—[a,b]
2-t-Bu-4,6-di-NO$_2$-phenol	4.10	3.54	—[b]
4-t-Bu-2,6-di-NO$_2$-phenol	3.79	3.21	—[b]
2-Me-phenol	2.45	—	—[a]
3-Me-phenol	2.34	—	—[a]
4-Me-phenol	2.42	—	—[a]
2-Et-phenol	2.81	—	—[a]
4-Et-phenol	2.88	—	—[a]
2-Pr-phenol	3.13	—	—[a]
4-Pr-phenol	3.09	—	—[a]
2-s-Bu-phenol	3.47	—	—[a]
4-s-Bu-phenol	3.43	—	—[a]
2-t-Bu-phenol	3.51	—	—[a]
3-t-Bu-phenol	3.25	—	—[a]
4-t-Bu-phenol	3.43	—	—[a]
2-Ph-phenol	3.40	—	—[a]
4-Ph-phenol	3.24	—	—[a]
4-t-Pent-phenol	3.64	—	—[a]
2,6-di-Me-phenol	2.47	—	—[a]
2,6-di-Et-phenol	2.73	—	—[a]
3-Me-4-Cl-phenol	3.29	—	—[a]
4-SO$_2$Me-phenol	1.27	—	—[a]
4-CN-phenol	2.11	—	—[a]
4-CF$_3$-phenol	3.25	—	—[a]
3-NO$_2$-phenol	2.56	—	—[a]

TABLE 5.4 (*Continued*)

Compound	$\log P_{mem}^N$	$\log P_{mem}^{SIP}$	Note/Ref.
2-Et-4,6-di-NO$_2$-phenol	3.02	—	—[a]
2-i-Pr-4,6-di-NO$_2$-phenol	3.14	—	—[a]
Benzenesulphonamide	0.82	—	—[c]
Aniline	1.04	—	—[c]
Nitrobenzene	1.71	—	—[c]
Naphthylamide	1.99	—	—[c]
4-Cl-1-naphthol	2.88	—	—[c]
Naphthalene	2.78	—	—[c]
2-Me-anthracene	3.75	—	—[c]
1,2,5,6-Dibenzanthracene	3.09	—	—[c]
Benzamide	0.21	—	—[c]
Methylphenylsulphone	0.88	—	—[c]
Hydrochlorothiazide	0.91	—	—[c]
Methylphenylsulphoxide	0.98	—	—[c]
Phenylurea	1.04	—	—[c]
Phenylbenzamide	1.05	—	—[c]
Phenol	1.32	—	—[c]
Dimethylphenylsulfonamide	1.60	—	—[c]
Acetophenone	1.76	—	—[c]
Benzonitrile	1.81	—	—[c]
Benzaldehyde	1.90	—	—[c]
Methylnaphthylsulphone	1.91	—	—[c]
Naphthylsulphonamide	2.01	—	—[c]
Anisole	2.10	—	—[c]
Methylbenzoate	2.20	—	—[c]
Triphenylphosphineoxide	2.21	—	—[c]
3-(2-Naphthoxy)propylmethylsulphoxide	2.60	—	—[c]
Chrysene	2.60	—	—[c]
Fluoroanthrene	2.61	—	—[c]
Toluene	2.71	—	—[c]
Phenanthrene	2.75	—	—[c]
Atenolol	—	1.36	—[c]
Xamoterol	—	1.46	—[c]
Proxichromil	—	1.50	—[c,d]
Amlodipine	3.75	3.75	—[e]
5-Phenylvaleric acid	2.95	0.50	—[e]
4-Phenylbutylamine	2.41	2.12	—[e]
5-Hydroxyquinoline	1.85	—	—[e]

[a]Temperature 25°C, equilibrium dialysis, small unilamellar vesicles (lecithin) [381].
[b]Temperature 20°C, equilibrium dialysis, small unilamellar vesicles (DOPC), 0.1 M KCl [382].
[c]Centrifugation method (15 min, 150,000 g), brush-border membrane vesicles [433].
[d]Ionic strength 0.015 M (NaCl).
[e]Temperature 37°C, 0.02 M ionic strength, ultrafiltration method, small unilamellar vesicles (DMPC) [441].

say, by zeta-potential measurements [395,397,398], then we could predict what pK_a^{mem} should be. This is an important consideration, since membranes often bear (negative) surface charge.

5.15   PREDICTION OF ABSORPTION FROM LIPOSOME PARTITION STUDIES?

It is clear that charged species partition more strongly into liposomes than anticipated from octanol properties (Figs. 5.7 and 5.11). Although octanol has been a useful model system, it cannot address the role of ionic forces evident in biological membranes. In addition, it is apparent that certain hydrophilic species such as acyclovir, famotidine, atenolol, and morphine partition into liposomes as neutral species more strongly than suggested by octanol measurements (Fig. 2.6). Hydrogen bonding is certain to be a part of this. If amphiphilic charged or H-bonding species have such a strong affinity for membranes, can passive absorption of charged species be facilitated? What does it mean that acyclovir indicates a log P_{mem} that is 3.5 units higher than log P_{oct}? These questions are revisited at the end of the book.

5.16   log P_{mem}^N, log P_{mem}^{SIP} "GOLD STANDARD" FOR DRUG MOLECULES

Table 5.3 lists a carefully selected collection of log P_{mem}^N and log P_{mem}^{SIP} values of drug molecules taken from various literature and some unpublished sources. Table 5.4 contains a similar collection of values for a series of substituted phenols and a variety of mostly uncharged compounds.

CHAPTER 6

SOLUBILITY

The treatise by Grant and Higuchi [37] comprehensively covers pre-1990 solubility literature. In this chapter, we present a concise, multimechanistic [1] solubility equilibrium model ("not just a number"; see Section 1.6) and stress what is new since 1990 [39]; we also cite some important classic works. Many protocols have been described in the literature for measuring solubility–pH profiles, using various detection systems [12,26,37–39,459–503]. Classical approaches are based on the saturation shake-flask method [37–39]. New methods are usually validated against it. The classical techniques are slow and not easily adapted to the high-throughput needs of modern drug discovery research. At the early stages of research, candidate compounds are stored as DMSO solutions, and solubility measurements need to be performed on samples introduced in DMSO, often as 10 mM solutions. It is known that even small quantities of DMSO (<5%) in water can increase the apparent solubility of molecules, and that it is a challenge to determine the true aqueous solubility of compounds when DMSO is present. To this end, a new method has been developed which extracts true aqueous solubility from DMSO-elevated values [26].

The accurate prediction of the solubility of new drug candidates still remains an elusive target [1,12,502]. Historical solubility databases used as "training sets" for prediction methods contain a large portion of oil substances, and not enough crystalline, drug-like compounds. Also, the quality of the historical data in the training sets is not always easy to verify. Such methods, for reasons of uncertain training

Absorption and Drug Development: Solubility, Permeability, and Charge State. By Alex Avdeef
ISBN 0-471-423653.   Copyright © 2003 John Wiley & Sons, Inc.

data, often perform poorly in predicting solubilities of crystalline drug compounds [504–506].

6.1 SOLUBILITY–pH PROFILES

The basic relationships between solubility and pH can be derived for any given equilibrium model. In this section simple monoprotic and diprotic molecules are considered [26,472–484,497].

6.1.1 Monoprotic Weak Acid, HA (or Base, B)

The protonation reactions for ionizable molecules have been defined in Section 3.1. When a solute molecule, HA (or B), is in equilibrium with its precipitated form, HA(s) (or B(s)), the process is denoted by the equilibrium expression

$$HA(s) \rightleftarrows HA \qquad (\text{or } B(s) \rightleftarrows B) \qquad (6.1)$$

and the corresponding equilibrium constant is defined as

$$S_0 = \frac{[HA]}{[HA(s)]} = [HA] \qquad \left[\text{or } S_0 = \frac{[B]}{[B(s)]} = [B]\right] \qquad (6.2)$$

By convention, $[HA(s)] = [B(s)] = 1$. Eqs. (6.1) represent the precipitation equilibria of the uncharged species, and are characterized by the intrinsic solubility equilibrium constant, S_0. The zero subscript denotes the zero charge of the precipitating species. In a saturated solution, the *effective* (total) solubility S, at a particular pH is defined as the sum of the concentrations of all the compound species dissolved in the aqueous solution:

$$S = [A^-] + [HA] \qquad [\text{or } S = [B] + [BH^+]] \qquad (6.3)$$

where [HA] is a constant (intrinsic solubility) but $[A^-]$ is a variable. It's convenient to restate the equation in terms of only constants and with pH as the only variable. Substitution of Eqs. (3.1) [or (3.2)] into (6.3) produces the desired equation.

$$
\begin{aligned}
S &= \frac{[HA]\, K_a}{[H^+]} + [HA] & \left(\text{or } S = [B] + \frac{[B][H^+]}{K_a}\right) \\
&= [HA]\left(\frac{K_a}{[H^+]} + 1\right) & \left(\text{or } = [B]\left\{\frac{[H^+]}{K_a} + 1\right\}\right) \\
&= S_0(10^{-pK_a+pH} + 1) & \left(\text{or } = S_0\left\{10^{+pK_a-pH} + 1\right\}\right)
\end{aligned}
\qquad (6.4)
$$

Figure 6.1a shows a plot of log S versus pH for the weak-acid case (indomethacin, pK_a 4.42, log S_0 − 5.58, log mol/L [$pION$]) and Fig. 6.2a shows that of a weak base (miconazole, pK_a 6.07, log S_0 − 5.85 [$pION$]). As is evident from the acid curve, for pH $\ll$ pK_a [i.e., $10^{-pK_a+pH} \ll 1$ in Eq. (6.4)], the function reduces to the horizontal line log $S = $ log S_0. For pH $\gg$ pK_a (i.e., $10^{-pK_a+pH} \gg 1$), log S is a straight line as a function of pH, exhibiting a slope of +1. The base shows a slope of −1. The pH at which the slope is half-integral equals the pK_a. Note the mirror relationship between the curve for an acid (Fig. 6.1a) and the curve for a base (Fig. 6.2a).

6.1.2 Diprotic Ampholyte, XH_2^+

In a saturated solution, the three relevant equilibria for the case of a diprotic ampholyte are Eqs. (3.3) and (3.4), plus

$$XH(s) \rightleftarrows XH \qquad S_0 = \frac{[XH]}{[XH(s)]} = [XH] \qquad (6.5)$$

Note that [XH(s)] by convention is defined as unity. For such a case, effective solubility is

$$S = [X^-] + [XH] + [XH_2^+] \qquad (6.6)$$

where [HX] is a constant (intrinsic solubility) but [X$^-$] and [XH$_2^+$] are variables. As before, the next step involves conversions of all variables into expressions containing only constants and pH:

$$S = S_0(1 + 10^{-pK_{a2}+pH} + 10^{+pK_{a1}-pH}) \qquad (6.7)$$

Figure 6.3a shows the plot of log S versus pH of an ampholyte (ciprofloxacin, pK_a values 8.62 and 6.16, log S_0 − 3.72 [$pION$]). In Figs. 6.1b, 6.2b, and 6.3b are the log–log speciation profiles, analogous to those shown in Figs. 4.2b, 4.3b, and 4.4b. Note the discontinuities shown for the solubility speciation curves. These are the transition points between a solution containing some precipitate and a solution where the sample is completely dissolved. These log-log solubility curves are important components of the absorption model described in Section 2.1 and illustrated in Fig. 2.2.

6.1.3 Gibbs pK_a

Although Figs. 6.1a, 6.2a, and 6.3a properly convey the shapes of solubility–pH curves in saturated solutions of uncharged species, the indefinite ascendency (dotted line) in the plots can be misleading. It is not possible to maintain saturated solutions over 10 orders of magnitude in concentration! At some point long before the solubilities reach such high values, salts will precipitate, limiting further

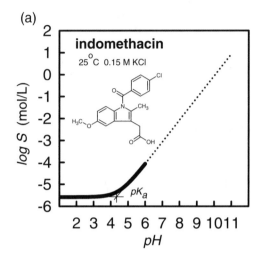

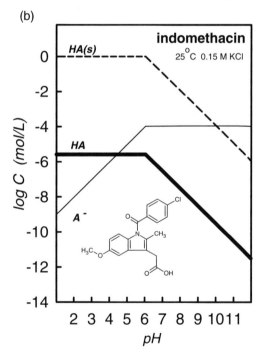

Figure 6.1 Solubility–pH profile (a) and a log–log speciation plot (b) for a weak acid (indomethacin, pK_a 4.42, log S_0 −5.58 [pION]). [Avdeef, A., *Curr. Topics Med. Chem.*, **1**, 277–351 (2001). Reproduced with permission from Bentham Science Publishers, Ltd.]

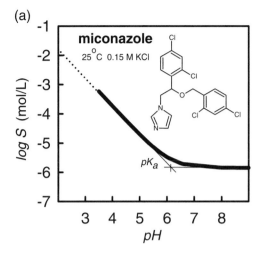

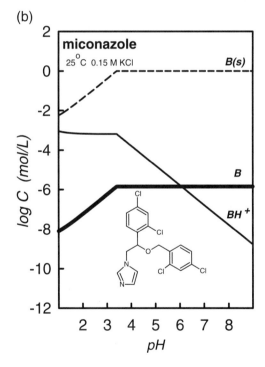

Figure 6.2 Solubility–pH profile (a) and a log–log speciation plot (b) for a weak base (miconazole, pK_a 6.07, log S_0 −5.85 [pION]). [Avdeef, A., *Curr. Topics Med. Chem.*, **1**, 277–351 (2001). Reproduced with permission from Bentham Science Publishers, Ltd.]

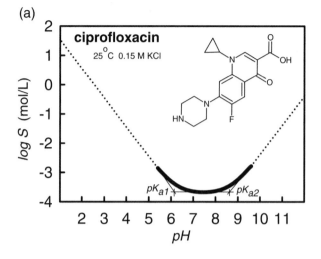

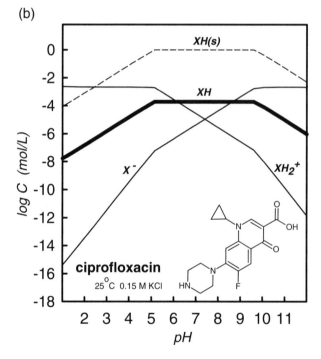

Figure 6.3 Solubility–pH profile (a) and a log–log speciation plot (b) for an ampholyte (ciprofloxacin, pK_a 8.62, 6.16, log S_0 −3.72 [pION]). [Avdeef, A., *Curr. Topics Med. Chem.*, **1**, 277–351 (2001). Reproduced with permission from Bentham Science Publishers, Ltd.]

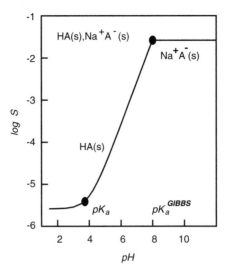

Figure 6.4 Solubility–pH profile of a weak acid, with salt precipitation taken into account. [Avdeef, A., *Curr. Topics Med. Chem.*, **1**, 277–351 (2001). Reproduced with permission from Bentham Science Publishers, Ltd.]

increases. Although precipitation of salts is not covered in detail in this chapter, it is nevertheless worthwhile to consider its formation in this limiting sense. As the pH change raises the solubility, at some value of pH the solubility product of the salt will be reached, causing the shape of the solubility–pH curve to change from that in Fig. 6.1a to that in Fig. 6.4, an example of a weak acid exhibiting salt precipitation.

As a new rule of thumb [473], in 0.15 M NaCl (or KCl) solutions titrated with NaOH (or KOH), acids start to precipitate as salts above $\log (S/S_0) \approx 4$ and bases above $\log (S/S_0) \approx 3$. It is exactly analogous to the *diff* 3–4 rule; let us call the solubility equivalent the "*sdiff* 3–4" rule [473]. Consider the case of the monoprotic acid HA, which forms the sodium salt (in saline solutions) when the solubility product K_{sp} is exceeded. In additions to Eqs. (3.1) and (6.1), one needs to add the following reaction/equation to treat the case:

$$Na^+A^-(s) \rightleftarrows Na^+ + A^- \qquad K_{sp} = \frac{[Na^+][A^-]}{[Na^+A^-(s)]} = [Na^+][A^-] \qquad (6.8)$$

Effective solubility is still defined by Eq. (6.3). However, Eq. (6.3) is now solved under three limiting conditions with reference to a special pH value:

1. If the solution pH is below the conditions leading to salt formation, the solubility–pH curve has the shape described by Eq. (6.4) (curve in Fig. 6.1a).
2. If pH is above the characteristic value where salt starts to form (given high enough a sample concentration), Eq. (6.3) is solved differently. Under

this circumstance, $[A^-]$ becomes the constant term and $[HA]$ becomes variable.

$$S = [A^-] + \frac{[H^+][A^-]}{K_a}$$

$$= [A^-]\left(1 + \frac{[H^+]}{K_a}\right)$$

$$= \frac{K_{sp}}{[Na^+]}(1 + 10^{+pK_a-pH})$$

$$= S_i(1 + 10^{+pK_a-pH}) \tag{6.9}$$

where S_i refers to the solubility of the conjugate base of the acid, which depends on the value of $[Na^+]$ and is hence a conditional constant. Since $pH \gg pK_a$ and $[Na^+]$ may be assumed to be constant, Eq. (6.9) reduces to that of a horizontal line in Fig. 6.4: $\log S = \log S_i$ for $pH > 8$.

3. If the pH is exactly at the special point marking the onset of salt precipitation, the equation describing the solubility-pH relationship may be obtained by recognizing that both terms in Eq. 6.3 become constant, so that

$$S = S_0 + S_i \tag{6.10}$$

Consider the case of a very concentrated solution of the acid hypothetically titrated from low pH ($<pK_a$) to the point where the solubility product is first reached (high pH). At the start, the saturated solution can only have the uncharged species precipitated. As pH is raised past the pK_a, the solubility increases, as more of the free acid ionizes and some of the solid HA dissolves, as indicated by the solid curve in Fig. 6.1a. When the solubility reaches the solubility product, at a particular elevated pH, salt starts to precipitate, but at the same time there may be remaining free acid precipitate. The simultaneous presence of the solid free acid and its solid conjugate base invokes the Gibbs phase rule constraint, forcing the pH and the solubility to constancy, as long as the two interconverting solids are present. In the course of the thought-experiment titration, the alkali titrant is used to convert the remaining free acid solid into the solid salt of the conjugate base. During this process, pH is absolutely constant (a "perfect" buffer system). This special pH point has been designated the Gibbs pK_a, that is, pK_a^{gibbs} [472,473]. The equilibrium equation associated with this phenomenon is

$$HA(s) \rightleftarrows A^-(s) + H^+ \qquad K_a^{gibbs} = \frac{[H^+][A^-(s)]}{[HA(s)]} = [H^+] \tag{6.11}$$

Note that pK_a^{gibbs} is the conceptual equivalent of pK_a^{oct} and pK_a^{mem} [(see. Eq. (5.1)]. We should not be surprised that this is a *conditional* constant, depending on the value of the background salt.

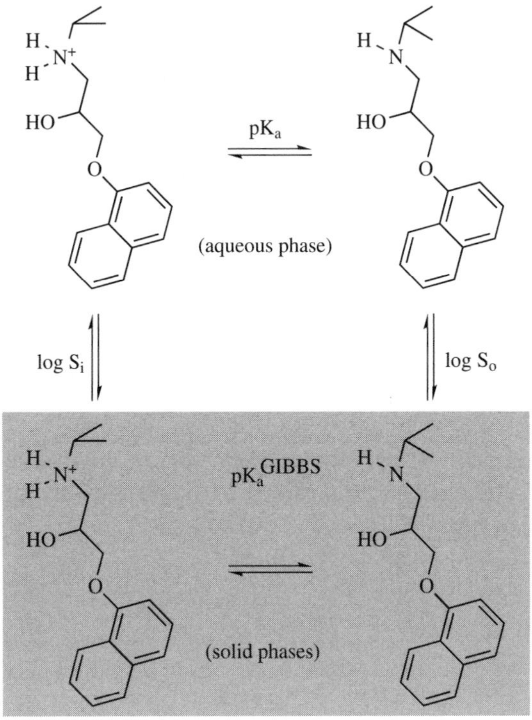

Figure 6.5 Solubility tetrad equilibria. [Avdeef, A., *Curr. Topics Med. Chem.*, **1**, 277–351 (2001). Reproduced with permission from Bentham Science Publishers, Ltd.]

At this point we bring in the now familiar tetrad diagram, Fig. 6.5, and conclude that

$$sdiff \ (\log S^{I-N}) = \log \ S_i - \log \ S_0 = |pK_a^{\text{gibbs}} - pK_a| \qquad (6.12)$$

Figure 6.4 shows a hypothetical solubility–pH profile with *sdiff* = 4, as typical as one finds with simple acids in the presence of 0.15 M Na$^+$ or K$^+$ [473]. Compare Eq. (6.12) with Eq. (4.6).

In principle, all the curves in Figs. 6.1a, 6.2a, and 6.3a would be expected to have solubility limits imposed by the salt formation. Under conditions of a constant counterion concentration, the effect would be indicated as a point of discontinuity (pK_a^{gibbs}), followed by a horizontal line of constant solubility S_i.

6.2 COMPLICATIONS MAY THWART RELIABLE MEASUREMENT OF AQUEOUS SOLUBILITY

There are numerous experimental complications in the measurement of solubility. Solid phases, formed incipiently, are often metastable with respect to a thermodynamically more stable phase, especially with highly insoluble compounds. An "active" form of a solid, a very fine crystalline precipitate with a disordered lattice, can drop out of a strongly oversaturated solution, which then "ages" only slowly into a more stable "inactive" form [465]. Hence, if measurements are done following initial precipitation, higher solubilities are observed. Amorphism [464] and polymorphism [466] can be troubling complications. Various solvates of a solid (either water or cosolvent in the crystal lattice) have different solubilities [43].

Certain surface-active compounds [499], when dissolved in water under conditions of saturation, form self-associated aggregates [39,486–488] or micelles [39,485], which can interfere with the determination of the true aqueous solubility and the pK_a of the compound. When the compounds are very sparingly soluble in water, additives can be used to enhance the rate of dissolution [494,495]. One can consider DMSO used in this sense. However, the presence of these solvents can in some cases interfere with the determination of the true aqueous solubility. If measurements are done in the presence of simple surfactants [500], bile salts [501], complexing agents such as cyclodextrins [489–491,493], or ion-pair-forming counterions [492], extensive considerations need to be applied in attempting to extract the true aqueous solubility from the data. Such corrective measures are described below.

6.3   DATABASES AND THE "IONIZABLE MOLECULE PROBLEM"

Two sensibly priced commercial databases for solubility exist [366,507]. An article in the journal *Analytical Profiles of Drug Substances* carries solubility data [496]. Abraham and Le [508] published a list of intrinsic aqueous solubilities of 665 compounds, with many ionizable molecules. It is difficult to tell from published lists what the quality of the data for *ionizable* molecules is. Sometimes, it is not clear what the listed number stands for. For example, S_w, water solubility, can mean several different things: either intrinsic value, or value determined at a particular pH (using buffers), or value measured by saturating distilled water with excess compound. In the most critical applications using ionizable molecules, it may be necessary to scour the original publications in order to be confident of the quality of reported values.

6.4   EXPERIMENTAL METHODS

Lipinski et al. [12] and Pan et al. [463] compared several commonly used methods of solubility measurement in early discovery, where samples are often introduced as 10 mM DMSO solutions. Turbidity-based and UV plate scanner-based detections systems were found to be useful. The methods most often used in discovery and in preformulation will be briefly summarized below.

6.4.1 Saturation Shake-Flask Methods

Solubility measurement at a single pH [37–39] under equilibrium conditions is largely a labor-intensive procedure, requiring long equilibration times (12 h–7 days). It's a simple procedure. The drug is added to a standard buffer solution (in a flask) until saturation occurs, indicated by undissolved excess drug. The thermostated saturated solution is shaken as equilibration between the two phases is established. After microfiltration or centrifugation, the concentration of the substance in the supernatant solution is then determined using HPLC, usually with UV detection. If a solubility–pH profile is required, then the measurement needs to be performed in parallel in several different pH buffers.

6.4.2 Turbidimetric Ranking Assays

Turbidity-detection-based methods [12,459–463], popularized by Lipinski and others, in part have met some high-throughput needs of drug discovery research. The approach, although not thermodynamically rigorous, is an attempt to rank molecules according to expected solubilities. Usually, the measurements are done at one pH. Various implementations of the basic method are practiced at several pharmaceutical companies, using custom-built equipment. Detection systems based on 96-well microtiter plate nephelometers are well established. An automated solubility analyzer incorporating such a detector usually requires the user to develop an appropriate chemistry procedure and to integrate a robotic fluidic system in a customized way. It is important that turbidity methods using an analate addition strategy be designed to keep the DMSO concentration in the buffer solution constant in the course of the additions. The shortcomings of the turbidity methodology are (1) poor reproducibility for very sparingly water-soluble compounds, (2) use of excessive amounts ($\leq 5\%$ v/v) of DMSO in the analate addition step, and (3) lack of standardization of practice.

6.4.3 HPLC-Based Assays

In an effort to increase throughput, several pharmaceutical companies have transferred the classical saturation shake-flask method to 96-well plate technology using a robotic liquid dispensing system [463]. Analyses are performed with fast generic gradient reverse-phase HPLC. In some companies, the DMSO is eliminated by a freeze-drying procedure before aqueous buffers are added. This adds to the assay time and can be problematic with volatile samples (e.g., coumarin). Still, the serial chromatographic detection systems are inherently slow. Data handling and report generation are often the rate-limiting steps in the operations.

6.4.4 Potentiometric Methods

Potentiometric methods for solubility measurement have been reported in the literature [467–471]. A novel approach, called *dissolution template titration* (DTT), has been introduced [472–474]. One publication called it the "gold standard" [509].

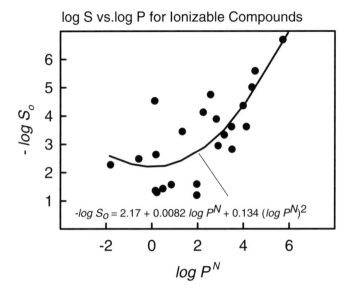

Figure 6.6   Empirical relationship between intrinsic solubility of ionizable molecules and their octanol–water log P [pION]. [Avdeef, A., *Curr. Topics Med. Chem.*, **1**, 277–351 (2001). Reproduced with permission from Bentham Science Publishers, Ltd.]

The procedure takes as input parameters the measured (or calculated) pK_a and the measured (or calculated) octanol–water partition coefficient, log P. The latter parameter is used to estimate the intrinsic solubility S_0, using the Hansch-type expression [38], log $S_0 = 1.17 - 1.38$ log P, or an improved version for ionizable molecules of moderate lipophilicity (Fig. 6.6):

$$\log S_0 = -2.17 - 0.0082 \ \log P - 0.134(\log P)^2 \qquad (6.13)$$

Using the pK_a and the estimated S_0, the DTT procedure simulates the entire titration curve before the assay commences. Figure 6.7 shows such a titration curve of propoxyphene. The simulated curve serves as a template for the instrument to collect individual pH measurements in the course of the titration. The pH domain containing precipitation is apparent from the simulation (filled points in Fig. 6.7). Titration of the sample suspension is done in the direction of dissolution (high to low pH in Fig. 6.7), eventually well past the point of complete dissolution (pH < 7.3 in Fig. 6.7). The rate of dissolution of the solid, described by the classical Noyes–Whitney expression [37], depends on a number of factors, which the instrument takes into account. For example, the instrument slows down the rate of pH data taking as the point of complete dissolution approaches, where the time needed to dissolve additional solid substantially increases (between pH 9 and 7.3 in Fig. 6.7). Only after the precipitate completely dissolves, does the instrument collect the remainder of the data rapidly (unfilled circles in Fig. 6.7). Typically, 3–10 h is required for the entire equilibrium solubility data taking. The more insoluble the

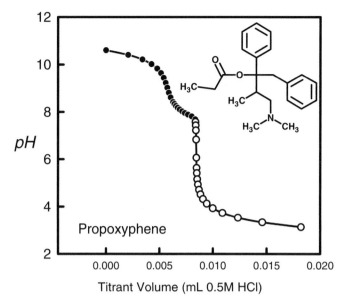

Figure 6.7 Dissolution template titration (DTT) curve of propoxyphene: 0.51 mg of the hydrochloride salt was dissolved in 5.1 mL of 0.15 M KCl solution, with 0.0084 mL of 0.5 M KOH used to raise the pH to 10.5.

compound is anticipated to be (based on the template) the longer the assay time. An entire solubility–pH profile is deduced from the assay.

A graphical analysis follows, based on Bjerrum plots (see Sections 3.3.1 and 4.14). The Bjerrum difference plots are probably the most important graphical tools in the initial stages of solution equilibrium analysis in the pH-metric method. The difference curve is a plot of $\bar{n}_H$, the average number of bound protons (i.e., the hydrogen ion binding capacity), versus p_cH ($-\log$ [H$^+$]). Such a plot can be obtained by subtracting a titration curve containing no sample ("blank" titration) from a titration curve with sample; hence the term "difference" curve. Another way of looking at it is as follows. Since it is known how much strong acid [HCl] and strong base [KOH] have been added to the solution at any point and since, it is known how many dissociable protons n the sample substance brings to the solution, the *total* hydrogen ion concentration in solution is known, regardless of what equilibrium reactions are taking place (model independence). By measuring the pH, and after converting it into p_cH [116], the *free* hydrogen ion concentration is known. The difference between the total and the free concentrations is equal to the concentration of the *bound* hydrogen ions. The latter concentration divided by that of the sample substance C gives the average number of bound hydrogen ions per molecule of substance $\bar{n}_H$

$$\bar{n}_H = \frac{([\text{HCl}] - [\text{KOH}] + nC - [\text{H}^+] + K_w/[\text{H}^+])}{C} \tag{6.14}$$

where K_w is the ionization constant of water (1.78×10^{-14} at 25°C, 0.15 M ionic strength).

Figure 6.8 shows the Bjerrum plots for an weak acid (benzoic acid, pK_a 3.98, log S_0 − 1.55, log mol/L [474]), a weak base (benzydamine, pK_a 9.26, log S_0 −3.83, log mol/L [472]), and an ampholyte (acyclovir, pK_a 2.34 and 9.23, log S_0 − 2.16, log mol/L [pION]). These plots reveal the pK_a and pK_a^{app} values as the p_cH values at half-integral $\bar{n}_H$ positions. By simple inspection of the dashed curves in Fig. 6.8, the pK_a values of the benzoic acid, benzydamine, and acyclovir are 4.0, 9.3, and (2.3, 9.2), respectively. The pK_a^{app} values depend on the concentrations used, as is evident in Fig. 6.8. It would not have been possible to deduce the constants by simple inspection of the titration curves (pH vs. volume of titrant, as in Fig. 6.7). The difference between pK_a and pK_a^{app} can be used to determine log S_0, the intrinsic solubility, or log K_{sp}, the solubility product of the salt, as will be shown below.

In addition to revealing constants, Bjerrum curves are a valuable diagnostic tool that can indicate the presence of chemical impurities and electrode performance problems [165]. Difference curve analysis often provides the needed "seed" values for refinement of equilibrium constants by mass-balance-based nonlinear least squares [118].

As can be seen in Fig. 6.8, the presence of precipitate causes the apparent pK_a, pK_a^{app}, to shift to higher values for acids and to lower values for bases, and in opposite but equal directions for ampholytes, just as with octanol (Chapter 4) and liposomes (Chapter 5). The intrinsic solubility can be deduced by inspection of the curves and applying the relationship [472]

$$\log S_0 = \log \frac{C}{2} - |pK_a^{app} - pK_a| \tag{6.15}$$

where C is the sample concentration. To simplify Eq. (6.15), Fig. 6.9 shows characteristic Bjerrum plots taken at 2 M concentration of an acid (ketoprofen, log S_0 − 3.33 [473]), a base (propranolol, log S_0 − 3.62 [473]), and an ampholyte (enalapril maleate, log S_0 − 1.36 [474]). In Fig. 6.9, all examples are illustrated with $C = 2\,M$, so that the difference between true pK_a and the apparent pK_a is directly read off as the log S_0 value.

In an ideally designed experiment, only a *single* titration is needed to determine the solubility constant *and* the aqueous pK_a. This is possible when the amount of sample, such as a weak base, added to solution is such that from the start of an alkalimetric titration (pH $\ll pK_a$) to the midbuffer region (pH $= pK_a$) the compound stays in solution, but from that point to the end of titration (pH $\gg pK_a$), precipitation occurs. (The idea is similar to that described by Seiler [250] for log P determinations by titration.) After each titrant addition, pH is measured. The curve represented by unfilled circles in Fig. 6.8b is an example of such a titration of a weak base whose pK_a is 9.3, with precipitation occurring above pH 9.3, with onset indicated by the "kink" in the curve at that pH. In practice, it is difficult to know *a priori* how much compound to use in order to effect such a special condition. So, two or more titrations may be required, covering a probable range of concentrations,

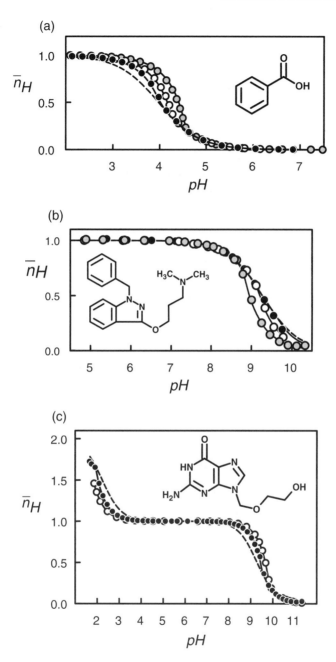

Figure 6.8 Bjerrum plots for (a) benzoic acid (black circle = 87 mM, unfilled circle = 130 mM, gray cicle = 502 mM), (b) benzydamine (black circle = 0.27 mM, unfilled circle = 0.41 mM, gray circle = 0.70 mM), and (c) acyclovir (black circle = 29 mM, unfilled circle = 46 mM). The dashed curves correspond to conditions under which no precipitation takes place.

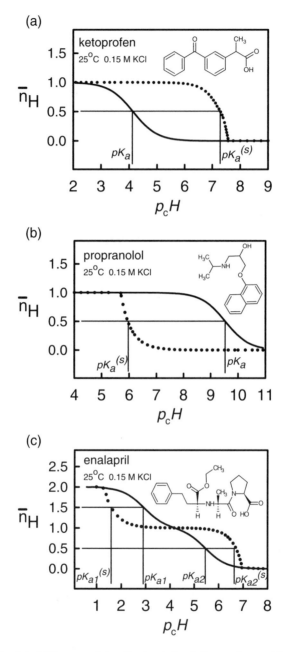

Figure 6.9 Simulated Bjerrum plots of saturated solutions of an acid, a base, and an ampholyte. The sample concentration was chosen as 2 M, a special condition where the difference between the true pK_a and the apparent pK_a is equal to $-\log S_0$. [Avdeef, A., *Curr. Topics Med. Chem.*, **1**, 277–351 (2001). Reproduced with permission from Bentham Science Publishers, Ltd.]

using as little sample as necessary to cause precipitation near the mid-point. For compounds extremely insoluble in water, cosolvents such as methanol, ethanol, DMSO, or acetonitrile may be used, with the solubility constant determined by extrapolation to zero cosolvent [43].

Usually, the solubility of the salt is determined from separate, more concentrated solutions. To conserve on sample, the titration of the salt may be performed with an excess of the counterion concentration [479]. Also, some amount of sample salt may be conserved by titrating in cosolvent mixtures, where salts are often less soluble.

The graphically deduced constants are subsequently refined by a weighted non-linear least squares procedure [472]. Although the potentiometric method can be used in discovery settings to calibrate high-throughput solubility methods and computational procedures, it is too slow for HTS applications. It is more at home in a preformulation lab.

6.4.5 Fast UV Plate Spectrophotometer Method

A high-throughput method using a 96-well microtiter plate format and plate UV spectrophotometry has been described [26]. Solubilities at a single pH, or at ≤ 12 pH values can be determined, using one of two methods.

6.4.5.1 *Aqueous Dilution Method*

A known quantity of sample is added to a known volume of a universal buffer solution of sufficient capacity and of known pH. The amount of sample must be sufficient to cause precipitation to occur in the formed saturated solution. After waiting for a period of time to allow the saturated solution to reach the desired steady state, the solution is filtered to remove the solid and obtain a clear solution, whose spectrum is then taken by the UV spectrophotometer. Mathematical treatment of the spectral data yields the area-under-the-curve of the filtered sample solution, AUC_S.

A reference solution is prepared by a dilution method. A known quantity of sample is dissolved in a known volume of the system buffer of known pH; the amount of sample is X times less than in the above case in order to avoid precipitation in the formed solution. The spectrum is immediately taken by the UV spectrophotometer, to take advantage of the possibility that solution may be "supersaturated" (i.e., solid should have precipitated, but because not enough time was allowed for the solid to precipitate, the solution was temporarily clear and free of solid). Mathematical treatment of the spectral data yields the AUC of the reference sample solution, AUC_R. The ratio $R = AUC_R/AUC_S$ is used to automatically recognize the right conditions for solubility determination: when the reference has no precipitate, and the sample solution is saturated with precipitate. Under these conditions, solubility is determined from the expression

$$S = \frac{C_R}{R} \tag{6.16}$$

TABLE 6.1 Intrinsic Solubility S_0, Corrected for the Drug DMSO/Drug Aggregation Effects

Compound	pK_a	S_0^{APP} (µg/mL)	Corrected S_0 (µg/mL)	pSOL S_0 (µg/mL)	Shake-Flask S_0 (µg/mL)
Amitriptyline	9.45[a]	56.9	3.0	2.0[a]	2.0[a]
Chlorpromazine	9.24[a]	19.4	3.4	3.5[a]	0.1[a]
Diclofenac	3.99[b]	22.6	3.8	0.8[b]	0.6[b]
Furosemide	10.63, 3.52[b]	29.8	2.9	5.9[b]	12.0[b](2.9[c])
Griseofulvin	Nonionizable	37.6	20.2	—	9[d]
Indomethacin	4.42[a]	7.2	4.1	2.0[a]	2.0[a], 1[e]
Miconazole	6.07[f]	11.1	1.6	0.7[f]	—
2-Naphthoic acid	4.16[f]	33.3	20.2	—	22.4[g]
Phenazopyridine	5.15[f]	12.2	12.2	14.3[f]	—
Piroxicam	5.07, 2.33[h]	10.5	1.1	—	9.1[i] (3.3[c]), 8–16[j] (2.2–4.4[c])
Probenecid	3.01[f]	4.6	0.7	0.6[f]	
Terfenadine	9.53[f]	4.4	0.1	0.1[f]	

[a]M. A. Strafford, A. Avdeef, P. Artursson, C. A. S. Johansson, K. Luthman, C. R. Brownell, and R. Lyon, Am. Assoc. Pharm. Sci. Ann. Mtng. 2000, poster presentation.
[b]Ref. 433.
[c]Corrected for aggregate formation: unpublished data.
[d]J. Huskonen, M. Salo, and J. Taskinen, J. Chem. Int. Comp. Soc. **38**, 450–456 (1998).
[e]Ref. 507.
[f]pION, unpublished data.
[g]K. G. Mooney, M. A. Mintun, K. J. Himmestein, and V. J. Stella, J. Pharm. Sci. **70**, 13–22 (1981).
[h]Ref. 162.
[i]C. R. Brownell, FDA, private correspondence, 2000.
[j]Ref. 500 (24 h).

where C_R is the calculated concentration of the reference solution. Some results are presented in Table 6.1. The apparent intrinsic solubilities S_0^{app}, determined in this way (eq. 6.16) are listed in column 3, for the compounds used in one study. All the S_0^{app} values reported in Table 6.1 were determined in the presence of 0.5% v/v DMSO, except for phenazopyridine, where 0.26% was used.

The results of a pH 4–9.5 solubility assay of chlorpromazine are shown in Fig. 6.10. The horizontal line represents the upper limit of measurable solubility (e.g., 125 µg/mL), which can be set by the instrument according to the requirements of the assay. When the measured concentration reaches the line, the sample is completely dissolved, and solubility cannot be determined. This is automatically determined by the instrument, based on the calculated value of R. When measured points fall below the line, the concentration corresponds to the apparent solubility S^{app}.

6.4.5.2 Cosolvent Method

The sample plate is prepared as in the preceding method. But before the spectra are taken, a volume Y of a water-miscible cosolvent is added to a volume Z of sample

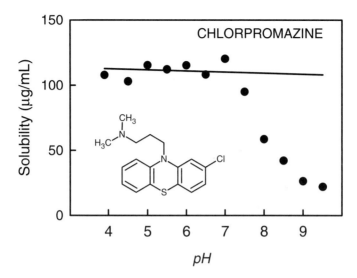

Figure 6.10 High-throughput solubility–pH determination of chlorpromazine. The horizontal line indicates the set upper limit of solubility, where the compound completely dissolves and solubility cannot be specified. The points below the horizontal line are measured in the presence of precipitation and indicate solubility. The solubility–pH curve was collected in the presence of 0.5 vol% DMSO, and is affected by the cosolvent (see text). [Avdeef, A., *Curr. Topics Med. Chem.*, **1**, 277–351 (2001). Reproduced with permission from Bentham Science Publishers, Ltd.]

solution to produce a new solution, in which the compound is now diluted by $Z/(Y + Z)$. Suitable cosolvents are ones with the lowest vapor pressure, the greatest capability in dissolving a solute (i.e., highest solubilizing power) and the lowest UV absorption. The spectrum of the solution is then immediately taken by the UV spectrophotometer. Mathematical treatment of the spectral data yields the area under the curve of the filtered cosolvent sample solution, $\text{AUC}_S^{\text{COS}}$.

The reference plate is prepared differently. A known quantity of sample is added to a known volume of system solution of known pH with the amount of sample *comparable* to that found in the sample plate, and no effort is made in this step to suppress precipitation in the formed solution. A volume Y of the cosolvent is immediately added to a volume Z of reference solution to produce a new solution, in which the compound is now diluted by $Z/(Y + Z)$. The spectrum of the solution is then immediately taken by the UV spectrophotometer. Mathematical treatment of the spectral data yields the area under the curve of the cosolvent reference solution, $\text{AUC}_R^{\text{COS}}$. Define $R^{\text{COS}} = \text{AUC}_R^{\text{COS}}/\text{AUC}_S^{\text{COS}}$. The solubility of the sample compound then is

$$S = \frac{(1 + Y/Z)C_R^{\text{COS}}}{R^{\text{COS}}} \tag{6.17}$$

where C_R^{COS} is the calculated concentration of the compound in the reference solution.

(a)

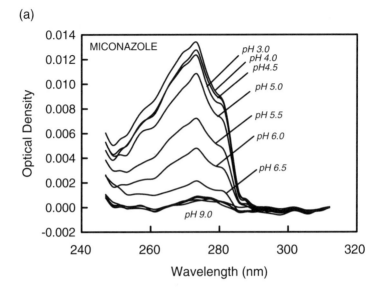

(b)

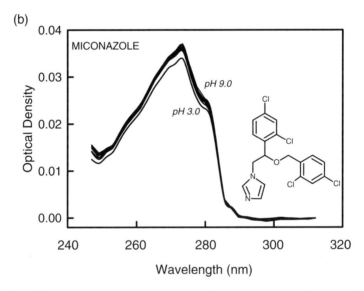

Figure 6.11 UV spectra of saturated solutions of miconazole as a function of pH: (a) sample; (b) reference. [Avdeef, A., *Curr. Topics Med. Chem.*, **1**, 277–351 (2001). Reproduced with permission from Bentham Science Publishers, Ltd.]

Figure 6.11 shows the measured absorption spectra of miconazole (reference and sample). As precipitation takes place to varying degrees at different pH values, the spectra of the sample solutions change in optical densities, according to Beer's law. This can be clearly seen in Fig. 6.12 for the sample spectra, where the sample spectra have the lowest OD values at pH 9.0 and systematically show higher OD values

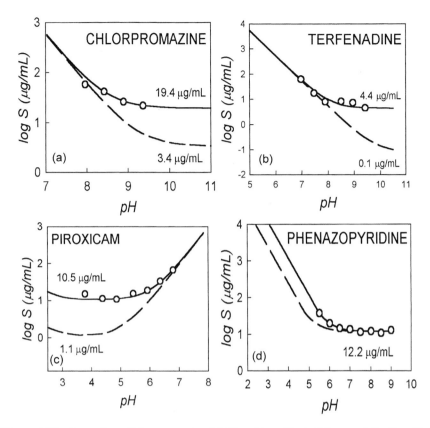

Figure 6.12 Correction of the apparent solubility–pH profile (solid curves) for the effect of DMSO and/or aggregation: (a) chlorpromazine; (b) terfenadine; (c) piroxicam; (d) phenazopyridine. [Avdeef, A., *Curr. Topics Med. Chem.*, **1**, 277–351 (2001). Reproduced with permission from Bentham Science Publishers, Ltd.]

as pH is lowered, a pattern consistent with that of an weak base. The changing OD values indicate that solubility changes with pH.

6.5 CORRECTION FOR THE DMSO EFFECT BY THE Δ-SHIFT METHOD

6.5.1 DMSO Binding to the Uncharged Form of a Compound

It was found that the log S/pH curves were altered in the presence of as little as 0.5% v/v DMSO, in that the apparent pK_a values, pK_a^{app}, derived from log S versus pH [481], were different from the true pK_a values by about one log unit. The pK_a^{app} values were generally higher than the true pK_a values for weak acids (positive shift), and lower than those for weak bases (negative shift). This has been called the "Δ shift" [Avdeef, unpublished]. It is thought to be caused in some cases by DMSO binding to the drugs. Just as the equilibrium model in Section 6.1.3 was expanded to allow for the salt solubility equilibrium, Eq. (6.4), the same can be done with a binding equation based on DMSO (e.g., in 0.5% v/v);

$$HA + nDMSO \rightleftarrows HA(DMSO)_n \qquad (6.18)$$

Such a reaction can cause a shift in the apparent ionization constant. It was discovered that the Δ shift, when subtracted from the logarithm of the apparent (DMSO-distorted) solubility S_0^{APP}, yields the true aqueous solubility constant:

$$\log S_0 = \log S_0^{APP} \pm \Delta \qquad (6.19)$$

where $\pm$ includes $-$ for acids and $+$ for bases. For an amphoteric molecule (which has both acid and base functionality) with two pK_a values either sign may be used, depending on which of the two values is selected. DMSO makes the compound appear more soluble, but the true aqueous solubility can be determined from the apparent solubility by subtracting the pK_a difference. Figure 6.12 illustrates the apparent solubility–pH curve (solid line) and the true aqueous solubility–pH curve (dashed line), correcting for the effect of DMSO for several of the molecules considered.

6.5.2   Uncharged Forms of Compound–Compound Aggregation

Shifts in pK_a can also be expected if water-soluble aggregates form from the uncharged monomers. This may be expected with surface-active molecules or molecules such as piroxicam [500]. Consider the case where no DMSO is present, but aggregates form, of the sort

$$mHA \rightleftarrows (HA)_m \qquad (6.20)$$

The working assumption is that the aggregates are water soluble, that they effectively make the compound appear more soluble. If ignored, they will lead to erroneous assessment of intrinsic solubility. It can be shown that Eq. (6.19) also applies to the case of aggregation.

6.5.3   Compound–Compound Aggregation of Charged Weak Bases

Consider the case of a weak base, where the *protonated*, positively charged form self-associates to form aggregates, but the uncharged form does not. This may be the case with phenazopyridine (Fig. 6.12). Phenazopyridine is a base that consistently shows *positive* shifts in its apparent pK_a, the opposite of what's expected of uncharged compound DMSO or aggregation effects. A rationalization of this effect can be based on the formation of partially protonated aggregates (perhaps micelles). Assume that one of the species is $(BH^+)_n$.

$$nBH^+ \rightleftarrows (BH^+)_n \qquad (6.21)$$

It can be shown that for such a case, the observed solubility–pH curve is shifted horizontally, not vertically, as with uncharged-compound DMSO/aggregation effects, and that the apparent intrinsic solubility is not affected by the phenomenon.

6.5.4 Ionizable Compound Binding by Nonionizable Excipients

It can be postulated that a number of phenomena, similar to those of reactions in Eqs. (6.17), (6.19), and (6.20), will shift the apparent pK_a in a manner of the discussions above. For example, the additives in drug formulations, such as surfactants, bile salts, phospholipids, ion-pair-forming counterions, cyclodextrins, or polymers may make the drug molecule appear more soluble. As long as such excipients do not undergo a change of charge state in the pH range of interest (i.e., the excipients are effectively non-ionizable), and the drug molecule is ionizable in this range, the difference between the apparent pK_a, pK_a^{app}, and the true pK_a will reveal the true aqueous solubility, as if the excipient were not present. Table 6.2 summarizes some of the relationships developed between solubility, pK_a, and pK_a^{app}.

6.5.5 Results of Aqueous Solubility Determined from Δ Shifts

Since the pK_a values of the compounds studied are reliably known (Table 6.1), it was possible to calculate the Δ shifts (Table 6.2). These shifts were used to calculate the corrected aqueous intrinsic solubilities S_0, also listed in Table 6.2.

TABLE 6.2 True Aqueous Solubility Determined from pK_a Shifts of Monoprotic Compounds

Ionizable Compound Type	$\Delta = pK_a^{APP} - pK_a$	True Aqueous log S_0	Examples
Acid	$\Delta > 0$	log $S_0^{APP} - \Delta$	Diclofenac, furosemide, indomethacin, probenecid, naphthoic acid
Acid	$\Delta \leq 0$	log S_0^{APP}	Prostaglandin F2a [485]
Base	$\Delta \geq 0$	log S_0^{APP}	Phenazopyridine
Base	$\Delta < 0$	log $S_0^{APP} + \Delta$	Amitriptyline, chlorpromazine, miconazole, terfenadine

Table 6.3   Solubility Constants of Drug Molecules, Measured by the Dissolution Template Titration Method[a]

Compound	$-\log S_0$ (log mol L^{-1})	Ref.
Acyclovir	2.24	506
Amiloride	3.36	506
Amiodarone	8.10	pION
Amitriptyline	5.19	506,—[b]
Amoxicilin	2.17	506
Ampicillin	1.69	pION
Atenolol	1.30	473
Atropine	1.61	506
Benzoic acid	1.59	474
Benzydamine	3.83	472
Bromocriptine	4.70	509
Cephalexin	1.58	pION
Chlorpromazine	5.27	506,—[b]
Cimetidine	1.43	474
Ciprofloxacin	3.73	506
Clozapine	3.70	509
Desipramine	3.81	506
Diclofenac	5.59	473
Diltiazem	2.95	474
Doxycycline	2.35	506
Enalapril	1.36	474
Erythromycin	3.14	506
Ethinyl estradiol	3.95	506
Famotidine	2.48	473
Flurbiprofen	4.36	473
Furosemide	4.75	473
Hydrochlorothiazide	2.63	473
Ibuprofen	3.62	473
Indomethacin	5.20	506,—[b]
Ketoprofen	3.33	473
Labetolol	3.45	473
Lasinavir	4.00	509
Methotrexate	4.29	506
Metoprolol	1.20	474
Miconazole	5.85	25
Metolazone	4.10	509
Nadolol	1.57	474
Nalidixic acid	4.26	pION
2-Naphthoic acid	3.93	25
Naproxen	4.21	473
Norfloxacin	2.78	pION
Nortriptyline	4.18	pION

Table 6.3   (*Continued*)

Compound	$-\log S_0$ (log mol L^{-1})	Ref.
Phenazopyradine	4.24	506,—[b]
Phenytoin	4.13	473
Pindolol	3.70	*p*ION
Piroxicam	5.48	*p*ION
Primaquine	2.77	506
Probenecid	5.68	25
Promethazine	4.39	506
Propoxyphene	5.01	474
Propranolol	3.62	473
Quinine	2.82	474
Rufinamide	3.50	509
Tamoxifen	7.55	506
Terfenadine	6.69	474
Theophylline	1.38	506
Trovafloxacin	4.53	474
Valsartan	4.20	509
Verapamil	4.67	506
Warfarin	4.74	506
Zidovudine	1.16	506

[a]Temperature 25°C, 0.15 M ionic strength (KCl).
[b]M. A. Strafford, A. Avdeef, P. Artursson, C. A. S. Johansson, K. Luthman, C. R. Brownell, and R. Lyon, Am. Assoc. Pharm. Sci. Ann. Mtng. 2000, poster presentation.

6.6   LIMITS OF DETECTION

The HTS method of Section 6.4.5 can reproduce to 0.1 μg/mL. Turbidity-based methods have sensitivities well above 1 μg/mL. The pH-metric method can decrease to 5 ng/mL [Avdeef, unpublished]. Reports of such a low limit of detection can be found in the literature [495].

6.7   log S_0 "GOLD STANDARD" FOR DRUG MOLECULES

Table 6.3 lists a set of reliably determined log S_0 solubility constants for a series of ionizable drugs determined by the pH-metric DTT solubility method.

CHAPTER 7

PERMEABILITY

7.1 PERMEABILITY IN THE GASTROINTESTINAL TRACT AND AT THE BLOOD–BRAIN BARRIER

Measured permeability (especially when combined with solubility and charge state) can be viewed as a surrogate property for predicting oral (gastrointestinal) absorption of preclinical drug candidate molecules. This chapter considers the transport of molecules by passive diffusion through phospholipid bilayers. The emphasis is on (1) the current state-of-the-art measurement of permeabilities by the so-called PAMPA method and (2) the theoretical physicochemical models that attempt to rationalize the observed transport properties. Such models are expected to lead to new insights into the in vivo absorption processes. In oral absorption predictions, the established in vitro assay to assess the permeability coefficients is based on Caco-2 cultured-cell confluent monolayers [48,510–515]. We refer to this topic in various places, drawing on the biophysical aspects of the work reported in the literature. We also consider some physicochemical properties of the blood–brain barrier (BBB), insofar as they contrast to those of the gastrointestinal tract (GIT). Our main focus, however, is on results derived from simpler in vitro systems based on artificial membranes.

In order to rationalize membrane permeability and oral absorption in terms of physicochemical drug properties, good experimental data and sound theoretical

Absorption and Drug Development: Solubility, Permeability, and Charge State. By Alex Avdeef
ISBN 0-471-423653. Copyright © 2003 John Wiley & Sons, Inc.

models are needed. Since lipophilicity is such an important concept in ADME (absorption, distribution, metabolism, excretion) predictions, models that address the permeability–lipophilicity relationships are expected to provide important insights. In the simplest models, permeability is linearly related to the membrane–water partition coefficient [Eq. (2.3)], but in practice, linearity is not generally observed over a wide range of lipophilicities. To explain this, different theoretical models for passive membrane diffusion have been described in the literature.

In assays based on synthetic membranes, the nonlinearity may be caused by (1) unstirred water layer; (2) aqueous pores in oily membranes; (3) membrane retention of lipophilic solute; (4) excessive lipophilicity (non-steady-state conditions, long acceptor-side solute desorption times); (5) transmembrane pH gradients; (6) effects of buffers (in the unstirred water layer); (7) precipitation of solute in the donor side; (8) aggregation of solutes in the donor side (slowing diffusion); (9) specific hydrogen bonding, electrostatic, and hydrophobic/lipophilic interactions with membrane constituents; (10) solute charge state (pK_a effects) and membrane surface charge (Gouy–Chapman effects); and (11) the use of inappropriate permeability equations (e.g., neglecting membrane retention of lipophilic drugs).

In vitro systems based on cultured cells are subject to all the above mentioned nonlinear effects, plus those based on the biological nature of the cells. The apical and basolateral membranes have different lipid components, different surface charge domains, and different membrane-bound proteins. Active transporters abound. Some enhance permeability of drugs, others retard it. A very important efflux system, P-gp (where "P" denotes permeability), prevents many potentially useful drugs from passing into the cells. P-gp is particularly strongly expressed in the BBB and in cancer cells. The junctions between barrier cells can allow small molecules to permeate through aqueous channels. The tightness of the junctions varies in different parts of the GIT. The junctions are particularly tight in the endothelial cells of the BBB. The GIT naturally has a pH gradient between the apical and basolateral sides of the epithelial cell barrier. Metabolism plays a critical role in limiting bioavailability of drugs.

In penetrating biological barriers, drugs may have simultaneous access to several different mechanisms of transport. To develop an integrated model for the biological processes related to oral absorption is a daunting challenge, since many of these processes are not entirely understood. Most practical efforts have been directed to deriving sufficiently general core models for *passive* membrane transport (both transcellular and paracellular), addressing many of the effects observed in artificial membrane studies, as listed above. Components of the active transport processes, derived from more complex in vitro cultured-cell models, can then be layered on top of the core passive models.

In the bulk of this chapter we will focus on the rapidly emerging new in vitro technology based on the use of immobilized artificial membranes, constructed of phospholipid bilayers supported on lipophilic filters. One objective is to complete the coverage of the components of the transport model explored in Chapter 2, by considering the method for determining the top curve (horizontal line) in the plots in Fig. 2.2 (i.e., intrinsic permeabilities P_0 of drugs). Also, a new model for gastrointestinal (oral) absorption based on permeability measurements using artificial membranes will be presented.

Approximately 1400 measurements of permeability are presented in tables and figures in this chapter. Most of the data are original, not published previously. Unless otherwise noted, the permeability and membrane retention data are from pION's laboratories, based on the permeation cell design developed at pION. Cells of different designs, employing different filter and phospholipid membrane materials, produce different permeability values for reasons discussed below. Although the analysis of the measurements is the basis of the presentation in this chapter, much of the data can be further mined for useful quantitative structure–property information, and the reader is encouraged to do so. First-person references in this chapter, such as "our laboratory," refer specifically to pION's laboratory, and "our results" are those of several colleagues who have contributed to the effort, covering a period of >4 years, as cited in the acknowledgment section. Where possible, comparisons to published permeability results from other laboratories will be made.

The survey of over 50 artificial lipid membrane models (pION) in this chapter reveals a new and very promising in vitro GIT model, based on the use of levels of lecithin membrane components higher than those previously reported, the use of negatively charged phospholipid membrane components, pH gradients, and artificial sink conditions. Also, a novel direction is suggested in the search for an ideal in vitro BBB model, based on the salient differences between the properties of the GIT and the BBB.

We return to using the K_p and K_d symbols to represent the partition coefficient and the apparent partition (distribution) coefficient, respectively. The effective, apparent, membrane, and intrinsic permeability coefficients are denoted P_e, P_a, P_m, and P_0, respectively, and D refers to the diffusivity of molecules.

The coverage of permeability in this book is more comprehensive than that of solubility, lipophilicity, and ionization. This decision was made because permeability is not as thoroughly treated in the pharmaceutical literature as the other topics, and also because much emphasis is placed on the PAMPA in this book, which is indeed a very new technique [547] in need of elaboration.

7.2 HISTORICAL DEVELOPMENTS IN ARTIFICIAL-MEMBRANE PERMEABILITY MEASUREMENT

7.2.1 Lipid Bilayer Concept

The history of the development of the bilayer membrane model is fascinating, and spans at least 300 years, beginning with studies of soap bubbles and oil layers on water [517–519].

In 1672 Robert Hooke observed under a microscope the growth of "black" spots on soap bubbles [520]. Three years later Isaac Newton [521], studying the "images of the Sun very faintly reflected [off the black patched on the surface of soap bubbles]," calculated the thickness of the black patches to be equivalent to 95 Å. (Anders Jonas Ångström, 'father of spectroscopy,' who taught at the University of Uppsala, after whom the Å unit is named, did not appear until about 150 years later.)

Ben Franklin, a self-trained scientist of eclectic interests, but better known for his role in American political history, was visiting England in the early 1770s. He published in the Philosophical Transactions of the Royal Society in 1774 [552]:

> At length being at Clapham where there is, on the common, a large pond, which I observed to be one day very rough with the wind, I fetched out a cruet of oil, and dropt a little of it on the water . . . and there the oil, though not more than a tea spoonful, . . . spread amazingly, and extended itself gradually till it reached the lee side, making all that quarter of the pond, perhaps half an acre, as smooth as a looking glass . . . so thin as to produce prismatic colors . . . and beyond them so much thinner as to be invisible.

Franklin mentioned Pliny's account of fisherman pouring oil on troubled waters in ancient times, a practice that survives to the present. (Franklin's experiment was reenacted by the author at the pond on Clapham Common with a teaspoon of olive oil. The spreading oil covered a surface not larger than that of a beach towel–it appears that technique and/or choice of oil is important. The olive oil quickly spread out in circular patterns of brilliant prismatic colors, but then dissolved from sight. Indeed, the pond itself has shrunken considerably over the intervening 230 years.)

More than 100 years later, in 1890, Lord Rayleigh, a professor of natural philosophy at the Royal Institution of London, was conducting a series of quantitative experiments with water and oil, where he carefully measured the area to which a volume of oil would expand. This led him to calculate the thickness of the oil film [517,518]. A year after publishing his work, he was contacted by a German woman named Agnes Pockels, who had done extensive experiments in oil films in her kitchen sink. She developed a device for carefully measuring the exact area of an oil film. Lord Rayleigh helped Agnes Pockels in publishing her results in scientific journals (1891–1894) [517,518].

Franklin's teaspoon of oil (assuming a density 0.9 g/mL and average fatty-acid molecular weight 280 g/mol) would contain 10^{+22} fatty-acid tails. The half-acre pond surface covered by the oil, $\sim$2000 m^2, is about $2 \times 10^{+23}$ Å^2. So, each tail would be expected to occupy about 20 Å^2, assuming that a single monolayer (25 Å calculated thickness) of oil formed on the surface of the pond.

Pfeffer in 1877 [523] subjected plant cell suspensions to different amounts of salt and observed the cells to shrink under hypertonic conditions and swell in hypotonic conditions. He concluded there was a semipermeable membrane separating the cell interior from the external solution, an invisible (under light microscope) plasma membrane.

Overton in the 1890s at the University of Zürich carried out some 10,000 experiments with more than 500 different chemical compounds [518,524]. He measured the rate of absorption of the compounds into cells. Also, he measured their olive oil–water partition coefficients, and found that lipophilic compounds readily entered the cell, whereas hydrophilic compounds did not. This lead him to conclude that the cell membrane must be oil-like. The correlation that the greater the lipid solubility of a compound, the greater is the rate of penetration of the plasma membrane became known as *Overton's rule*. Collander confirmed these observations but noted that some small hydrophilic molecules, such as urea and glycerol, could also pass into cells. This could be explained if the plasma membrane contained water-filled pores. Collander and Bärlund concluded that molecular size and lipophilicity are two important properties for membrane uptake [525].

Fricke measured resistance of solutions containing suspensions of red blood cells (RBCs) using a Wheatstone bridge [518]. At low frequencies the impedance of the suspensions of RBC was very high. But at high frequencies, the impedance decreased to a low value. If cells were surrounded by a thin membrane of low dielectric material, of an effective resistance and a capacitance in parallel to the resistor, then current would flow around the cells at low frequencies, and "through" the cells (shunting through the capacitor) at high frequencies. Hober in 1910 evaluated the equivalent electrical circuit model and calculated the thickness of the RBC membrane to be 33 Å if the effective dielectric constant were 3 and 110 Å if the effective dielectric constant were 10 [518].

In 1917 Langmuir [526], working in the laboratories of General Electric, devised improved versions of apparatus (now called the Langmuir trough) originally used by Agnes Pockels, to study properties of monolayers of amphiphilic molecules at the air–water interface. The technique allowed him to deduce the dimensions of fatty acids in the monolayer. He proposed that fatty acid molecules form a monolayer on the surface of water by orienting themselves vertically with the hydrophobic hydrocarbon chains pointing away from the water and the lipophilic carboxyl groups in contact with the water.

Gorter and Grendel in 1925 [527], drawing on the work of Langmuir, extracted lipids from RBC ghosts and formed monolayers. They discovered that the area of the monolayer was twice that of the calculated membrane surface of intact RBC, indicating the presence of a "bilayer." This was the birth of the concept of a lipid bilayer as the fundamental structure of cell membranes (Fig. 7.1).

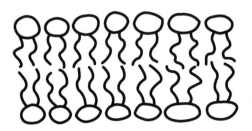

Figure 7.1   Lipid bilayer.

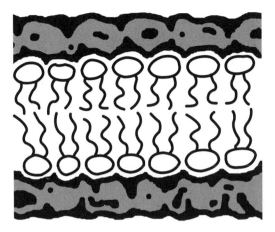

Figure 7.2 Danielli–Davson membrane model. A layer of protein was thought to sandwich a lipid bilayer.

The first membrane model to be widely accepted was that proposed by Danielli and Davson in 1935 [528]. On the basis of the observation that proteins could be adsorbed to oil droplets obtained from mackerel eggs and other research, the two scientists at University College in London proposed the "sandwich" of lipids model (Fig. 7.2), where a bilayer is covered on both sides by a layer of protein. The model underwent revisions over the years, as more was learned from electron microscopic and X-ray diffraction studies. It was eventually replaced in the 1970s by the current model of the membrane, known as the fluid mosaic model, proposed by Singer and Nicolson [529,530]. In the new model (Fig. 7.3), the lipid bilayer was retained, but the proteins were proposed to be globular and to freely float within the lipid bilayer, some spanning the entire bilayer.

Mueller, Rudin, Tien, and Wescott in 1961, at the Symposium of the Plasma Membrane [531] described for the first time how to reconstitute a lipid bilayer

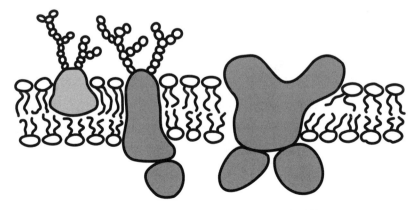

Figure 7.3 Fluid mosaic modern model of a bilayer.

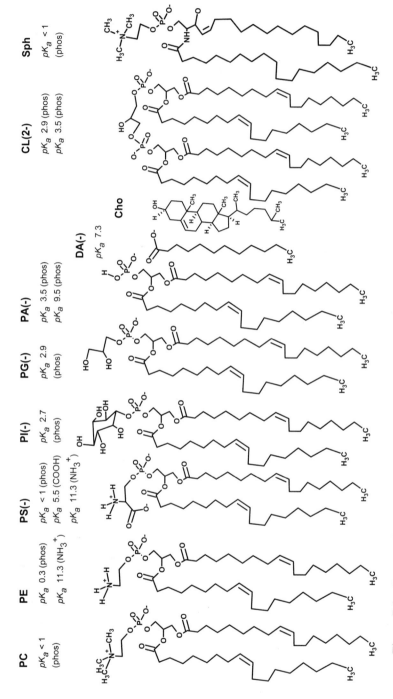

Figure 7.4 Common lipid components of biological membranes. For simplicity, all acyl chains are shown as oleyl residues.

in vitro. It is considered the seminal work on the self-assembly of planar lipid bilayers [516,518,519,531,532]. Their research led them to the conclusion that a soap film in its final stages of thinning has a structure of a single bilayer, with the oily tails of detergent molecules pointing to the side of air, and the polar heads sandwiching a layer of water. Their experimental model drew on three centuries of observations, beginning with the work of Hooke. The membranes prepared by the method of Rudin's group became known as *black lipid membranes* (BLMs). Soon thereafter, vesicles with walls formed of lipid bilayers, called *liposomes*, were described by Bangham [533].

7.2.2 Black Lipid Membranes (BLMs)

Mueller et al. [516,531,532] described in 1961 that when a small quantity of a phospholipid (1-2% wt/vol n-alkane or squalene solution) was carefully placed over a small hole (0.5 mm diameter) in a thin sheet of teflon or polyethylene (10–25 μm thick), a thin film gradually forms at the center of the hole, with excess lipid flowing toward the perimeter (forming a "plateau–Gibbs border"). Eventually, the central film turns optically black as a single (5-nm-thick) bilayer lipid membrane (BLM) forms over the hole. Suitable lipids for the formation of a BLM are mostly isolated from natural sources, including phosphatidylcholine (PC), phosphatidylethanolamine (PE), phosphatidylserine (PS), phosphatidylinositol (PI), and sphingomyelin (Sph). Such membranes have been viewed as useful models of the more complex natural membranes [516,532–544]. Figure 7.4 shows the most common membrane components. Sphingomyelin is an example of a broad class of sphingolipids, which include cerebrosides (carbohydrates attached to the head groups) and gangliosides (found in plasma membrane of nerve cells).

However, a serious drawback in working with BLMs as a model system is that they are extremely fragile (requiring a vibration-damping platform and a Faraday cage during measurements of electrical properties), and tedious to make [536–542]. That notwithstanding, Walter and Gutknecht [537] studied the permeation of a series of simple carboxylic acids across eggPC/decane BLMs. Intrinsic permeability coefficients, P_0, were determined from tracer fluxes. A straight-line relationship was observed between log P_0 and hexadecane–water partition coefficients, log K_p, for all but the smallest carboxylic acid (formic): log $P_0 = 0.90$ log $K_p + 0.87$. Using a similar BLM system, Xiang and Anderson [538] studied the pH-dependent transport of a series of α-methylene-substituted homologs of p-toluic acid. They compared the eggPC/decane permeabilities to partition coefficients determined in octanol–, hexadecane–, hexadecene–, and 1,9-decadiene–water systems. The lowest correlation was found from comparisons to the octanol–water scale. With the hexadecane–water system, log $P_0 = 0.85$ log $K_p - 0.64$ (r^2 0.998), and with decadiene–water system, log $P_0 = 0.99$ log $K_p - 0.17$ (r^2 0.996). Corrections for the unstirred water layer were key to these analyses. Figure 7.5 shows the linear correlation between the logarithms of the permeability coefficients and the partition coefficients for the five lipid systems mentioned above.

BLACK LIPID MEMBRANE (BLM) PERMEABILITY - LIPOPHILICITY

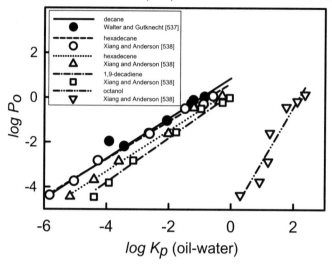

Figure 7.5   Intrinsic permeabilities of ionizable acids versus oil–water partition coefficients.

7.2.3 Microfilters as Supports

Efforts to overcome the limitations of the fragile membranes (as delicate as soap bubbles) have evolved with the use of membrane supports, such as polycarbonate filters (straight-through pores) [543] or other more porous microfilters (sponge-like pore structure) [545–548].

Thompson et al. [543] explored the use of polycarbonate filters, and performed experiments to make the case that just single bilayer membranes formed in each of the straight-through pores. Several possible pore-filling situations were considered: lipid–solvent plug, lipid–solvent plug plus BLM, multilamellar BLM, and unilamellar BLM. The key experiment in support of a single-bilayer disposition involved the use of amphotericin B (Fig. 7.6), which is an amphiphilic polyene zwitterionic molecule, not prone to permeate bilayers, but putatively forming tubular membrane-spanning oligomers if the molecules are first introduced from both sides of a bilayer, as indicated schematically in Fig. 7.6. Once a transmembrane oligomer forms, small ions, such as Na^+ or K^+, are able to permeate through the pore formed. The interpretation of the voltage–current curves measured supported such a single-bilayer membrane structure when polycarbonate microfilters are used as a scaffold support.

Cools and Janssen [545] studied the effect of background salt on the permeability of warfarin through octanol-impregnated membranes (Millipore ultrafiltration filters, VSWP, 0.025-μm pores). At a pH where warfarin was in its ionized form, it was found that increasing background salt increased permeability (Fig. 7.7). This

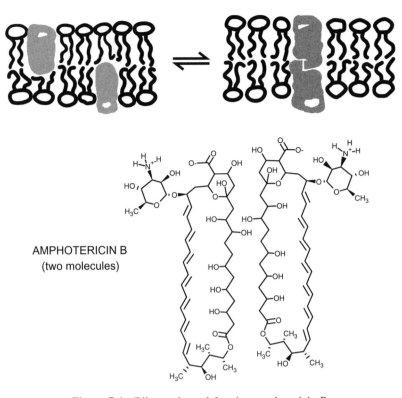

AMPHOTERICIN B
(two molecules)

Figure 7.6 Bilayer channel-forming amphotericin B.

ION-PAIR PERMEABILITY

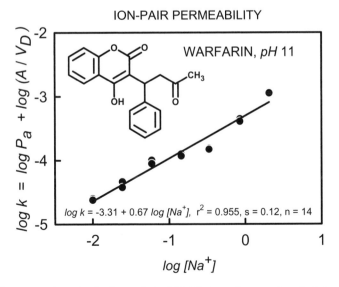

WARFARIN, *pH* 11

$\log k = -3.31 + 0.67 \log [Na^+]$, $r^2 = 0.955$, $s = 0.12$, $n = 14$

$\log k = \log P_a + \log (A / V_D)$

$\log [Na^+]$

Figure 7.7 Permeation of anionic warfarin (pH 11) through octanol-soaked (impregnated) microfilter as a function of sodium ion concentration.

observation was thought to support an ion pair mechanism of transport of charged drugs across real biological membranes. However, current understanding of the structure of wet octanol (Fig. 2.7), suggests that this isotropic solvent system may not be a suitable model for passive diffusion of *charged* drugs across phospholipid bilayers, since the water clusters in octanol may act as 'shuttles' for the transport of ion pairs. This would not be expected under in vivo conditions.

Camenisch et al. [546] measured the pH 7.4 permeabilities of a diverse group of drugs across octanol-and isopropylmyristate-impregnated artificial membranes (Millipore GVHP mixed cellulose ester filters, 0.22 µm pores), and compared them to permeabilities of the Caco-2 system, and octanol–water apparent partition coefficients, log $K_{d(7.4)}$. The uncharged drug species diffused passively, in accordance with the pH partition hypothesis. (When the GVHP membrane was not impregnated with a lipid, the permeabilities of all the tested drugs were high and largely undifferentiated, indicating only the unstirred water layer resistance.) Over the range of lipophilicities, the curve relating the effective permeability, log P_e, to log $K_{d(7.4)}$ was seen as sigmoidal in shape, and only linear in midrange; between log $K_{d(7.4)} - 2$ and 0, log P_e values correlated with the apparent partition coefficients (Fig. 7.8). However, outside that range, there was no correlation between permeabilities and the octanol–water partition coefficients. At the high end, the permeabilities of very lipophilic molecules were limited by the unstirred water layer. At the other end, very hydrophilic molecules were observed to be more permeable than predicted by octanol, due to an uncertain mechanism.

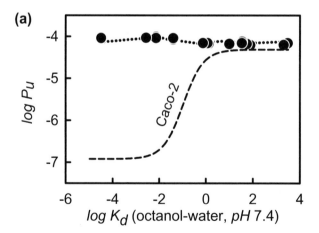

Figure 7.8 Permeation of drugs through oil-soaked microfilters; comparisons to Caco-2 permeabilities (dashed curves) [546]: (a) oil-free (untreated hydrophilic) filters; (b) unstirred water layer permeability versus log MW; (c) octanol-soaked (impregnated) filters; (d) isopropylmyristate-soaked filters.

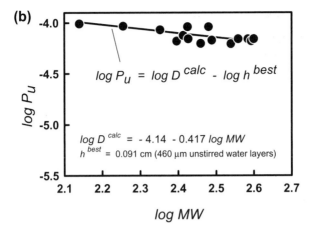

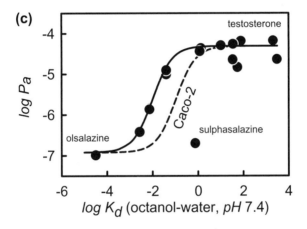

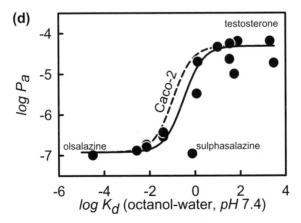

Figure 7.8 (*Continued*)

7.2.4 Octanol-Impregnated Filters with Controlled Water Pores

Ghosh [548] used cellulose nitrate microporous filters (500 μm thick) as scaffold material to deposit octanol into the pores and then under controlled pressure conditions, displace some of the oil in the pores with water, creating a membrane with parallel oil and water pathways. This was thought to serve as a possible model for some of the properties of the outermost layer of skin, the stratum corneum. The relative proportions of the two types of channel could be controlled, and the properties of 5–10% water pore content were studied. Ibuprofen (lipophilic) and antipyrine (hydrophilic) were model drugs used. When the filter was filled entirely with water, the measured permeability of antipyrine was 69 (in 10^{-6} cm/s); when 90% of the pores were filled with octanol, the permeability decreased to 33; 95% octanol content further decreased permeability to 23, and fully octanol-filled filters indicated 0.9 as the permeability.

7.3 PARALLEL ARTIFICIAL-MEMBRANE PERMEABILITY ASSAY (PAMPA)

7.3.1 Egg Lecithin PAMPA Model (Roche Model)

Kansy et al. [547,550] from Hoffmann-La Roche published a widely read study of the permeation of drugs across phospholipid-coated filters, using a technique they coined as "PAMPA," which stands for parallel artificial-membrane permeability assay. Their report could not have come at a better time—just when the paradigm was shifting into screening for biopharmaceutic properties at high speeds, along side the biological screening.

In the commercial version of the PAMPA assay, a "sandwich" (Fig. 7.9) is formed from a specially-designed 96-well microtiter plate [*p*ION] and a 96-well microfilter plate [several sources], such that each composite well is divided into two chambers: donor at the bottom and acceptor at the top, separated by a 125-μm-thick microfilter disk (0.45 μm pores, 70% porosity, 0.3 cm^2 cross-sectional area), coated with a 10% wt/vol dodecane solution of egg lecithin (a mixed lipid containing mainly PC, PE, a slight amount of PI, and cholesterol), under conditions that multilamellar bilayers are expected to form inside the filter channels when the system contacts an aqueous buffer solution [543].

The Roche investigators were able to relate their measured fluxes to human absorption values with a hyperbolic curve, much like that indicated in Caco-2

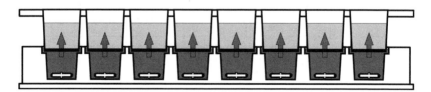

Figure 7.9 Cross section of a *p*ION 96-well microtitre plate PAMPA sandwich assembly.

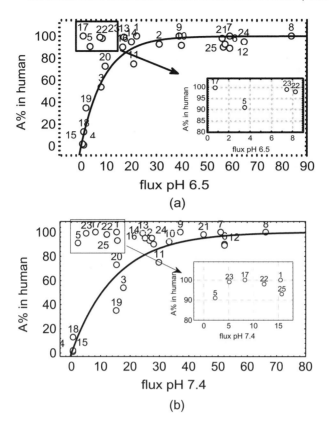

Figure 7.10 Absorption% versus PAMPA flux [547]: (a) pH 6.5; (b) pH 7.4. [Reprinted from Kansy, M.; Senner, F.; Gubernator, K. *J. Med. Chem.,* **41**, 1070–1110 (1998), with permission from the American Chemical Society.]

screening [48,82,91,97,108–110,510–515,551–553]. The outliers in their assays, inset in Fig. 7.10, were molecules known to be actively transported. Since the artificial membranes have no active transport systems and no metabolizing enzymes, the assay would not be expected to model actively transported molecules. What one sees with PAMPA is pure passive diffusion, principally of the uncharged species.

More recently, several publications have emerged, describing PAMPA-like systems. [25–28,509, 554–565] The PAMPA method has attracted a lot of favorable attention, and has spurred the development of a commercial instrument, [25–28,556] and the organization of the first international symposium on PAMPA in 2002 [565].

7.3.2 Hexadecane PAMPA Model (Novartis Model)

Faller and Wohnsland [509,554] developed the PAMPA assay using phospholipid-free hexadecane, supported on 10-μm thick polycarbonate filters(20% porosity,

0.3 cm^2 cross-sectional area), and were able to demonstrate interesting predictions. Their PAMPA method appeared to be a satisfactory substitute for obtaining alkane-water partition coefficients, which are usually very difficult to measure directly, due to the poor solubility of drug molecules in alkanes. They applied the pH-based methods of Walter and Gutknecht [537] to extract the intrinsic permeability coefficients, P_0, of the molecules they studied. A plot of log P_0 vs. hexadecane-water log K_d is a straight line with a slope of 0.86 (r^2 0.96), as shown in Fig. 7.11. Apparently, membrane retention was not measured in the original version of the method. A later measurement in our laboratory, where retention was considered, indicated a slope of 1.00, albeit with a slightly poorer fit (r^2 0.92), as shown by the open circles in Fig. 7.11.

7.3.3 Brush-Border Lipid Membrane (BBLM) PAMPA Model (Chugai Model)

Sugano et al. [561,562] explored the lipid model containing several different phospholipids, closely resembling the mixture found in reconstituted brush border lipids [433,566] and demonstrated dramatically improved property predictions. The best-performing lipid composition consisted of a 3% wt/vol lipid solution in 1,7-octadiene (lipid consisting of 33% wt/wt cholesterol, 27% PC, 27% PE, 7% PS, 7% PI). The donor and acceptor compartments were adjusted in the pH interval between 5.0 and 7.4 [562]. With such a mixture, membrane retention is expected to be extensive when lipophilic drugs are assayed. The use of 1,7-octadiene in the assay was noted to require special safety precautions.

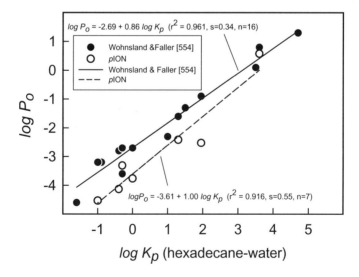

Figure 7.11 Intrinsic permeabilities versus alkane–water partition coefficients for drugs: PAMPA filters soaked with alkane [509].

7.3.4 Hydrophilic Filter Membrane PAMPA Model (Aventis Model)

Zhu et al. [563] found the use of hydrophilic filters (low-protein-binding PVDF) as an advantage in lowering the permeation time to 2 h. Egg lecithin, 1% wt/vol in dodecane, was used as the membrane medium. Over 90 compounds were characterized at pH 5.5 and 7.4. For each molecule, the greater P_e value of the two measured at different pH [509,554] was used to compare to Caco-2 permeabilities reported in the literature. It is noteworthy that many ionizable molecules did not follow the permeability-pH dependency expected from the pH partition hypothesis. It may be that water channels were contributing to the unexpected permeability-pH trends. Solute retention by the membrane was not considered. They tried using the Chugai five-component model, but found difficulties in depositing the lipid mixture on hydrophilic filters. Human intestinal absorption (HIA) values were compared to PAMPA measurements, Caco-2 permeabilities, partition coefficients (log K_p/ log K_d), polar surface area (PSA) and quantitative structure-property relations (QSPRs) developed by Winiwarter et al. [56] It was concluded that PAMPA and Caco-2 measurements best predicted HIA values.

7.3.5 Permeability–Retention–Gradient–Sink PAMPA Models (pION Models)

The system reported by Avdeef and co-workers [25–28,556–560] is an extension of the Roche approach, with several novel features described, including a way to assess membrane retention [25–28,556,557] and a way to quantify the effects of iso-pH [558] and gradient pH [559] conditions applied to ionizable molecules. A highly pure synthetic phospholipid, dioleoylphosphatidylcholine (DOPC), was initially used to coat the filters (2% wt/vol DOPC in dodecane). Other lipid mixtures were subsequently developed, and are described in detail in this chapter.

7.3.6 Structure of Phospholipid Membranes

The structure of the filter-immobilized artificial membranes is not known with certainty. Thompson et al. [543] hypothesized that polycarbonate filters had a single bilayer per pore, based largely on the behavior of amphotericin B in the pore-forming oligomerization reaction. Hennesthal and Steinem [568], using scanning force microscopy, estimated that a single bilayer spans exterior pores of porous alumina. These observations may be incomplete, as there is considerable complexity to the spontaneous process of the formation of BLMs (Section 7.2.1). When 2% phosphatidylcholine (PC)–dodecane solution is suspended in water, where the water exceeds 40 wt%, the lipid solution takes on the inverted hexagonal (H_{II}) structure, where the polar head groups of the PC face water channels in a cylindrical structure [569]. Such structures can alter transport properties, compared to those of normal phases [570]. (It may be possible to model the paracellular transport mechanism, should the presence of aqueous pores be established.) Suspensions of 2% PC–dodecane have been titrated potentiometrically from pH 10 down to pH 3. Along the way, at about pH 4, the pH electrode stopped functioning and appeared to be choked by a clear gelatinous coating, suggesting that some sort of phase transition had taken place then [Avdeef, unpublished].

7.4 THE CASE FOR THE IDEAL IN VITRO ARTIFICIAL MEMBRANE PERMEABILITY MODEL

7.4.1 Lipid Compositions in Biological Membranes

Different tissues have different lipid compositions. The most common lipid components of membranes are PC and PE. Lipid extracts from brain and lung are also rich in PS; heart tissue is rich in PG, and liver is rich in PI [567]. Human blood cells, as "ghost" erythrocytes (with cytoplasm contents removed), are often used as membrane models. These have different compositions between the inner and outer leaflets of the bilayer membrane. Phospholipids account for 46% of the outer leaflet membrane constituents, with PC and Sph about equal in amount. The inner leaflet is richer in phospholipids (55%), with the mix: 19% PE, 12% PS, 7% PC and 5% Sph [567].

Proulx [571] reviewed the published lipid compositions of brush-border membranes (BBM) isolated from epithelial cells from pig, rabbit, mouse, and rat small intestines. Table 7.1 shows the lipid makeup for the rat, averaged from five reported studies [571]. Krämer et al. [572,573] reported Madin–Darby canine kidney (MDCK) and BBB lipid composition profiles, listed in Table 7.1, for comparative purposes. Also shown are typical compositions of soy- and egg-derived lecithin extracts. Sugano's composition [561,562] is an attempt to mimic the BBLM. Table 7.1 lists the anionic-to-zwitterionic lipid weight ratios. On a molar basis, cholesterol accounts for about 50% of the total lipid content (37% on a weight basis) in the BBLM. The cholesterol content in BBLM is higher than that found in kidney epithelial (MDCK) and cultured brain endothelial cells (Table 7.1). (Slightly different BBLM lipid composition was reported by Alcorn et al. [433].) The outer (luminal) leaflet of the BBLM is rich in sphingomyelin, while the inner leaflet (cytosol) is rich in PE and PC. Apical (brush border) and basolateral lipids are different in the epithelium. The basolateral membrane content (not reported by Proulx) is high in PC, whereas the BBM has nearly the same PC as PE content. It appears that the BBB has the highest negative lipid content, and the BBM has the lowest negative lipid content of the three systems listed in the table. Cholesterol content follows the opposite pattern.

7.4.2 Permeability–pH Considerations

The effective permeability of ionizable molecules depends on pH, and the shapes of the permeability–pH profiles can be theoretically predicted when the pK_a of the molecule is known, the pH partition hypothesis are valid, and the resistance of

TABLE 7.1   Lipid Compsitions (%w/w) of Biological Membranes[a]

Lipid[b] [Ref.]	BBM[c] [30]	MDCK[d] [38]	BBB[e] [37]	Sugano BBM Model[f]	Soy "20% Extract" Lecithin[g]	Egg "60% Extract" Lecithin[g]
PC(±)	20	22	18	27	24	73
PE(±)	18	29	23	27	18	11
PS(−)	6	15	14	7	—	—
PI(−)	7	10	6	7	12	1
Sph(±)	7	10	8	—	—	—
FA(−)	—	1	3	—	—	—
PA(−)	—	—	—	—	4	—
LPI(−)	—	—	—	—	—	2
CL(2−)	—	—	2	—	—	—
LPC(±)	—	—	—	—	5	—
CHO + CE	37	10	26	33	—	—
TG	—	1	1	—	37[h]	13[h]
Negative : zwitterionic lipid ratio (exclusive of CHO and TG)	1 : 3.5	1 : 2.3	1 : 1.8	1 : 3.9	1 : 2.9	1 : 28.0

[a]The %w/w values in this table for BBB and MDCK are conversions from the originally reported %mol/mol units.
[b]PC = phosphatidylcholine, PE = phosphatidylethanolamine, PS = phosphatidylserine, PI = phosphatiosphatidylinositol, Sph = sphingomyelin, FA = fatty acid, PA = phosphatidic acid, LPI = lyso-PI, CL = cardiolipin, LPC = Iyso-PC, CHO = cholesterol, CE = cholesterol ester, TG = triglycerides.
[c]BBM = reconstituted brush-border membrane, rat (average of four studies).
[d]MDCK = Madin–Darby canine kidney cultured epithelial cells [563].
[e]BBB = blood–brain barrier lipid model, RBE4 rat endothelial immortalized cell line.
[f]Refs. 561 and 562.
[g]From Avanti Polar Lipids, Alabaster, AL.
[h]Unspecified neutral lipid, most likely asymmetric triglycerides.

the unstirred water layer (UWL; see Section 7.7.6) may be neglected (as, e.g., in the GIT and the BBB) [536,558,559]. The pH effects of ionizable molecules is illustrated in Fig. 7.12, for a series of weak acids and bases [562]. It is clear that if the 'wrong' pH is used in screening the permeabilities of molecules, highly promising molecules, such as furosemide or ketoprofen (Fig. 7.12), may be characterized as false negatives. The ideal pH to use for in vitro screening ought to reflect the in vivo pH conditions.

Said et al. [78] directly measured the "acid microclimate" on the surface of gastrointestinal tract (GIT) epithelial cells (intact with mucus layer) in rats. The pH on the apical (donor) side of the cells varied from 6.0 to 8.0, while the pH on the basolateral (acceptor) side was 7.4. Furthermore, the pH gradient between

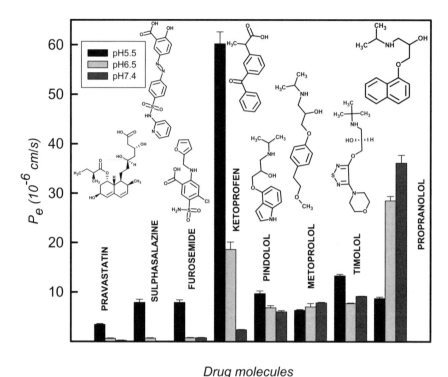

Drag molecules

Figure 7.12 Chugai model PAMPA permeabilities as a function of pH for several drug molecules [561].

the donor and acceptor sides varied with position in the GIT, as indicated in Table 7.2. Others have measured microclimate pH as low as 5.2 [73].

Yamashita et al. [82] determined drug permeabilities by performing Caco-2 assays under two pH conditions: pH 6.0_{donor}–$7.4_{acceptor}$ and pH 7.4_{donor}–$7.4_{acceptor}$. These choices adequately span the microclimate range in the GIT. Weak acids were more permeable under the gradient-pH condition, compared to the iso-pH condition. Weak bases behaved in the opposite way. Uncharged molecules showed the same permeabilities under the two conditions. The gradient-pH set of permeability measurements better predicted human absorption than the iso-pH set ($r^2 = 0.85$ vs. 0.50, respectively). The authors may have underestimated some of the permeabilities, by using equations which implied 'iso-pH' conditions (see, Section 7.5).

In designing the ideal screening strategy, it appears important to consider using pH gradients. If the *in vivo* conditions are to be mimicked, at least two effective permeability measurements should be attempted, as suggested by the above mentioned researchers: pH 6.0_{donor}–$7.4_{acceptor}$ (gradient pH) and pH 7.4_{donor}–$7.4_{acceptor}$ (iso-pH), the microclimate pH range spanned in the GIT.

TABLE 7.2 **Microclimate pH on the Apical Side of Epithelial Cells in the GIT in Rats**

Position in the GIT	Microclimate pH
Stomach	8.0
Proximal duodenum	6.4
Distal duodenum	6.3
Proximal jejunum	6.0
Midjejunum	6.2
Distal jejunum	6.4
Proximal ileum	6.6
Midileum	6.7
Distal ileum	6.9
Proximal colon	6.9
Distal colon	6.9

Source: Refs. 52 and 70.

7.4.3 Role of Serum Proteins

Sawada et al. [574–576] characterized the iso-pH 7.4 MDCK permeabilities of very lipophilic molecules, including chlorpromazine (CPZ) [574]. They included 3% wt/vol bovine serum albumin (BSA) on the apical (donor) side, and 0.1–3% BSA on the basolateral (acceptor) side, and found that plasma protein binding greatly affected the ability of molecules to permeate cellular barriers. They observed cell tissue retention of CPZ ranging within 65–85%, depending on the amount of BSA present in the receiving compartment. They concluded that the rapid rate of disappearance of lipophilic compounds from the donor compartment was controlled by the unstirred water layer (UWL; see Section 7.7.6), a rate that was about the same for most lipophilic compounds; however, the very slow appearance of the compounds in the receiving compartment depended on the rate of desorption from the basolateral side of the membranes, which was strongly influenced by the presence of serum proteins in the receiving compartment. They recommended the use of serum proteins in the receiving compartment, so as to better mimic the in vivo conditions when using cultured cells as in vitro assays.

Yamashita et al. [82] also studied the effect of BSA on transport properties in Caco-2 assays. They observed that the permeability of highly lipophilic molecules could be rate limited by the process of desorption off the cell surface into the receiving solution, due to high membrane retention and very low water solubility. They recommended using serum proteins in the acceptor compartment when lipophilic molecules are assayed (which is a common circumstance in discovery settings).

7.4.4 Effects of Cosolvents, Bile Acids, and Other Surfactants

Figure 7.13 shows some of the structures of common bile acids. In low ionic strength solutions, sodium taurocholate forms tetrameric aggregates, with critical

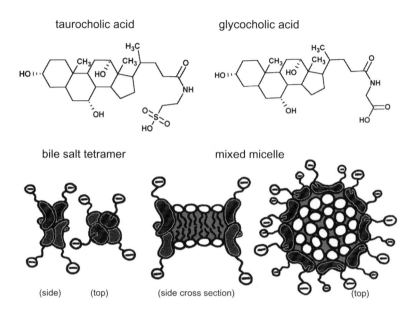

taurocholic acid   glycocholic acid

bile salt tetramer   mixed micelle

(side)   (top)   (side cross section)   (top)

Figure 7.13   Examples of bile salts and aggregate structures formed in aqueous solutions.

micelle concentration (CMC) 10–15 mM. Sodium deoxycholate can have higher levels of aggregation, with lower cmc (4–6 mM) [577]. Mixed micelles form in the GIT, where the edges of small sections of planar bilayer fragments are surrounded by a layer of bile salts (Fig. 7.13).

Yamashita et al. [82] added up to 10 mM taurocholic acid, cholic acid (cmc 2.5 mM), or sodium laurel sulfate (SLS; low ionic strength cmc 8.2 mM) to the donating solutions in Caco-2 assays. The two bile acids did not interfere in the transport of dexamethasone. However, SLS caused the Caco-2 cell junctions to become leakier, even at the sub-CMC 1 mM level. Also, the permeability of dexamethasone decreased at 10 mM SLS.

These general observations have been confirmed in PAMPA measurements in our laboratory, using the 2% DOPC–dodecane lipid. With very lipophilic molecules, glycocholic acid added to the donor solution slightly reduced permeabilities, taurocholic acid increased permeabilities, but SLS arrested membrane transport altogether in several cases (especially cationic, surface-active drugs such as CPZ).

Yamashita et al. [82] tested the effect of PEG400, DMSO, and ethanol, with up to 10% added to solutions in Caco-2 assays. PEG400 caused a dramatic decrease (75%) in the permeability of dexamethasone at 10% cosolvent concentration; DMSO caused a 50% decrease, but ethanol had only a slight decreasing effect.

Sugano et al. [562] also studied the effect of PEG400, DMSO, and ethanol, up to 30%, in their PAMPA assays. In general, water-miscible cosolvents are expected to decrease the membrane-water partition coefficients. In addition, the decreased dielectric constants of the cosolvent-water solutions should give rise to a higher proportion of the ionizable molecule in the *uncharged* state [25]. These two effects oppose each other. Mostly, increasing levels of cosolvents were observed to lead to decreasing permeabilities. However, ethanol made the weak-acid ketoprofen (pK_a 4.12) more permeable with increasing cosolvent levels, an effect consistent with the increasing pK_a with the decreasing dielectric constant of the cosolvent mixtures (leading to a higher proportion of uncharged species at a given pH). But the same reasoning cannot be used to explain why the weak-base propranolol (pK_a 9.5) decreased in permeability with increasing amounts of ethanol. This may be due to the increased solubility of propranolol in water with the added ethanol in relation to the solubility in the membrane phase. This leads to a lowered membrane/mixed-solvent partition coefficient, hence lowering flux due to a diminished sample concentration gradient in the membrane (Fick's law) [25]. DMSO and PEG400 dramatically reduced permeabilities for several of the molecules studied. Cosolvent use is discussed further in Section 7.7.9.

7.4.5 Ideal Model Summary

The literature survey in this section suggests that the ideal in vitro permeability assay would have pH 6.0 and 7.4 in the donor wells, with pH 7.4 in the acceptor wells. (Such a two-pH combination could differentiate acids from bases and nonionizables by the differences between the two P_e values.) Furthermore, the acceptor side would have 3% wt/vol BSA to maintain a sink condition (or some sink-forming equivalent). The donor side may benefit from having a bile acid (i.e., taurocholic or glycocholic, 5–15 mM), to solubilize the most lipophilic sample molecules. The ideal lipid barrier would have a composition similar to those in Table 7.1, with the membrane possessing substantial negative charge (mainly from PI and PS). Excessive DMSO or other cosolvents use requires further research, due to their multimechanistic effects. *In vitro* assays where permeabilities of lipophilic molecules are diffusion-limited [574–576], the role of the unstirred water layer (UWL; see Section 7.7.6) needs to be accounted, since under *in vivo* conditions, the UWL is nearly absent, especially in the BBB.

7.5 DERIVATION OF MEMBRANE-RETENTION PERMEABILITY EQUATIONS (ONE-POINT MEASUREMENTS, PHYSICAL SINKS, IONIZATION SINKS, BINDING SINKS, DOUBLE SINKS)

The equations used to calculate permeability coefficients depend on the design of the in vitro assay to measure the transport of molecules across membrane barriers. It is important to take into account factors such as pH conditions (e.g., pH gradients), buffer capacity, acceptor sink conditions (physical or chemical), any precipitate of the solute in the donor well, presence of cosolvent in the donor compartment, geometry of the compartments, stirring speeds, filter thickness, porosity, pore size, and tortuosity.

In PAMPA measurements each well is usually a one-point-in-time (single-timepoint) sample. By contrast, in the conventional multitimepoint Caco-2 assay, the acceptor solution is frequently replaced with fresh buffer solution so that the solution in contact with the membrane contains no more than a few percent of the total sample concentration at any time. This condition can be called a "physically maintained" sink. Under pseudo–steady state (when a practically linear solute concentration gradient is established in the membrane phase; see Chapter 2), lipophilic molecules will distribute into the cell monolayer in accordance with the effective membrane–buffer partition coefficient, even when the acceptor solution contains nearly zero sample concentration (due to the physical sink). If the physical sink is maintained indefinitely, then eventually, all of the sample will be depleted from both the donor and membrane compartments, as the flux approaches zero (Chapter 2). In conventional Caco-2 data analysis, a very simple equation [Eq. (7.10) or (7.11)] is used to calculate the permeability coefficient. But when combinatorial (i.e., lipophilic) compounds are screened, this equation is often invalid, since a considerable portion of the molecules partitions into the membrane phase during the multitimepoint measurements.

The extra timepoint measurements make the traditional Caco-2 assay too slow for high-throughput applications. Since the PAMPA assay was originally developed for high-throughout uses, there is no continuous replacement of the acceptor compartment solution. Some technical compromises are necessary in order to make the PAMPA method fast. Consequently, care must be exercised, in order for the single-timepoint method to work reliably. If the PAMPA assay is conducted over a long period of time (e.g., >20 h), the system reaches a state of equilibrium, where the sample concentration becomes the same in both the donor and acceptor compartments (assuming no pH gradients are used) and it becomes impossible to determine the permeability coefficient. Under such conditions, the membrane will also accumulate some (but sometimes nearly all) of the sample, according to the membrane–buffer partition coefficient. In the commonly practiced PAMPA assays it is best to take the single timepoint at 3–12 h, before the system reaches a state of equilibrium. Since the acceptor compartment is not assumed to be in a sink state, the permeability coefficient equation takes on a more complicated form [Eq. (7.20) or (7.21)] than that used in traditional Caco-2 assays.

For ionizable sample molecules, it is possible to create an effective sink condition in PAMPA by selecting buffers of different pH in the donor and acceptor compartments. For example, consider salicylic acid (pK_a 2.88; see Table 3.1). According to the pH partition hypothesis, only the free acid is expected to permeate lipophilic membranes. If the donor pH < 2 and the acceptor pH is 7.4, then as soon as the free acid reaches the acceptor compartment, the molecule ionizes, and the concentration of the free acid becomes effectively zero, even though the total concentration of the species in the acceptor compartment may be relatively high. This situation may be called an 'ionization-maintained' sink.

Another type of nonphysical sink may be created in a PAMPA assay, when serum protein is placed in the acceptor compartment and the sample molecule that passes across the membrane then binds strongly to the serum protein. Consider phenazopyridine (pK_a 5.15; see Table 3.1) in a pH 7.4 PAMPA assay, where the acceptor solution contains 3% wt/vol BSA (bovine serum albumin). As soon as the free base reaches the acceptor compartment, it binds to the BSA. The unbound fraction becomes very low, even though the total concentration of the base in the acceptor compartment may be relatively high. This may be called a *binding-maintained* sink.

In this chapter we use the term "sink" to mean any process that can significantly lower the concentration of the neutral form of the sample molecule in the acceptor compartment. Under the right conditions, the ionization and the binding sinks serve the same purpose as the physically maintained sink often used in Caco-2 measurements. We will develop several transport models to cover these "chemical" sink conditions. When both of the chemical sink conditions (ionization and binding) are imposed, we will use the term "double sink" in this chapter.

The chemical sink may be thought of as a method used to increase the volume of distribution of species in the acceptor solution beyond the geometric volume of the receiving compartment. As such, this extension of terminology should be clear to traditional Caco-2 users. The use of the chemical sinks in PAMPA is well suited to automation, and allows the assay to be conducted at high-throughput speeds. As mentioned above, the one-point-in-time (single-timepoint) sampling can lead to errors if not properly executed. We will show that when multitimepoint PAMPA is done (see Fig. 7.15), the equations developed in this chapter for high-speed single-timepoint applications are acceptably good approximations.

7.5.1 Thin-Membrane Model (without Retention)

Perhaps the simplest Fick's law permeation model consists of two aqueous compartments, separated by a very thin, pore-free, oily membrane, where the unstirred water layer may be disregarded and the solute is assumed to be negligibly retained in the membrane. At the start ($t = 0$ s), the sample of concentration $C_D(0)$, in mol/cm^3 units, is placed into the donor compartment, containing a volume (V_D, in cm^3 units) of a buffer solution. The membrane (area A, in cm^2 units) separates the donor compartment from the acceptor compartment. The acceptor compartment also contains a volume of buffer (V_A, in cm^3 units). After a permeation time, t (in seconds), the experiment is stopped. The concentrations in the acceptor and donor compartments, $C_A(t)$ and $C_D(t)$, respectively, are determined.

Two equivalent flux expressions define such a steady-state transport model [41]

$$J(t) = P[C_D(t) - C_A(t)] \tag{7.1}$$

and

$$J(t) = \frac{-V_D}{A} \frac{dC_D(t)}{dt} \tag{7.2}$$

where P denotes either the effective or the apparent permeability, P_e or P_a, depending on the context (see later), in units of cm/s. These expressions may be equated to get the differential equation

$$\frac{dC_D(t)}{dt} = -\left(\frac{A}{V_D}\right)P[C_D(t) - C_A(t)] \tag{7.3}$$

It is useful to factor out $C_A(t)$ and solve the differential equation in terms of just $C_D(t)$. This can be done by taking into account the mass balance, which requires that the total amount of sample be preserved, and be distributed between the donor and the acceptor compartments (disregarding the membrane for now). At $t = 0$, all the solute is in the donor compartment, which amounts to $V_D C_D(0)$ moles. At time t, the sample distributes between two compartments:

$$V_D C_D(0) = V_D C_D(t) + V_A C_A(t) \tag{7.4}$$

This equation may be used to replace $C_A(t)$ in Eq. (7.3) with donor-based terms, to get the simplified differential equation

$$\frac{dC_D(t)}{dt} + aC_D(t) + b = 0 \tag{7.5}$$

where $a = AP/[(V_A V_D)/(V_A + V_D)] = \tau_{eq}^{-1}$, τ_{eq} is the time constant, and $b = APC_D(0)/V_A$. Sometimes, τ_{eq}^{-1} is called the *first-order rate constant, k* [in s^{-1} units (reciprocal seconds)]. The ordinary differential equation may be solved by standard techniques, using integration limits from 0 to t, to obtain an exponential solution, describing the *disappearance* of solute from the donor compartment as a function of time

$$\frac{C_D(t)}{C_D(0)} = \frac{m_D(t)}{m_D(0)} = \frac{V_A}{V_A + V_D}\left[\frac{V_D}{V_A} + \exp(-t/\tau_{eq})\right] \tag{7.6}$$

where $m_D(t)$ refers to the moles of solute remaining in the donor compartment at time t. Note that when $V_A \gg V_D$, Eq. (7.6) approximately equals $\exp(-t/\tau_{eq})$. Furthermore, $\exp(-t/\tau_{eq}) \approx 1 - t/\tau_{eq}$ when t is near zero. Using the mole balance relation [Eq. (7.4)], the exponential expression above [Eq. (7.6)] may be converted into another one, describing the *appearance* of solute in the acceptor compartment.

$$\frac{C_A(t)}{C_D(0)} = \frac{V_D}{V_A + V_D}\left(1 - \exp\frac{-t}{\tau_{eq}}\right) \tag{7.7}$$

In mole fraction units, this is

$$\frac{m_A(t)}{m_D(0)} = \frac{V_A}{V_A + V_D}\left(1 - \exp\frac{-t}{\tau_{eq}}\right) \tag{7.8}$$

Note that when $V_A \gg V_D$, Eq. (7.8) approximately equals $1 - \exp(-t/\tau_{eq})$. Furthermore, $1 - \exp(-t/\tau_{eq}) \approx t/\tau_{eq}$ when t is near zero. Figure 7.14 shows the forms of Eqs. (7.6) and (7.8) under several conditions. When less than $\sim$10% of the compound has been transported, the reverse flux due to $C_A(t)$ term in Eq. (7.1) is nil. This is effectively equivalent to a sink state, as though $V_A \gg V_D$. Under these conditions, Eq. (7.8) can be simplified to

$$\frac{m_A(t)}{m_D(0)} \approx \frac{t}{\tau_{eq}} \approx \frac{APt}{V_D} \tag{7.9}$$

and the apparent permeability coefficient can be deduced from this "one-way flux" equation,

$$P_a = \frac{V_D}{At}\frac{m_A(t)}{m_D(0)} \tag{7.10}$$

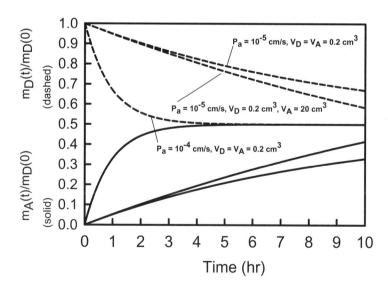

Figure 7.14 Relative concentrations of accetor and donor compartments as a function of time for the thin-membrane model.

We define this permeability as "apparent," to emphasize that there are important but hidden assumptions made in its derivation. This equation is popularly (if not nearly exclusively) used in culture cell in vitro models, such as Caco-2. The sink condition is maintained by periodically moving a detachable donor well to successive acceptor wells over time. At the end of the total permeation time t, the mass of solute is determined in each of the acceptor wells, and the mole sum $m_A(t)$ is used in Eq. (7.10). Another variant of this analysis is based on evaluating the slope in the early part of the appearance curve (e.g., solid curves in Fig. 7.14):

$$P_a = \frac{V_D}{A} \frac{\Delta m_A(t)/\Delta t}{m_D(0)} \tag{7.11}$$

It is important to remember that Eqs. (7.10) and (7.11) are both based on assumptions that (1) sink conditions are maintained, (2) data are taken early in the transport process (to further assure sink condition), and (3) *there is no membrane retention of solute*. In discovery settings where Caco-2 assays are used, the validity of assumption 3 is often untested.

The more general solutions (but still neglecting membrane retention, hence still "apparent") are given by "two-way flux" in Eqs. 7.12 (disappearance kinetics) and (7.13) (appearance kinetics).

$$P_a = -\frac{2.303 \, V_D}{At} \left(\frac{1}{1 + r_V}\right) \cdot \log_{10}\left[-r_V + (1 + r_V) \cdot \frac{C_D(t)}{C_D(0)}\right] \tag{7.12}$$

$$= -\frac{2.303 \, V_D}{At} \left(\frac{1}{1 + r_V}\right) \cdot \log_{10}\left[1 - (1 + r_V^{-1}) \cdot \frac{C_A(t)}{C_D(0)}\right] \tag{7.13}$$

where the aqueous compartment volume ratio, $r_V = V_D/V_A$. Often, $r_V = 1$. From analytical considerations, Eq. (7.13) is better to use than (7.12) when only a small amount of the compound reaches the acceptor wells; analytical errors in the calculated P_a, based on Eq. (7.13), tend to be lower.

Palm et al. [578] derived a two-way flux equation which is equivalent to Eq. (7.13), and applied it to the permeability assessment of alfentanil and cimetidine, two drugs that may be transported by passive diffusion, in part, as charged species. We will discuss this apparent violation of the pH partition hypothesis (Section 7.7.7.1).

7.5.2   Iso-pH Equations with Membrane Retention

The popular permeability equations [(7.10) and (7.11)] derived in the preceding section presume that the solute does not distribute into the membrane to any appreciable extent. This assumption may not be valid in drug discovery research, since most of the compounds synthesized by combinatorial methods are very lipophilic and can substantially accumulate in the membrane. Neglecting this leads to underestimates of permeability coefficients. This section expands the equations to include membrane retention.

7.5.2.1 *Without Precipitate in Donor Wells and without Sink Condition in Acceptor Wells*

When membrane retention of the solute needs to be considered, one can derive the appropriate permeability equations along the lines described in the preceding section: Eqs. (7.1)–(7.3) apply (with P designated as the effective permeability, P_e). However, the mass balance would need to include the membrane compartment, in addition to the donor and acceptor compartments. At time t, the sample distributes (mol amounts) between three compartments:

$$V_D C_D(0) = V_A C_A(t) + V_D C_D(t) + V_M C_M(t) \qquad (7.14)$$

The partition coefficient is needed to determine the moles lost to the membrane, $V_M C_M(t)$. If ionizable compounds are considered, then one must decide on the types of partition coefficient to use $-K_p$ (true pH-independent partition coefficient) or K_d (pH-dependent apparent partition coefficient). If the permeability assay is based on the measurement of the *total* concentrations, $C_D(t)$ and $C_A(t)$, summed over all charge-state forms of the molecule, and only the uncharged molecules transport across the membrane to an appreciable extent, it is necessary to consider the apparent partition (distribution) coefficient, K_d, in order to explain the pH dependence of permeability.

The apparent membrane–buffer partition (distribution) coefficient K_d, defined at $t = \infty$, is

$$K_d = \frac{C_M(\infty)}{C_D(\infty)} = \frac{C_M(\infty)}{C_A(\infty)} \qquad (7.15)$$

since at equilibrium, $C_D(\infty) = C_A(\infty)$, *in the absence of a pH gradient and other sink conditions*. At equilibrium ($t = \infty$), the mole balance equation [Eq. (7.14)] can be expanded to factor in the partition coefficient, Eq. (7.15):

$$\begin{aligned} V_D C_D(0) &= V_D C_D(\infty) + V_A C_A(\infty) + V_M K_d C_D(\infty) \\ &= V_D C_D(\infty) + V_A C_D(\infty) + V_M K_d C_D(\infty) \\ &= C_D(\infty)(V_D + V_A + V_M K_d) \end{aligned} \qquad (7.16)$$

It is practical to make the approximation that $C_M(\infty) \approx C_M(t)$. This is justified if the membrane is saturated with the sample in a short period of time. This lag (steady-state) time may be approximated from Fick's second law as $\tau_{LAG} = h^2/(\pi^2 D_m)$, where h is the membrane thickness in centimeters and D_m is the sample diffusivity *inside the membrane*, in cm^2/s [40,41]. Mathematically, τ_{LAG} is the time at which Fick's second law has transformed into the limiting situation of Fick's first law. In the PAMPA approximation, the lag time is taken as the time when solute molecules first appear in the acceptor compartment. This is a tradeoff approximation to achieve high-throughput speed in PAMPA. With h = 125 μm and $D_m \approx 10^{-7}$ cm^2/s, it should take ~3 min to saturate the lipid membrane with sample. The observed times are of the order of 20 min (see below), short enough for our purposes. Cools and Janssen [545] reported 10–30-min lag times with octanol-impregnated filters. With thinner BLM membranes, the time to reach steady state under sink conditions was reported to be 3–6 min [537]. Times as short as 50 s have been reported in BLM membranes [84].

From Eq. (7.16), one can deduce $C_D(\infty)$, and apply it in the next step. Before equilibrium is reached, at time $t > \tau_{\text{LAG}}$, the moles of solute in the membrane may be estimated from

$$
\begin{aligned}
V_M C_M(t) \approx V_M C_M(\infty) &= V_M K_d C_D(\infty) \\
&= \frac{V_D C_D(0) V_M K_d}{V_A + V_D + V_M K_d}
\end{aligned}
\tag{7.17}
$$

At this point, we introduce the *retention fraction R*, which is defined as the mole fraction of solute "lost" to the membrane. Equation (7.16) is used in the steps leading to Eq. (7.18):

$$
\begin{aligned}
R &= 1 - \frac{m_D(t)}{m_D(0)} - \frac{m_A(t)}{m_D(0)} \\
&= 1 - \frac{C_D(\infty)}{C_D(0)} - \frac{V_A}{V_D} \cdot \frac{C_A(\infty)}{C_D(0)} \\
&= \frac{V_M K_d}{V_A + V_D + V_M K_d}
\end{aligned}
\tag{7.18}
$$

Note that from Eqs. (7.17) and (7.18), $R \approx V_M C_M(t)/V_D C_D(0)$ for $t > \tau_{\text{LAG}}$. The substitution of the apparent partition coefficient with the retention ratio allows us to state the mole balance at time t (provided $t > \tau_{\text{LAG}}$) in a much simplified form:

$$
V_A C_A(t) + V_D C_D(t) = V_D C_D(0)(1 - R)
\tag{7.19}
$$

Given this relationship between $C_A(t)$ and $C_D(t)$, where retention is factored in, we can proceed to convert Eq. (7.3) into Eq. (7.5), where a is the same as before, and b now needs to be multiplied by the partition-related factor, $1 - R$. The so-modified ordinary differential, Eq. (7.5), is solved by standard methods, using integration limits from τ_{LAG} to t (not 0 to t), and the desired effective permeability derived as

$$
P_e = -\frac{2.303 V_D}{A(t - \tau_{\text{LAG}})} \left(\frac{1}{1 + r_V} \right) \cdot \log_{10}\left[-r_V + \left(\frac{1 + r_V}{1 - R} \right) \cdot \frac{C_D(t)}{C_D(0)} \right]
\tag{7.20}
$$

$$
= -\frac{2.303 V_D}{A(t - \tau_{\text{LAG}})} \left(\frac{1}{1 + r_V} \right) \cdot \log_{10}\left[1 - \left(\frac{1 + r_V^{-1}}{1 - R} \right) \cdot \frac{C_A(t)}{C_D(0)} \right]
\tag{7.21}
$$

Note that Eqs. (7.20) and (7.21) are nearly identical to Eqs. (7.12) and (7.13); the differences are the $1 - R$ term (to reflect membrane retention) and the lag time offset, τ_{LAG} (the time needed to saturate the membrane with solute). These differences warrant the new equations to be denoted with the subscript e.

When using the 96-well microtiter plate format, typical metrics are $V_A = 200$–400 μL, $V_D = 200$–400 μL, $A = 0.3$ cm^2, $V_M = 4$–6 μL, h (filter thickness) = 125 μm, 70% porosity (ϵ), t (permeation time) = 3–15 h, $\tau_{\text{LAG}} = 0$–60 min. As noted

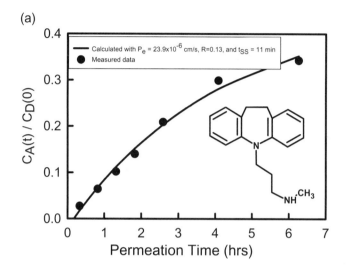

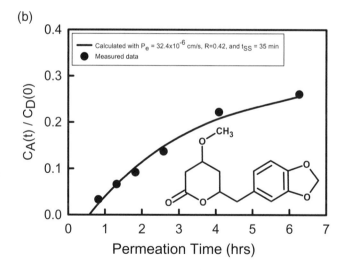

Figure 7.15 Kinetics of transport across a filter-immobilized artificial membrane: (a) desipramine and (b) dihydromethysticin concentrations in acceptor well. [Reprinted from Avdeef, A., in van de Waterbeemd, H.; Lennernäs, H.; Artursson, P. (Eds.). *Drug Bioavailability. Estimation of Solubility, Permeability, Absorption and Bioavailability.* Wiley-VCH: Weinheim, 2003 (in press), with permission from Wiley-VCH Verlag GmbH.]

above, the time constant for the kinetic process is defined as $\tau_{eq} = [(V_A V_D)/(V_A + V_D)]/(AP_e)$. For membranes made with 2% DOPC in dodecane, metoprolol at pH 7.4, has $\tau_{eq} = 4.8 \times 10^{-5}$ s or 134 h for the donor concentration to decay to $1/e$ (37%) from the final equilibrium value. For diltiazem, the time constant is 5.3 h. However, for membranes made with 20% soy lecithin in dodecane, under sink conditions created by an anionic surfactant in the acceptor wells, the metoprolol and diltiazem time constants decrease to 3.2 and 2.6 h, respectively, since the permeability coefficients increase in the soy-based membrane under artificially imposed sink conditions (as discussed in a later section).

Figure 7.15 shows the appearance curves of desipramine and dihydromethysticin [556] in the acceptor wells as a function of time. Because some of the material is lost to the membrane, the curves level off asymptotically at acceptor concentration fractions considerably less the 0.5 value expected in the thin-membrane model (Fig. 7.14). The solid curve for desipramine in Fig. 7.15a is a least-squares fit of the data points to Eq. (7.21), with the parameters: P_e 24×10^{-6} cm/s, R 0.13, and τ_{LAG} 11 min. The solid curve for dihydromethysticin in Fig. 7.15b is described by the parameters: P_e 32×10^{-6} cm/s, R 0.42, and τ_{LAG} 35 min.

Ketoprofen, a weak-acid drug, with a pK_a 4.12 (25°C, 0.01 M ionic strength), was selected to illustrate Eqs. (7.20) and (7.21) in a series of simulation calculations, as shown in Fig. 7.16. The membrane–buffer apparent partition coefficients $K_{d(pH)}$ were calculated at various pH values, using the measured constants from

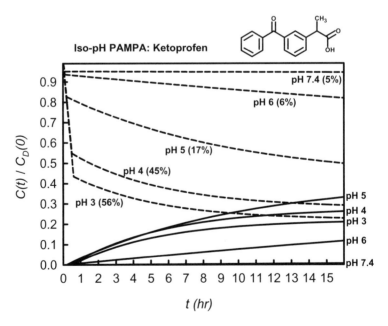

Figure 7.16 Relative concentrations of accetor and donor compartments as a function of time for the iso-pH ketoprofen model.

liposome–water partition studies: the surface ion pair (SIP) constant, $\log K_p^{SIP}$ 0.70, corresponding to the partitioning of the anionic form of the drug in bilayers at high pH, and the neutral-species partition coefficient, $\log K_p^N$ 2.14, evident at low pH [149]. For example, at pH 7.4, K_d is 5 and at pH 4.3, K_d is 58. Also used for the simulation calculation were the intrinsic permeability coefficient, P_0 1.7×10^{-4} cm/s, corresponding to the transport property of the uncharged form of ketoprofen, and the unstirred water layer permeability coefficient, P_u 2.2×10^{-5} cm/s. (These two types of permeability are described later in this chapter.)

At pH 3, ketoprofen is mostly in an uncharged state in solution. The dashed curve in Fig. 7.16 corresponding to pH 3 shows a rapid decline of the sample in the donor well in the first half-hour; this corresponds to the membrane loading up with the drug, to the extent of 56%. The corresponding appearance of the sample in the acceptor well is shown by the solid line at pH 3. The solid curve remains at zero for $t < \tau_{LAG}$. After the lag period, the acceptor curve starts to rise slowly, mirroring in shape the donor curve, which decreases slowly with time. The two curves nearly meet at 16 h, at a concentration ratio near 0.22, far below the value of 0.5, the expected value had the membrane retention not taken a portion of the material out of the aqueous solutions.

7.5.2.2 Sink Condition in Acceptor Wells

In Section 7.7.5.4, we discuss the effects of additives in the acceptor wells that create a sink condition, by strongly binding lipophilic molecules that permeate across the membrane. As a result of the binding in the acceptor compartment, the transported molecule has a reduced "active" (unbound) concentration in the acceptor compartment, $c_A(t)$, denoted by the lowercase letter c. The permeability equations in the preceding section, which describe the nonsink process, are inappropriate for this condition. In the present case, we assume that the reverse transport is effectively nil; that is, $C_A(t)$ in Eq. (7.1) may be taken as $c_A(t) \approx 0$. As a result, the permeability equation is greatly simplified:

$$P_a = -\frac{2.303\, V_D}{A(t - \tau_{LAG})} \cdot \log_{10}\left[\frac{1}{1-R} \cdot \frac{C_D(t)}{C_D(0)}\right] \tag{7.22}$$

Note that we call this the "apparent" permeability, since there is a hidden assumption (unbound concentration is zero).

7.5.2.3 Precipitated Sample in the Donor Compartment

When very insoluble samples are used, sometimes precipitate forms in the donor wells, and the solutions remain saturated during the entire permeation assay. Equations (7.20) and (7.21) would not appropriately represent the kinetics. One needs to consider the following modified flux equations [see, Eqs. (7.1) and (7.2)]

$$J(t) = P_e(S - C_A(t)) \tag{7.23}$$

and

$$J(t) = \frac{V_A}{A} \frac{dC_A(t)}{dt} \qquad (7.24)$$

The donor concentration becomes constant in time, represented by the solubility, $S = C_D(0) = C_D(t)$. Reverse flux can still occur, but as soon as the sample reaches the donor compartment, it would be expected to precipitate. Furthermore, the concentration in the acceptor compartment would not be expected to exceed the solubility limit: $C_A(t) \leq S$. After equating the two flux expressions, and solving the differential equation, we have the saturated-donor permeability equation

$$P_e = -\frac{2.303 \, V_A}{A(t - \tau_{\text{LAG}})} \cdot \log_{10}\left[1 - \frac{C_A(t)}{S}\right] \qquad (7.25)$$

Ordinarily it is not possible to determine the membrane retention of solute under the circumstances of a saturated solution, so no R terms appear in the special equation [Eq. (7.25)], nor is it important to do so, since the concentration gradient across the membrane is uniquely specified by S and $C_A(t)$. The permeability coefficient is "effective" in this case.

7.5.3 Gradient pH Equations with Membrane Retention: Single and Double Sinks

When the pH is different on the two sides of the membrane, the transport of ioniz-able molecules can be dramatically altered. In effect, sink conditions can be created by pH gradients. Assay improvements can be achieved using such gradients between the donor and acceptor compartments of the permeation cell. A three-com-partment diffusion differential equation can be derived that takes into account gradient pH conditions and membrane retention of the drug molecule (which clearly still exists—albeit lessened—in spite of the sink condition created). As before, one begins with two flux equations

$$J(t) = P_e^{(D \to A)} C_D(t) - P_e^{(A \to D)} C_A(t) \qquad (7.26)$$

and

$$J(t) = -\left(\frac{V_D}{A}\right) \frac{dC_D(t)}{dt} \qquad (7.27)$$

It is important to note that two different permeability coefficients need to be con-sidered, one denoted by the superscript $(D \to A)$, associated with donor (e.g., pH_D

5.0, 6.5, or 7.4)-to-acceptor (pH_A 7.4) transport, and the other denoted by the superscript ($A \rightarrow D$), corresponding to the reverse-direction transport. The two equivalent flux relationships can be reduced to an ordinary differential equation in $C_D(t)$, following a route similar to that in Section 7.5.2.1.

The gradient pH ($2\text{-}P_e$) model developed here implies that some backflux ($A \rightarrow D$) is possible. As far as we know, reported literature studies generally considered backflux to be nil under gradient pH conditions. That is, either Eq. (7.10) or (7.11) were used to interpret the membrane transport under a pH gradient conditions. If it can be assumed that $C_A(t)$ in Eq. (7.26) represents a fully charged (i.e., impermeable) form of the solute, then its contribution to backflux may be neglected, and an effective sink condition would prevail; that is, the concentration of the *uncharged* form of the solute, $c_A(t)$, is used in place of $C_A(t)$, where $c_A(t) \approx 0$. Under such circumstances, the generic sink equation, Eq. (7.22), may be used to determine an apparent permeability coefficient, P_a—"apparent" so as to draw attention to hidden assumptions (i.e., no reverse flux). However, valid use of Eq. (7.22) is restricted to strictly maintained sink conditions and presumes the absence of membrane retention of solute. This is a rather impractical constraint in high-throughput applications, where molecules with potentially diverse transport properties may be assayed at the same time.

A more general analysis requires the use of two effective permeability coefficients, one for each pH, each of which would be valid in the respective iso-pH conditions. Since fewer limiting assumptions are made, the more general method may be more suitable for high-throughput applications. We continue to derive the appropriate new model.

The donor–acceptor membrane mass balance is

$$\text{mol}_{\text{TOT}} = V_D C_D(0) = V_A C_A(\infty) + V_D C_D(\infty) + V_M C_M(\infty) \tag{7.28}$$

Each side of the barrier has a different membrane–buffer apparent partition coefficient K_d, defined at $t = \infty$ as

$$K_{d(A)} = \frac{C_M(\infty)}{C_A(\infty)} \tag{7.29}$$

and

$$K_{d(D)} = \frac{C_M(\infty)}{C_D(\infty)} \tag{7.30}$$

The moles lost to the membrane are derived from Eqs. (7.28)–(7.30):

$$\text{mol}_M = C_M(\infty) V_M = \frac{V_M V_D C_D(0)}{V_A / K_{d(A)} + V_D / K_{d(D)} + V_M} \tag{7.31}$$

The membrane retention fraction R may be defined as membrane-bound moles of sample, divided by the total moles of sample in the system:

$$R = \frac{\text{mol}_M}{\text{mol}_{TOT}} = \frac{V_M}{V_A/K_{d(A)} + V_D/K_{d(D)} + V_M} \tag{7.32}$$

The membrane saturates with solute early in the transport process. So, for $t \gg 20$ min, we may assume that $C_M(\infty) \approx C_M(t)$ is reasonably accurate. With this assumption, the acceptor concentration may be expressed in terms of the donor concentration as

$$C_A(t) = \frac{V_D}{V_A[C_D(0)(1 - R) - C_D(t)]} \tag{7.33}$$

A differential equation as a function of $C_D(t)$ only, similar to Eq. (7.5), can be derived, where the specific constants $a = A(P_e^{(A \rightarrow D)}/V_A + P_e^{(D \rightarrow A)}/V_D)$ and $b = C_D(0)(1 - R)AP_e^{(A \rightarrow D)}/V_A$. The solution to the ordinary differential equation is

$$P_e = -\frac{2.303\,V_D}{A(t - \tau_{SS})}\left(\frac{1}{1 + r_a}\right) \cdot \log_{10}\left[-r_a + \left(\frac{1 + r_a}{1 - R}\right) \cdot \frac{C_D(t)}{C_D(0)}\right] \tag{7.34}$$

where

$$r_a = \left(\frac{V_D}{V_A}\right)\frac{P_e^{(A \rightarrow D)}}{P_e^{(D \rightarrow A)}} = \frac{r_V P_e^{(A \rightarrow D)}}{P_e^{(D \rightarrow A)}} \tag{7.35}$$

is the sink asymmetry ratio (gradient-pH-induced). When the aqueous solution conditions are identical in the two chambers of the permeation cell (apart from the sample), $r_a = r_V$, and Eq. (7.34) becomes equivalent to Eq. (7.20). This presumes that the system is free of serum proteins or surfactants in the acceptor well. We discuss such assay extensions later.

7.5.3.1 Single Sink: Eq. (7.34) in the Absence of Serum Protein or Sink in Acceptor Wells

In general, Eq. (7.34) has two unknowns: $P_e^{(A \rightarrow D)}$ and $P_e^{(D \rightarrow A)}$. In serum protein-free assays, the following method is used to solve Eq. (7.34). At least two assays are done: one as gradient pH (e.g., pH 5.0_{donor}–7.4_{acceptor}) and the other as iso-pH (e.g., pH 7.4_{donor}–7.4_{acceptor}), with one pH common to the two assays. For iso-pH, $P_e^{(A \rightarrow D)} = P_e^{(D \rightarrow A)}$. This case can be solved directly using Eq. (7.20). Then, iteratively, Eq. (7.34) is solved for $P_e^{(D \rightarrow A)}$. Initially r_a is assumed to be r_V, but with each iteration, the r_a estimate is improved by using the calculated $P_e^{(D \rightarrow A)}$ and the $P_e^{(A \rightarrow D)}$ taken from the iso-pH case. The process continues until self-consistency is reached within the accuracy required.

In iso-pH serum protein- and surfactant-free solutions, the concentration of the sample in the acceptor wells cannot exceed that in the donor wells. With gradient-pH conditions, this limitation is lifted. At very long times, the concentrations in the donor and acceptor chambers reach equilibrium values, depending on the pH gradient

$$\frac{C_D(\infty)}{C_A(\infty)} = \frac{P_e^{(A \to D)}}{P_e^{(D \to A)}} \qquad (7.36)$$

or in terms of mole ratios

$$\frac{m_D(\infty)}{m_A(\infty)} = r_a \qquad (7.37)$$

This limiting ratio can be predicted for any gradient-pH combination, provided the pK_a values of the molecule, the unstirred water layer (UWL) P_u, and the intrinsic P_0 permeabilities were known [25]. (The topic of the UWL are discussed in greater detail in Section 7.7.6.) In gradient pH assays, it is sometimes observed that nearly all the samples move to the acceptor side, due to the sink conditions created, sometimes limiting the determination of concentrations. Shorter permeation times solve the problem, a welcome prospect in a high-throughput application. A 3–4-h period suffices, which is a considerable reduction over the original 15 h permeation time [547,550]. Shorter times would lead to greater uncertainties in the calculated permeability, since the approximate lag time τ_{LAG} can be as long as one hour for the most lipophilic molecules.

7.5.3.2 Double Sink: Eq. (7.34) in the Presence of Serum Protein or Sink in Acceptor Wells

If serum protein or surfactant is added to the acceptor wells, then, in general, $P_e^{(A \to D)}$ and $P_e^{(D \to A)}$ are not the same, even under iso-pH conditions. The acceptor-to-donor permeability needs to be solved by performing a separate iso-pH assay, where the serum protein or surfactant is added to the *donor* side, instead of the acceptor side. The value of P_e is determined, using Eq. (7.20), and used in gradient-pH cases in place of $P_e^{(A \to D)}$, as described in the preceding section. The gradient-pH calculation procedure is iterative as well.

Figure 7.17 shows the asymmetry ratios of a series of compounds (acids, bases, and neutrals) determined at iso-pH 7.4, under the influence of sink conditions created not by pH, but by anionic surfactant added to the acceptor wells (discuss later in the chapter). The membrane barrier was constructed from 20% soy lecithin in dodecane. All molecules show an upward dependence on lipophilicity, as estimated by octanol–water apparent partition coefficients, log $K_{d(7.4)}$. The bases are extensively cationic at pH 7.4, as well as being lipophilic, and so display the highest responses to the sink condition. They are driven to interact with the surfactant by both hydrophobic and electrostatic forces. The anionic acids are largely indifferent

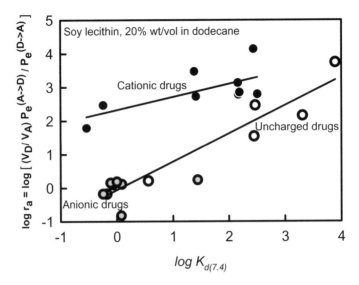

Figure 7.17 Surfactant-induced sink asymmetry ratio versus octanol–water apparent partition coefficient at pH 7.4.

to the presence of the anionic surfactant in the acceptor wells, with a slight suggestion of repulsion in one case (Fig. 7.17).

For ionizable lipophilic molecules, the right pH gradients can drive the solute in the acceptor compartment to the charged (impermeable) form; the uncharged fraction is then further diminished in concentration by binding to the serum protein or surfactant, in the double-sink assay.

7.5.3.3 Simulation Examples

Ketoprofen was selected to illustrate the properties of the gradient-pH permeability equation, Eq. (7.34), in a series of simulation calculations, as shown in Fig. 7.18. The membrane-buffer apparent partition coefficients, $K_{d(pH)}$, were calculated at various pH values, using the approach described in Section 7.5.2.1. The pH in the acceptor well was pH_A 7.4 in all cases, while that in the donor wells was pH_D 3–7.4. It is interesting to compare the transport properties of ketoprofen under iso-pH (Fig. 7.16) and gradient pH (Fig. 7.18) conditions. Under gradient pH conditions, at pH_D 3, ketoprofen is mostly in an uncharged state in solution. The dashed curve in Fig. 7.18 corresponding to pH_D 3 shows a rapid but not extensive decline of the sample in the donor well in the first few minutes; this corresponds to the membrane loading up with the drug, to the extent of only 9%, compared to 56% for iso-pH 3 conditions. The corresponding appearance of the sample in the acceptor well is shown by the solid line corresponding to pH_D 3, pH_A 7.4. After a short lag period, the acceptor curve starts to rise rapidly, mirroring in shape the donor curve, which decreases with time. The two curves cross at 7 h, whereas in the

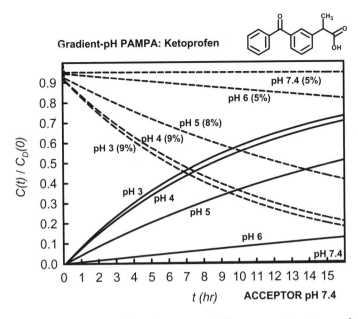

Figure 7.18 Relative concentrations of accetor and donor compartments as a function of time for the gradient–pH ketoprofen model.

iso-pH case, 16 h shows only near meeting. Also, the gradient pH curves cross slightly below the 0.5 concentration ratio, since membrane retention is only 9%.

7.5.3.4 Gradient pH Summary

The benefits of an assay designed under gradient pH conditions are (1) less retention and thus more analytical sensitivity, (2) shorter permeation times and thus higher throughput possible, and (3) more realistic modeling of the in vivo pH gradients found in the intestinal tract and thus better modeling. Time savings with increased sensitivity are important additions to an assay designed for high-throughput applications. A double-sink condition created by the combination of a pH gradient and serum protein (or an appropriate surfactant) in the acceptor compartment is an important component of the biophysical GIT transport model. In contrast, a no-sink condition may be more suitable for a BBB transport model. This is discussed in greater detail later.

7.6 PERMEABILITY–LIPOPHILICITY RELATIONS

7.6.1 Nonlinearity

In the introductory discussion in Chapter 2, it was indicated that the effective permeability P_e linearly depends on the apparent membrane–water partition

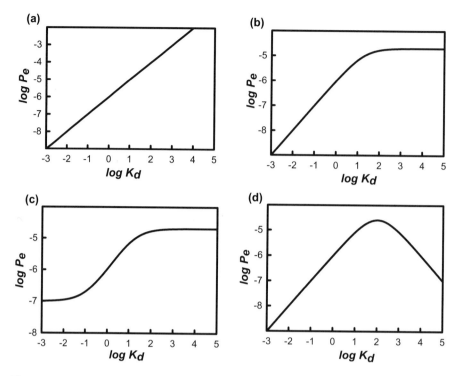

Figure 7.19   Permeability–lipophilicity relations: (a) linear; (b) hyperbolic; (c) sigmoidal; (d) bilinear.

coefficient, K_d [Eq. (2.3)]. The simple model system considered there assumed the membrane barrier to be a structureless homogeneous oil, free of aqueous pores, and also assumed the aqueous solutions on both sides of the barrier to be well mixed by convection, free of the UWL (Section 7.7.6) effect. A log P_e/log K_d plot would be a straight line. Real membrane barriers are, of course, much more complicated. Studies of permeabilities of various artificial membranes and culture-cell monolayers indicate a variety of permeability–lipophilicity relations (Fig. 7.19). These relationships have been the subject of two reviews [49,54]. Figure 7.19 shows linear [579], hyperbolic [580–582], sigmoidal [552,583,584], and bilinear [23,581,585,586] permeability–lipophilicity relations.

Early efforts to explain the nonlinearity were based on drug distribution (equilibrium) or transport (kinetic) in multicompartment systems [21,22]. In this regard, the 1979 review by Kubinyi is highly recommended reading [23]. He analyzed the transport problem using both kinetic and equilibrium models. Let us consider the simple three-compartment equilibrium model first. Imagine an organism reduced to just three phases: water (compartment 1), lipid (compartment 2), and receptor (compartment 3). The corresponding volumes are v_1, v_2, and v_3, respectively, and $v_1 \gg v_2 \gg v_3$. If all of the drug is added to the aqueous phase at time 0, concentration $C_1(0)$, then at equilibrium, the mass balance (see Section 7.5) would be

$v_1C_1(0) = v_1C_1(\infty) + v_2C_2(\infty) + v_3C_3(\infty)$. Two partition coefficients need to be defined: $K_{p2} = C_2(\infty)/C_1(\infty)$ and $K_{p3} = C_3(\infty)/C_1(\infty)$. With these, the mass balance may be rewritten as $v_1C_1(0) = v_1C_1(\infty) + v_2K_{p2}C_1(\infty) + v_3K_{p3}C_1(\infty) = C_1(\infty)(v_1 + v_2K_{p2} + v_3K_{p3})$. If the organic : aqueous volume ratios are r_2 and r_3, then the equilibrium concentrations in the three phases can be stated as

$$C_1(\infty) = \frac{C_1(0)}{(1 + r_2K_{p2} + r_3K_{p3})} \tag{7.38}$$

$$C_2(\infty) = \frac{C_1(0)K_{p2}}{(1 + r_2K_{p2} + r_3K_{p3})} \tag{7.39}$$

$$C_3(\infty) = \frac{C_1(0)K_{p3}}{(1 + r_2K_{p2} + r_3K_{p3})} \tag{7.40}$$

Further reduction is possible. To a good approximation, partition coefficients from different organic solvents may be interrelated by the so-called Collander equation [364,587]: $\log K_{p3} = a \log K_{p2} + c$, or $K_{p3} = 10^c K_{p2}^a$, where a and c are constants. Equations (7.38)–(7.40) can be expressed in log forms as a function of just one partition coefficient (i.e., $K_p = K_{p2}$):

$$\text{Water}: \quad \log\frac{C_1(\infty)}{C_1(0)} = -\log(1 + r_2K_p + r_310^c K_p^a) \tag{7.41}$$

$$\text{Lipid}: \quad \log\frac{C_2(\infty)}{C_1(0)} = \log K_p - \log(1 + r_2K_p + r_310^c K_p^a) \tag{7.42}$$

$$\text{Receptor}: \quad \log\frac{C_3(\infty)}{C_1(0)} = a\log K_p - \log(1 + r_2K_p + r_310^c K_p^a) + c \tag{7.43}$$

Figure 7.20 is a sample plot of relative equilibrium concentrations, Eqs. (7.41)–(7.43). In the example, the three phases were picked to be water, octanol, and phosphatidylcholine-based liposomes (vesicles consisting of a phospholipid bilayer), with the volumes $v_1 = 1$ mL (water), $v_2 = 50$ µL (octanol), and $v_3 = 10$ µL (liposomes). The Collander equation was deduced from Fig. 5.6: $\log K_{p,\text{liposome}} = 0.41 \log K_{p,\text{oct}} + 2.04$. Figure 7.20 suggests that when very hydrophilic molecules (with $\log K_{p,\text{oct}} < -6$) are placed into this three-phase mixture, most of them distribute into the water phase (solid curve), with only minor liposome phase occupation (dashed-dotted curve), but virtually no octanol phase occupation (dashed curve). In the example, molecules with $\log K_{p,\text{oct}}$ of -4 to $+3$, mostly reside in the liposome fraction, schematically modeling the lipophilic property of a hypothetical receptor site, reaching maximum occupancy for compounds with $\log K_{p,\text{oct}}$ at about $+1.5$. Very lipophilic molecules, with $\log K_{p,\text{oct}} > 5$ preferentially concentrate in the (more lipophilic) octanol compartment, becoming unavailable to the receptor region.

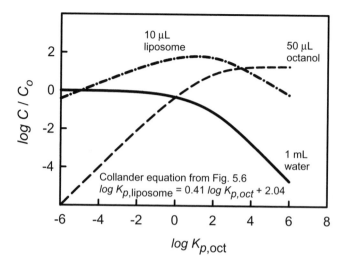

Figure 7.20 Three-compartment equilibrium distribution model (after Kubinyi [23]).

Kubinyi [23] showed that the bilinear equation (7.43) can be approximated by a general form

$$\log C = a \, \log \, K_p + c - b \, \log \, (rK_p + 1) \tag{7.44}$$

where a, b, c, r are empirical coefficients, determined by regression analysis, and C is the concentration in the intermediate phase. Equation (7.44) was used to calculate the curve in Fig. 7.19d.

Our present topic is the relationship between *permeability* and lipophilicity (kinetics), whereas we just considered a *concentration* and lipophilicity model (thermodynamics). Kubinyi demonstrated, using numerous examples taken from the literature, that the kinetics model, where the thermodynamic partition coefficient is treated as a ratio of two reaction rates (forward and reverse), is equivalent to the equilibrium model [23]. The liposome curve shape in Fig. 7.20 (dashed-dotted line) can also be the shape of a permeability-lipophilicity relation, as in Fig. 7.19d.

This relationship was further clarified by van de Waterbeemd in the "two-step distribution" model [588–590]. Later, the model was expanded by van de Waterbeemd and colleagues to include the effects of ionization of molecules, with the use of log K_d, in place of log K_p, as well as the effects of aqueous pores [49,54].

7.7 PAMPA: 50+ MODEL LIPID SYSTEMS DEMONSTRATED WITH 32 STRUCTURALLY UNRELATED DRUG MOLECULES

In the rest of the chapter, we describe over 50 specific PAMPA lipid models developed at *p*ION, identified in Table 7.3. The lipid models are assigned a two

TABLE 7.3 PAMPA Lipid Models

Lipid Type	Composition	pH_{DON}/pH_{ACC}	No Sink	Sink
			\multicolumn{2}{c}{Model Number}	
Neutral	2% DOPC[a]	7.4	1.0	1.1
	2% DOPC + 0.5% Cho	7.4	1A.0	—[d]
	Olive oil	7.4	2.0	—
	Octanol	7.4	3.0	—
	Dodecane	7.4	4.0	—
2-Component anionic[b]	2% DOPC + 0.6% DA	7.4	5.0	5.1
	2% DOPC + 1.1% DA	7.4	6.0	6.1
	2% DOPC + 0.6% PA	7.4	7.0	7.1
	2% DOPC + 1.1% PA	7.4	8.0	8.1
	2% DOPC + 0.6% PG	7.4	9.0	9.1
	2% DOPC + 1.1% PG	7.4	10.0	10.1
5-Component, anionic[c]	0.8% PC + 0.8% PE + 0.2% PS + 0.2% PI + 1.0% Cho	7.4	11.0	—[d]
Lecithin extracts[e] (anionic)	10% egg (Avanti)	7.4	12.0	12.1
	10% egg (Avanti) + 0.5% Cho	7.4	13.0	—[d]
	10% egg (Sigma)	7.4	14.0	14.1
	10% egg (Sigma) + 0.5% Cho	7.4	15.0	15.1
	10% soy	7.4	16.0	16.1
	20% soy	7.4	17.0	17.1
	20% soy + 0.5% Cho	7.4	18.0	18.1
	35% soy	7.4	19.0	19.1
	50% soy	7.4	—	20.1
	68% soy	7.4	21.0	—
	74% soy	7.4	—	22.1
Sink asymmetry	20% soy	7.4	—	23.2
Iso-pH	20% soy	6.5 / 6.5	—	24.1
	20% soy	5.0 / 5.0	—	25.1
Gradient–pH (corr UWL)	20% soy	6.5 / 7.4	—	26.1
	20% soy	6.0 / 7.4	—	27.1
	20% soy	5.5 / 7.4	—	28.1
	20% soy	5.0 / 7.4	—	29.1
	20% soy	4.5 / 7.4	—	30.1

[a]20 mg DOPC + 1 mL dodecane.

[b]20 mg DOPC + 6 (or 11) mg negative lipid (DA = dodecylcarboxylic acid, PA = phosphatidic acid, PG = phosphatidylglycerol) + 1 mL dodecane.

[c]Based on Sugano's formula, but using dodecane in place of 1,7-octadiene.

[d]Acceptor well solutions turn turbid in the presence of surfactant sink.

[e]Egg lecithin was "60% extract" grade. The products from Avanti and Sigma behaved differently. Soy lecithin was "20% extract" grade, from Avanti. The model number digit after the decimal point indicates 0 = no sink in system, 1 = sink in acceptor, 2 = sink in donor.

part serial number (Table 7.3). The first index is simply a serial designation and the second index indicates whether an artificial sink condition is in effect in the assay ($0 = $ no, $1 = $ yes). Special cases (e.g., cosolvent, cyclodextrin, bile salt, or mixed-micelle assays) will employ other values of the second index. We have selected 32 unrelated drug molecules, whose structures are shown in Fig. 7.21, to illustrate the properties of the PAMPA lipid models.

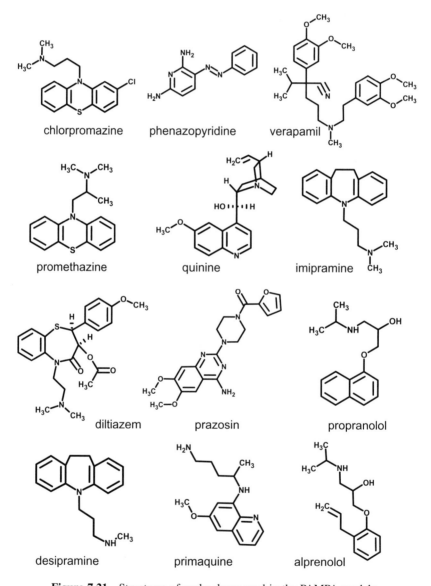

Figure 7.21   Structures of probe drugs used in the PAMPA models.

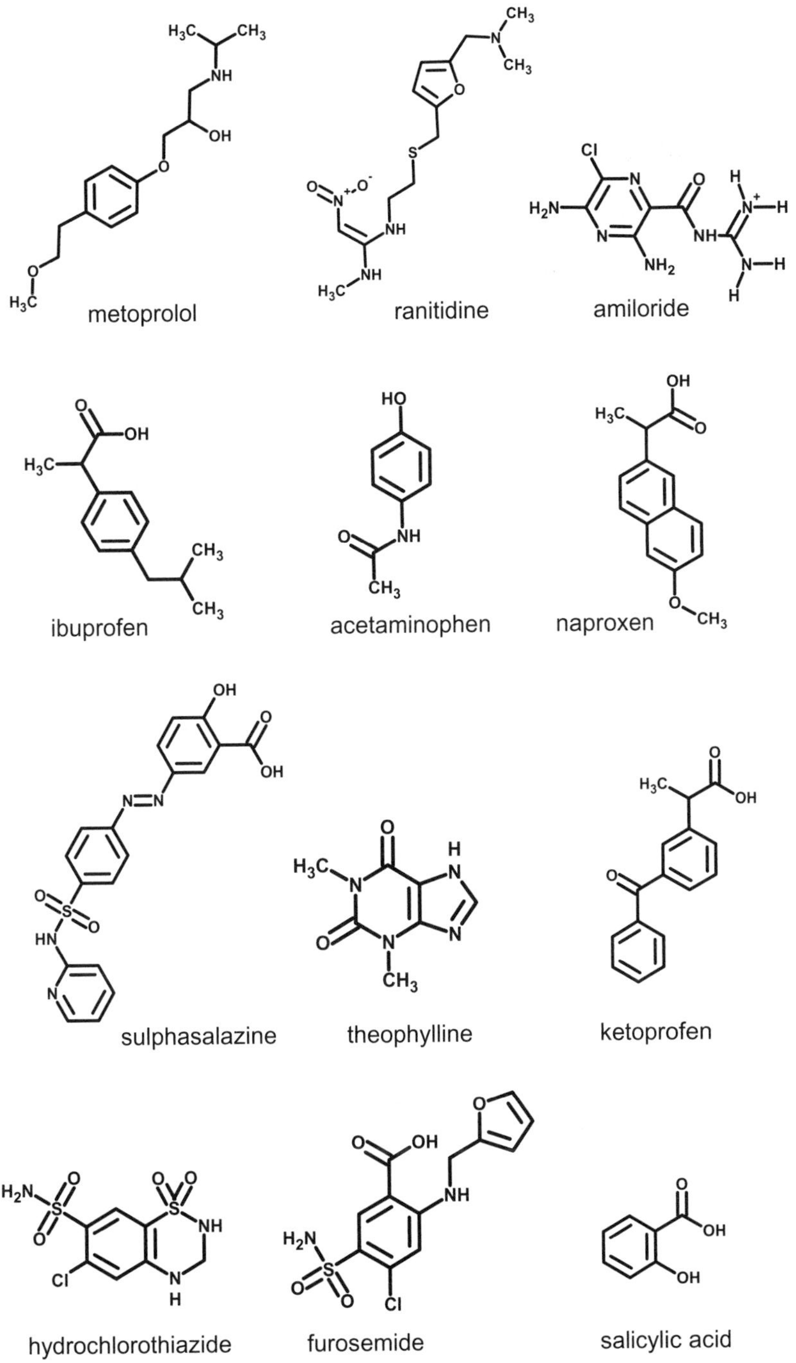

metoprolol

ranitidine

amiloride

ibuprofen

acetaminophen

naproxen

sulphasalazine

theophylline

ketoprofen

hydrochlorothiazide

furosemide

salicylic acid

Figure 7.21 (*Continued*)

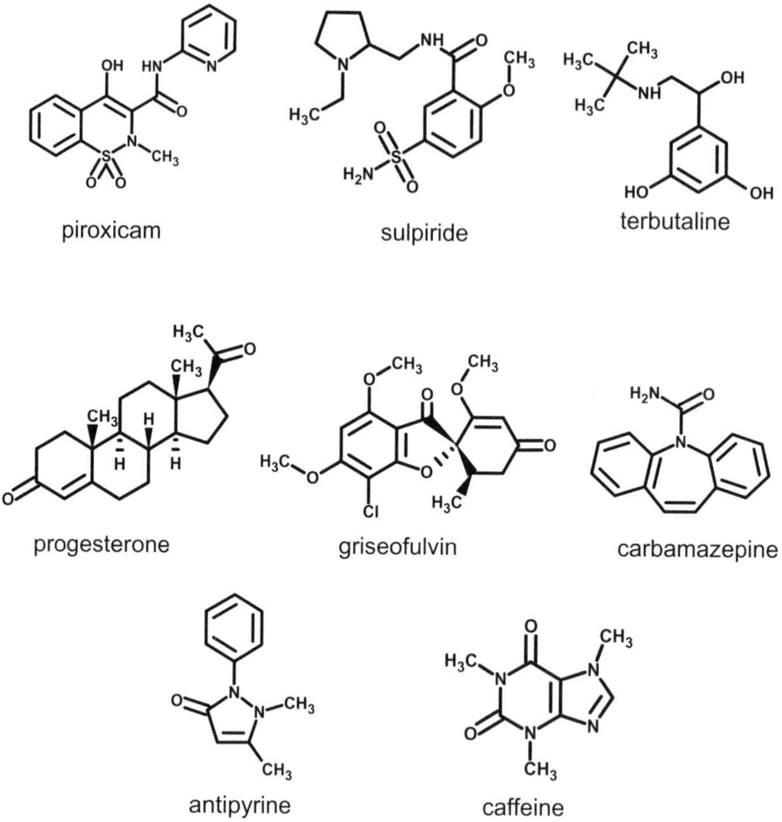

piroxicam

sulpiride

terbutaline

progesterone

griseofulvin

carbamazepine

antipyrine

caffeine

Figure 7.21   (*Continued*)

Table 7.4 summarizes the key pharmacokinetic (PK) and physicochemical properties of the selected probe molecules, consisting of bases, acids, and neutral species.

7.7.1 Neutral Lipid Models at pH 7.4

Four neutral lipid models were explored at pH 7.4: (1) 2% wt/vol DOPC in dodecane, (2) olive oil, (3) octanol, and (4) dodecane. Table 7.5 lists the effective permeabilities P_e, standard deviations (SDs), and membrane retentions of the 32 probe molecules (Table 7.4). The units of P_e and SD are 10^{-6} cm/s. Retentions are expressed as mole percentages. Figure 7.22a is a plot of log P_e versus log K_d (octanol–water apparent partition coefficients, pH 7.4) for filters loaded with 2% wt/vol DOPC in dodecane (model 1.0, filled-circle symbols) and with phospholipid-free dodecane (model 4.0, open-circle symbols). The dashed line in the plot was calculated assuming a UWL permeability (see Section 7.7.6) P_u, 16×10^{-6} cm/s (a typical value in an unstirred 96-well microtiter plate assay), and P_e of 0.8×10^{-6} cm/s

TABLE 7.4 Pharmacokinetic and Physicochemical Properties of Selected Probe Drugs[a]

Sample	% HIA	$P_e (10^{-4}$ cm/s)	$\log K_{d(7.4)}$	$\log K_p$	pK_{a1}	pK_{a2}	pK_{a3}	Charge Profile	f_u(pH 7.4)	Type
Chlorpromazine	100	—	3.45	5.40	9.24	—	—	$+ > o$	0.01	Base
Phenazopyridine	—	—	3.31	3.31	5.15	—	—	$+ > o$	0.99	Base
Verapamil	95	6.7	2.51	4.44	9.07	—	—	$+ > o$	0.02	Base
Promethazine	80	—	2.44	4.05	9.00	—	—	$+ > o$	0.02	Base
Quinine	80	—	2.19	3.50	4.09	8.55	—	$\ddagger > + > o$	0.07	Base
Imipramine	99	—	2.17	4.39	9.51	—	—	$+ > o$	0.008	Base
Diltiazem	99	—	2.16	2.89	8.02	—	—	$+ > o$	0.19	Base
Prazosin	50	—	2.00	2.18	7.11	—	—	$+ > o$	0.66	Base
Propranolol	99	2.9	1.41	3.48	9.53	—	—	$+ > o$	0.007	Base
Desipramine	95	4.4	1.38	3.79	10.16	—	—	$+ > o$	0.002	Base
Primaquine	100	—	1.17	3.00	3.55	10.03	—	$\ddagger > + > o$	0.002	Base
Alprenolol	93	—	0.86	2.99	9.51	—	—	$+ > o$	0.008	Base
Metoprolol	95	1.3	−0.24	1.95	9.56	—	—	$+ > o$	0.007	Base
Ranitidine	50	0.43	−0.53	1.28	1.96	8.31	—	$\ddagger > + > o$	0.11	Base
Amiloride	50	—	−0.60	−0.26	8.65	—	—	$+ > \pm$	0.000	Base
Ibuprofen	80	—	1.44	4.13	4.59	—	—	$o > -$	0.002	Acid
Acetaminophen	100	—	0.34	0.34	9.78	—	—	$o > -$	1.00	Acid
Naproxen	99	8.3	0.09	3.24	4.32	—	—	$o > -$	0.001	Acid
Sulfasalazine	13	—	0.08	3.61	2.80	8.25	10.96	$o > - - > = > \equiv$	0.000	Acid
Theophylline	98	—	0.00	0.00	8.70	—	—	$o > -$	0.95	Acid

TABLE 7.4 (*Continued*)

Sample	% HIA	P_e (10^{-4} cm/s)	log $K_{d(7.4)}$	log K_p	pK_{a1}	pK_{a2}	pK_{a3}	Charge Profile	f_u(pH 7.4)	Type
Ketoprofen	100	8.4	−0.11	3.16	4.12	—	—	$o > -$	0.001	Acid
Hydrochlorothiazide	67	0.04	−0.18	−0.03	8.91	10.25	—	$o > - >=$	0.97	Acid
Furosemide	61	0.05	−0.24	2.56	3.67	10.93	—	$o > - >=$	0.000	Acid
Salicyclic Acid	100	—	−1.68	2.19	3.02	—	—	$o > -$	0.000	Acid
Piroxicam	100	7.8	0.00	1.98	2.33	5.22	—	$+ > o > -$	0.007	Amphorphous
Sulpiride	35	—	−1.15	1.31	9.12	10.14	—	$+ > o > -$	0.05	Amphorphous
Terbutaline	60	0.3	−1.35	−0.08	8.67	10.12	11.32	$+ > ± > - >=$	0.02	Zwitterionic
Progesterone	91	—	3.89	3.89	—	—	—	o	1.00	Neutral
Griseofulvin	28	—	2.18	2.18	—	—	—	o	1.00	Neutral
Carbamazepine	100	4.3	2.45	2.45	—	—	—	o	1.00	Neutral
Antipyrine	100	4.5	0.56	0.56	—	—	—	o	1.00	Neutral
Caffeine	100	—	−0.07	−0.07	—	—	—	o	1.00	Neutral

[a] %HIA human intestinal absorption fraction, oral dose administration; P_e is human jejunal permeability [56]; log $K_{d(7.4)}$ apparent octanol–water partition coefficient; log K_p octanol–water partition coefficient; pK_a are ionization constants, at 0.01 M ionic strength; charge profile: the order in which charges on molecules change as pH is raised by 2–10. For example, for terbutaline at pH < 8.67 (pK_{a1}), the main species in solution is a cation (+); for pH 8.67–10.12, a zwitterion exists (±); between pH 10.12 and 11.32, an anion forms (−); and for pH > 11.32, the dianion predominates (=). The symbol > denotes transition in charge state when pH is increased. The fraction of the molecule in the uncharged form at pH 7.4 is represented by f_u.

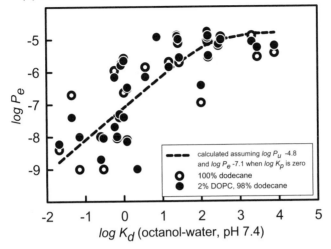

(a) Permeability and Lipophilicity

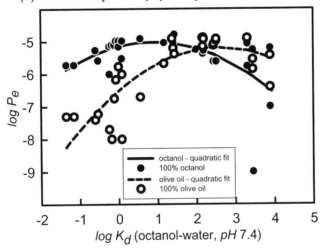

(b) Permeability and Lipophilicity

Figure 7.22 Lipophilic nature of membrane retention, $\log(\%R)$ versus octanol–water apparent partition coefficient, pH 7.4, neutral lipid models.

corresponding to where $\log K_d$ is zero (approximately equal to the P_e of metoprolol in 2% DOPC). Although the scatter of points is considerable, the pattern of the relationship between $\log P_e$ and $\log K_d$ best resembles the hyperbolic plot in Fig. 7.19b, with diffusion-limited (UWL) permeabilities for $\log K_d > 2$ and membrane-limited permeabilities for $\log K_d < 1$. (We discuss the UWL further Section 7.7.6.)

TABLE 7.5 Neutral Lipid PAMPA Models, pH 7.4[a]

Sample	2%DOPC (Model 1.0) P_e (SD)	%R	Olive Oil (Model 2.0) P_e(SD)	%R	Octanol (Model 3.0) P_e (SD)	%R	Dodecane (Model 4.0) P_e (SD)	%R
Chlorpromazine	5.5 (0.4)	85	1.4 (0.2)	82	0.000 (0.005)	90	2.8 (1.2)	89
Phenazopyridine	8.4 (1.1)	70	12.2 (2.5)	80	1.7 (0.2)	87	12.7 (1.6)	23
Verapamil	9.7 (1.0)	39	10.7 (2.8)	55	2.4 (0.3)	72	11.9 (1.0)	28
Promethazine	7.3 (0.7)	70	7.5	73	2.4 (0.2)	82	6.8 (1.4)	67
Quinine	3.1 (0.6)	1	11.9 (0.8)	21	5.2 (0.2)	63	2.8 (0.2)	10
Imipramine	11.1 (0.8)	56	7.0 (0.8)	75	4.2 (0.1)	76	8.5 (3.0)	55
Diltiazem	17.4 (1.8)	21	13.0 (2.6)	26	5.4 (1.2)	50	9.7 (0.1)	10
Prazosin	0.38 (0.07)	15	—	—	5.6 (0.2)	52	0.11 (0.03)	10
Propranolol	10.0 (0.5)	18	6.2 (0.9)	18	9.3 (0.2)	33	7.6 (0.1)	11
Desipramine	12.3 (0.4)	40	12.4 (1.8)	24	9.8 (0.6)	42	12.9 (1.1)	9
Primaquine	1.4 (0.1)	70	2.1	21	9.2 (0.1)	22	2.0 (0.1)	6
Alprenolol	11.8 (0.3)	16	—	—	—	—	—	—
Metoprolol	0.69 (0.04)	11	—	—	7.1 (0.1)	16	1.1 (0.1)	4
Ranitidine	0.009 (0.004)	2	0.06 (0.06)	0	2.6 (0.1)	13	0.000 (0.005)	2
Amiloride	0.002 (0.005)	0	0.04 (0.01)	2	5.4 (0.4)	14	0.01 (0.01)	2

	P_e (SD)							
Ibuprofen	2.7 (0.5)	38	4.1 (0.03)	11	16.6 (8.3)	34	1.9 (0.5)	0
Acetaminophen	0.001(0.005)	1	—	2	—	—	—	—
Naproxen	0.33 (0.03)	4	1.0 (0.1)	2	10.5 (0.7)	14	0.31 (0.12)	4
Sulfasalazine	0.007 (0.004)	1	0.01 (0.01)	0	3.0 (0.1)	11	0.008 (0.005)	3
Theophylline	0.04 (0.01)	1	—	—	10.5 (0.1)	12	0.23 (0.05)	1
Ketoprofen	0.05 (0.01)	4	0.18 (0.17)	2	7.8 (0.2)	13	0.04 (0.04)	1
Hydrochlorothiazide	0.01 (0.01)	1	0.01 (0.01)	3	7.8 (0.1)	14	0.009 (0.005)	1
Furosemide	0.02 (0.01)	1	0.02 (0.01)	2	1.0 (0.1)	12	0.02 (0.04)	0
Salicyclic acid	0.006 (0.004)	1	—	—	—	—	0.004 (0.006)	7
Piroxicam	2.2 (0.1)	3	1.7 (0.2)	4	6.9 (0.1)	18	2.6 (0.2)	1
Sulpiride	0.01 (0.01)	1	0.05 (0.05)	0	1.9 (0.1)	11	0.000 (0.005)	0
Terbutaline	0.04 (0.01)	6	0.05 (0.04)	2	1.9 (0.2)	14	0.20 (0.17)	0
Progesterone	6.3 (0.5)	84	0.4 (0.6)	83	0.1 (0.1)	91	3.7 (0.7)	82
Griseofulvin	12.8 (1.2)	18	13.2 (3.1)	9	6.5 (0.6)	62	9.2 (0.4)	10
Carbamazepine	7.1 (0.3)	10	8.3 (0.1)	7	9.4 (0.2)	40	6.4 (0.3)	6
Antipyrine	0.7 (0.1)	13	0.2 (0.2)	5	12.3 (0.9)	15	1.4 (0.1)	4
Caffeine	1.6 (0.1)	2	0.68 (0.08)	1	10.6 (0.2)	13	1.7 (0.1)	1

[a] All P_e and SD(P_e) are in units of 10^{-6} cm/s.

Figure 7.22b is a similar plot for the other two lipids considered: olive oil (unfilled symbols) and octanol (filled symbols). Both lipids can be described by a bilinear relationship, patterned after the case in Fig. 7.19d [Eq. (7.44)]. Octanol shows a declining log P_e relationship for very lipophilic molecules (log $K_d > 2$). The probe set of 32 molecules does not have examples of very hydrophilic molecules, with log $K_d < -2$, so the expected hydrophilic ascending part of the solid curve in Fig. 7.22b is not fully shown. Nevertheless, the shape of the plot is very similar to that reported by Camenisch et al. [546], shown in Fig. 7.8c. The UWL in the latter study (stirred solutions) is estimated to be ~ 460 μm (Fig. 7.8b), whereas the corresponding value in unstirred 96-well microtiter late assay is about 2300 μm. For this reason, the high point in Fig. 7.22b is $\sim 16 \times 10^{-6}$ cm/s, whereas it is $\sim 70 \times 10^{-6}$ cm/s in Fig. 7.8c.

Kansy et al. [550] reported the permeability–lipophilicity relationship for about 120 molecules based on the 10% wt/vol egg lecithin plus 0.5% wt/vol cholesterol in dodecane membrane lipid (model 15.0 in Table 7.3), shown in Fig. 7.23. The vertical axis is proportional to *apparent* permeability [see Eq. (7.9)]. For log $K_d > 1.5$, P_a *decreases* with increasing log K_d. In terms of characteristic permeability–lipophilicity plots of Fig. 7.19, the Kansy result in Fig. 7.23 resembles the bilinear case in Fig. (7.19d). Some of the P_a values may be underestimated for the most lipophilic molecules because membrane retention was not considered in the analysis.

7.7.1.1 DOPC

The 2% DOPC in dodecane (model 1.0, Table 7.3) was the first PAMPA model explored by the *p*ION group [25–28,556–558]. The lipid is commercially available in a highly purified preparation (in flame-sealed ampules packed under nitrogen), and is most like that used in the original BLM experiments [516,518,519, 523,532,542]. The lipid is completely charge neutral. It shows relatively low membrane retention for most molecules in Table 7.5, with the exception of chlorpromazine, phenazopyridine, primaquine, and progesterone. Our experience has been that as long as $R < 90\%$, most drug molecules have sufficient UV absorptivity to be adequately characterized when the initial concentrations are ~ 50 μm (a typical concentration in high-throughput applications). Lipid systems based on 10% or higher lecithin content can show very high membrane retention, in some cases preventing the assessment of permeability by UV spectrophotometry.

A few molecules have unexpectedly low permeability in 2% DOPC, not consistent with their octanol–water partition coefficients. Notably, metoprolol has a P_e value ~ 10 times lower in 2% DOPC, compared to 10% egg lecithin. Also, prazosin P_e appears to be significantly lower in DOPC, compared to other lipids.

The quality of the data collected from 2% DOPC membranes is unmatched by any other system we have explored. It's not uncommon to see interplate reproducibility $< 5\%$ and intraplate even better than that (1–3% SD). As will be seen later, lipid model 1.0 does not predict GIT absorption as well as some of the newer *p*ION models. However, this may not be the case when BBB models are explored in detail.

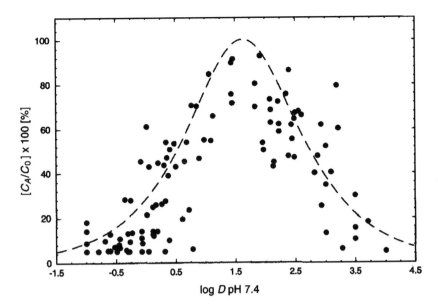

Figure 7.23 Relative acceptor compartment concentrations versus octanol–water apparent partition coefficients [550]. [Reprinted from Kansy, M.; Fischer, H.; Kratzat, K.; Senner, F.; Wagner, B.; Parrilla, I., in Testa, B.; van de Waterbeemd, H.; Folkers, G.; Guy, R. (Eds.). *Pharmacokinetic Optimization in Drug Research*, Verlag Helvetica Chimica Acta: Zürich and Wiley-VCH, Weinheim, 2001, pp. 447–464, with permission from Verlag Helvetica Chimica Acta AG.]

7.7.1.2 Olive Oil

Olive oil was the "original" model lipid for partition studies, and was used by Overton in his pioneering research [518,524]. It fell out of favor since the 1960s, over concerns about standardizing olive oil from different sources. At that time, octanol replaced olive oil as the standard for partition coefficient measurements. However, from time to time, literature articles on the use of olive oil appear. For example, Poulin et al. [264] were able to demonstrate that partition coefficients based on olive oil–water better predict the in vivo adipose-tissue distribution of drugs, compared to those from octanol–water. The correlation between in vivo log $K_{p\,\text{(adipose tissue–plasma)}}$ and log $K_{p\,\text{(olive oil–water)}}$ was 0.98 (r^2), compared to 0.11 (r^2) in the case of octanol. Adipose tissue is white fat, composed mostly of triglycerides. The improved predictive performance of olive oil may be due to its triglyceride content.

It was thus interesting for us to examine the permeability and membrane retention properties of olive oil. As Table 7.5 shows, most of the P_e values for olive oil are less than or equal to those of 2% DOPC, with notable exceptions; for instance, quinine is 4 times more permeable and progesterone is 16 times less permeable in olive oil than in DOPC. Both lipids show progesterone retention to be >80%, but quinine retention in olive oil is substantially greater than in DOPC.

7.7.1.3 Octanol

Octanol permeability is important to explore, since it is the principal basis for the lipophilicity scale in pharmaceutical research. Most interesting to us, in this light, is to address the question of ion pair partitioning and its meaning in the prediction of absorption of charged drugs. It has been discussed in the literature that quaternary ammonium drugs, when matched with lipophilic anions, show considerably increased octanol–water partition coefficients [291]. It has been hypothesized that with the right counterion, even charged drugs could be partly absorbed in the GIT. Given the structure of wet octanol, it could be argued that the 25 mol% water in octanol may be an environment that can support highly charged species, if lipophilic counterions are added. Unexpectedly high partition coefficients can be measured for ion pair forming drugs. But does this mean that ion pair transport takes place in vivo? This was addressed by the pION group by comparing permeability coefficients derived from DOPC and octanol lipid membrane models. For molecules showing very low permeabilities in DOPC (model 1.0) and very high permeabilities in octanol-impregnated membranes (Model 3.0), one could hypothesize that the water clusters in wet octanol act like "ion pair shuttles," an interesting effect, but perhaps with uncertain physiological interpretation [560].

Figure 7.22b shows that hydrophilic molecules, those with log $K_d < 1$, are much more permeable in octanol than in olive oil. The same may be said in comparison to 2% DOPC and dodecane. Octanol appears to enhance the permeability of hydrophilic molecules, compared to that of DOPC, dodecane, and olive oil. This is dramatically evident in Fig. 7.7, and is confirmed in Figs. 7.8c and 7.22b. The mechanism is not precisely known, but it is reasonable to suspect a "shuttle" service may be provided by the water clusters in octanol-based PAMPA (perhaps like an inverted micelle equivalent of endocytosis). Thus, it appears that charged molecules can be substantially permeable in the octanol PAMPA. However, do charged molecules permeate phospholipid bilayers to any appreciable extent? We will return to this question later, and will cite evidence at least for a partial answer.

Membrane retention of lipophilic molecules is significantly increased in octanol, compared to 2% DOPC. Chlorpromazine and progesterone show $R > 90\%$ in octanol. Phenazopyridine, verapamil, promethazine, and imipramine show $R > 70\%$.

7.7.1.4 Dodecane

Dodecane-coated filters were studied to determine what role hydrogen-bonding and electrostatic effects play in the 2% DOPC system. Measuring the differences between P_e deduced from 2% DOPC in dodecane and 0% DOPC in dodecane might indicate the extent of H-bonding and/or electrostatic interactions for specific probe molecules. Table 7.5 indicates that some molecules are retarded by the presence of DOPC (e.g., phenazopyridine, verapamil, metoprolol, theophylline, terbutaline, antipyrine), while most molecules are accelerated by DOPC (e.g., chlorpromazine, imipramine, diltiazem, prazosin, progesterone). The quantitative structure–permeability relationships for a much larger set of drug-like molecules are currently investigated in our laboratory (see Section 7.7.8).

It is also quite interesting that lipid model 4.0 may be used to obtain alkane partition coefficients at high-throughput speeds, as suggested by Faller and Wohnsland [509,554]. It is also interesting to note that since our P_e are corrected for membrane retention, the slope in Fig. 7.11 corresponding to the dashed line (our data) is 1.0, whereas the data not corrected for retention (solid line) show a lesser slope. This may not matter if the objective is to obtain alkane–water log K_p values at high speeds.

7.7.2 Membrane Retention (under Iso-pH and in the Absence of Sink Condition)

The membrane retention R is often stated as a mole percentage of the sample lost to the membrane. Its value can at times be very high, as high as 85% for chlorpromazine and 70% for phenazopyridine, with membranes made of 2% DOPC dissolved in dodecane. Regression analysis of log %R versus log $K_{d(7.4)}$, the octanol–water apparent partition coefficient, produces r^2 0.59. For DOPC-free dodecane, such analysis yields a higher r^2 (0.67). Olive oil and octanol further improve, with r^2 of 0.80 and 0.90, respectively. As far as %R representing lipophilicity as indicated by octanol–water partition coefficients is concerned, the order of "octanol-likeness" is octanol > olive oil > dodecane > DOPC in dodecane. Figure 7.24 shows the log %R/log K_d plot for octanol-impregnated membranes, at pH 7.4. It's clear that retention is due to the lipophilicity of molecules.

Culture-cell assays are also subject to sample retention by the monolayer. Sawada et al. [574] studied the transport of chlorpromazine across MDCK cell

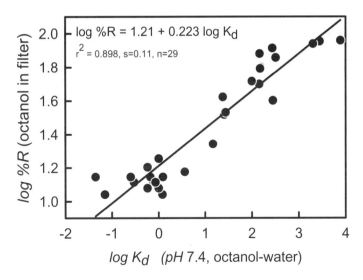

Figure 7.24 Membrane retention in octanol-soaked filters versus octanol–water apparent partition coefficients.

monolayers in the presence of various levels of serum proteins, and observed 65–85% retentions of the drug molecule by the MDCK cells. Wils et al. [591] reported retentions as high as 44% in Caco-2 cells. In a later publication, Sawada et al. [575] cited values as high as 89% for a homologous series of lipophilic molecules. Krishna et al. [551] more recently reported Caco-2 permeability results for lipophilic molecules, including progesterone and propranolol. They found retentions as high as 54%. It is undoubtedly a common phenomenon with research compounds, which are often very lipophilic. Yet in most reported assays, the effect is ignored, it appears. Ho et al. [514] derived an equation [similar to Eq. (7.22)] to describe the phenomenon in cultured cells, but its application in cultured-cell assays is scarce so far.

Retention may be a good predictor of the PK volume of distribution, of protein binding [264,592] or possibly even of conditions suitable for P-gp binding and extrusion of drugs. Apparently, these themes have not yet been adequately explored.

It is curious that the log of the expression for R, Eq. (7.18), produces a "Kubinyi-like" bilinear equation

$$\log R = \log K_d - \log(rK_d + 1) + \log r \qquad (7.45)$$

where the oil–water volume ratio, $r = V_M/(V_A + V_D)$. Its form is essentially that of Eq. (7.44). When 2% DOPC in dodecane is used for the PAMPA membrane lipid, V_M could be taken as the volume of dodecane (4–6 μL) or the volume of DOPC

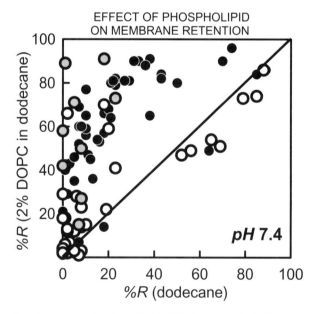

Figure 7.25 Membrane retention in 2%DOPC/dodecane-soaked filters versus dodecane-soaked filters.

(0.08–0.12 µL). The choice rests on the presumed structure of the membrane lipids (and where the drug preferentially partitions), which is not absolutely certain at present (see Section 7.3.6). It may be best to treat r as an empirical parameter, determined by regression against some lipophilicity model.

Figure 7.25 is a plot of %R (2%DOPC in dodecane) versus %R (100% dodecane). It shows that even 2% DOPC in dodecane can influence membrane retention to a considerable extent, compared to retentions observed in the absence of DOPC. Many molecules show retentions exceeding 70% in DOPC, under conditions where the retentions in dodecane are below 10%. However, it cannot be assumed that retention is always very low in dodecane, since several points in Fig. 7.25 are below the diagonal line, with values as high as 90% (chlorpromazine).

7.7.3 Two-Component Anionic Lipid Models with Sink Condition in the Acceptor Compartment

The use of simple single-component neutral lipids has played a valuable role in development of the PAMPA technique. Since it was an early objective of such work to predict GIT absorption, it became necessary to test the effect of phospholipid mixtures, where variable amounts of negative lipid could be introduced. Table 7.1 indicates that brush-border membrane (BBM) lipid mixture contains one negative phospholipid for every 3.5 zwitterionic lipids, and the blood–brain barrier (BBB) lipid has even a higher negative lipid content. The simplest model to simulate the BBM mixture could consist of two components: DOPC plus a negatively charged phospholipid: for example, phosphatidylserine, phosphatidylinositol, phosphatidylglycerol, phosphatidic acid, or cardiolipin (see Fig. 7.4). Even a fatty acid, such as dodecylcarboxylic acid (DA), could play the role of introducing negative charge to the mixture. Our design criterion was to begin with 2% DOPC and add the additional negatively charged lipid in the proportion consistent with BBM (0.6% added lipid) or BBB (1.1% added lipid) negative-zwitterionic proportions (Table 7.1).

Since there would be increased overall lipid concentration in the dodecane solution, we decided to create a sink condition in the acceptor wells, to lower the membrane retention. We discovered that the pH 7.4 buffer saturated with sodium laurel sulfate serves as an excellent artificial sink-forming medium. Since the new PAMPA membranes would possess substantial negative charge, the negatively charged micellar system was not expected to act as an aggressive detergent to the two-component artificial membrane infused in the microfilter.

Six two-component models were tested under sink conditions (models 5.1–10.1 in Table 7.3), employing three negatively charged lipids (dodecylcarboxylic acid, phosphatidic acid, and phosphatidylglycerol). These models were also tested in the absence of the sink condition (models 5.0–10.0 in Table 7.3).

Tables 7.6–7.8 list the P_e, SD, and %R of the 32 probe molecules in the thirteen new PAMPA lipid models, one of which is 2% DOPC assayed under sink conditions (model 1.1). The latter model served as a benchmark for assessing the effects of negative membrane charge.

TABLE 7.6 Two-Component Anionic Lipid PAMPA Models (with Sink), pH 7.4[a]

Sample	2%DOPC (Model 1.1)		+0.6%DA (Model 5.1)		+1.1%DA (Model 6.1)		+0.6%PA (Model 7.1)		+1.1%PA (Model 8.1)	
	P_e(SD)	%R	P_e(SD)	%R	P_e(SD)	%R	P_e(SD)	%R	P_e(SD)	%R
Chlorpromazine	21.3 (0.3)	16	31.4 (1.2)	30	35.5 (6.7)	31	31.2 (0.5)	27	34.3 (7.2)	31
Phenazopyridine	31.4 (1.6)	26	27.0 (1.6)	30	23.9 (2.8)	31	23.2 (1.3)	36	21.6 (1.1)	42
Verapamil	37.4 (4.2)	17	40.8 (0.9)	29	25.1 (1.9)	25	24.9 (2.1)	21	34.6 (3.8)	31
Promethazine	31.5 (1.7)	12	44.4 (7.4)	24	28.5 (0.6)	24	40.1 (4.3)	18	37.3 (2.0)	20
Quinine	2.7 (0.1)	13	27.7 (2.5)	38	30.7 (3.2)	31	19.1 (1.3)	27	24.7 (2.0)	33
Imipramine	22.2 (0.7)	21	29.9 (0.8)	33	26.8 (0.6)	32	25.0 (0.6)	32	29.0 (1.2)	32
Diltiazem	33.9 (8.7)	13	35.4 (9.1)	12	31.9 (5.8)	16	29.4 (1.3)	16	31.4 (1.4)	20
Prazosin	—	—	13.3 (1.5)	17	10.1 (0.7)	14	4.2 (0.4)	17	16.2 (0.6)	17
Propranolol	14.3 (0.1)	16	39.2 (1.0)	28	27.3 (2.6)	30	29.9 (1.5)	29	36.2 (1.3)	27
Desipramine	25.1 (6.3)	22	33.9 (6.8)	38	26.7 (4.6)	42	22.8 (3.1)	32	40.0 (7.0)	33
Primaquine	1.4 (0.1)	10	22.8 (0.4)	23	22.9 (0.3)	25	18.5 (3.0)	27	22.7 (0.7)	27
Metoprolol	0.41 (0.03)	3	5.8 (0.4)	16	8.7 (0.5)	17	4.7 (0.8)	13	16.9 (1.1)	18
Ranitidine	0.01 (0.02)	0	0.01 (0.02)	3	0.02 (0.03)	4	(nd)	1	0.02 (0.02)	1
Amiloride	(nd)	1	1.6 (0.3)	1	4.7 (0.1)	8	0.01 (0.02)	2	(nd)	1

Compound	P_e (SD)	n	P_e (SD)	n	P_e (SD)	n	P_e (SD)	n	P_e (SD)	n
Ibuprofen	2.4 (0.1)	7	11.9 (4.3)	14	10.1 (2.4)	28	3.8 (0.5)	10	(nd)	27
Naproxen	0.3 (0.1)	1	0.7 (0.1)	4	1.8 (0.6)	6	0.8 (0.2)	1	0.9 (0.5)	2
Sulfasalazine	0.03 (0.05)	3	(nd)	2	0.04 (0.06)	1	(nd)	3	(nd)	2
Theophylline	—	—	0.2 (0.1)	3	0.4 (0.4)	4	0.04 (0.06)	4	0.4 (0.3)	3
Ketoprofen	0.2 (0.2)	2	0.6 (0.2)	6	0.6 (0.1)	4	0.3 (0.1)	2	0.3 (0.1)	4
Hydrochlorothiazide	0.001 (0.005)	1	0.001 (0.005)	2	(nd)	3	0.002 (0.005)	1	0.01 (0.01)	2
Furosemide	0.04 (0.01)	1	(nd)	3	0.06 (0.08)	5	(nd)	1	0.1 (0.1)	2
Piroxicam	2.0 (0.1)	3	3.3 (0.1)	5	3.5 (0.1)	6	2.3 (0.1)	4	2.6 (0.1)	4
Sulpiride	0.1 (0.2)	1	(nd)	3	0.5 (0.1)	3	(nd)	2	0.03 (0.07)	4
Terbutaline	0.1 (0.1)	0	(nd)	4	0.1 (0.2)	7	0.1 (0.2)	3	0.5 (0.6)	4
Progesterone	22.7 (1.1)	34	30.3 (3.3)	44	21.2 (0.6)	32	35.7 (0.5)	34	29.2 (2.0)	37
Griseofulvin	18.2 (1.0)	10	19.4 (0.3)	17	18.9 (1.9)	24	19.4 (1.3)	21	21.2 (1.5)	22
Carbamazepine	6.4 (0.2)	13	17.3 (0.3)	9	18.0 (1.4)	19	9.4 (1.0)	12	13.7 (0.6)	15
Antipyrine	0.6 (0.12)	4	0.6 (0.1)	5	1.0 (0.3)	6	0.7 (0.4)	6	0.6 (0.2)	3
Caffeine	1.2 (0.1)	4	1.3 (0.2)	5	1.4 (0.2)	6	1.3 (0.1)	4	1.5 (0.1)	3

[a] All P_e and SD(P_e) are in units of 10^{-6} cm/s; (nd) = compound not detected in the acceptor compartment.

TABLE 7.7 Two-Component Anionic Lipid PAMPA Models (Only PG with Sink), pH 7.4[a]

Sample	+0.6%PG (Model 9.1) P_e(SD)	%R	+1.1%PG (Model 10.1) P_e(SD)	%R	+0.6%PA (Model 7.0) P_e(SD)	%R
Chlorpromazine	16.2 (2.1)	51	5.1 (1.6)	73	1.3 (1.5)	80
Phenazopyridine	17.2 (1.2)	53	5.4 (0.4)	56	3.6 (0.1)	56
Verapamil	21.1 (3.3)	37	8.4 (0.7)	53	4.8 (1.7)	56
Promethazine	35.4 (1.3)	45	13.1 (2.2)	62	(nd)	63
Quinine	2.4 (0.2)	38	5.2 (1.4)	60	7.2 (0.7)	54
Imipramine	24.3 (2.9)	49	7.6 (0.1)	60	1.8 (1.0)	56
Diltiazem	18.2 (3.7)	36	8.9 (2.8)	55	14.8 (0.1)	50
Prazosin	1.0 (0.5)	39	1.0 (0.2)	53	4.9 (1.2)	16
Propranolol	8.0 (0.6)	50	3.4 (1.4)	66	2.7 (0.2)	47
Desipramine	9.0 (2.0)	56	0.4 (0.6)	66	3.4 (2.9)	72
Primaquine	1.1 (0.2)	43	0.2 (0.2)	59	2.0 (0.3)	43
Alprenolol	—	—	—	—	7.9 (2.8)	42
Metoprolol	(nd)	22	(nd)	42	6.0 (0.8)	10
Ranitidine	(nd)	1	(nd)	2	0.1 (0.1)	1
Amiloride	0.03 (0.03)	2	(nd)	5	(nd)	0
Ibuprofen	18.9 (1.3)	0	(nd)	24	(nd)	28
Naproxen	(nd)	2	(nd)	4	2.2 (0.8)	0
Sulfasalazine	(nd)	1	0.004 (0.007)	2	0.03 (0.05)	3
Theophylline	(nd)	2	(nd)	3	(nd)	0
Ketoprofen	0.36 (0.04)	1	0.03 (0.04)	9	1.1 (0.6)	1
Hydrochlorothiazide	0.007 (0.007)	1	(nd)	3	0.04 (0.01)	0
Furosemide	(nd)	0	0.05 (0.08)	3	(nd)	0
Piroxicam	2.0 (0.2)	2	2.0 (0.1)	3	2.3 (0.1)	1
Sulpiride	(nd)	1	0.4 (0.3)	4	—	—
Terbutaline	(nd)	2	(nd)	3	(nd)	1
Progesterone	35.2 (2.2)	46	33.2 (0.9)	42	1.6 (0.1)	55
Griseofulvin	18.5 (2.7)	20	17.7 (1.5)	21	18.3 (0.9)	25
Carbamazepine	8.5 (0.7)	11	9.7 (0.5)	13	10.4 (1.4)	10
Antipyrine	1.1 (0.4)	3	0.7 (0.2)	5	1.4 (0.4)	1
Caffeine	1.5 (0.1)	4	2.1 (0.1)	4	2.1 (0.1)	0

[a]All P_e and SD(P_e) are in units of 10^{-6} cm/s; (nd) = compound not detected in the acceptor compartment.

TABLE 7.8 Two-Component Anionic Lipid PAMPA Models (without Sink), pH 7.4[a]

Sample	+0.6% DA (Model 5.0) P_e(SD)	%R	+1.1% DA (Model 6.0) P_e(SD)	%R	+1.1% PA (Model 8.0) P_e(SD)	%R	+0.6% PG (Model 9.0) P_e(SD)	%R	+1.1% PG (Model 10.0) P_e(SD)	%R
Chlorpromazine	0.6 (0.1)	59	0.2 (0.2)	60	0.28 (0.03)	62	0.4 (0.2)	64	0.3 (0.1)	68
Phenazopyridine	4.6 (0.3)	47	5.1 (0.2)	48	2.8 (0.1)	52	3.7 (0.1)	41	2.9 (0.1)	49
Verapamil	2.9 (1.3)	53	4.6 (0.4)	58	1.4 (0.2)	58	0.7 (0.3)	57	0.6 (0.6)	68
Promethazine	0.8 (0.3)	61	1.5 (0.3)	68	0.01 (0.01)	62	0.9 (0.4)	59	0.6 (0.3)	69
Quinine	6.0 (0.6)	37	7.4 (0.2)	47	3.4 (0.5)	53	0.5 (0.1)	28	0.2 (0.2)	51
Imipramine	2.3 (0.3)	53	2.7 (0.2)	59	2.0 (0.2)	51	2.5 (0.2)	37	0.8 (0.1)	46
Diltiazem	7.7 (0.3)	35	8.1 (1.0)	38	5.1 (0.4)	48	1.2 (0.2)	49	0.3 (0.4)	62
Prazosin	6.2 (0.4)	15	7.7 (0.5)	17	5.6 (1.3)	19	0.04 (0.01)	15	(nd)	36
Propranolol	4.6 (0.2)	47	6.9 (0.5)	57	1.9 (0.1)	50	0.9 (0.1)	34	0.4 (0.1)	53
Desipramine	2.8 (0.5)	54	2.1 (0.7)	61	1.2 (0.1)	55	2.3 (0.4)	40	1.6 (0.2)	58
Primaquine	7.7 (0.3)	45	5.2 (0.6)	48	1.4 (0.2)	48	(nd)	39	0.4 (0.2)	51
Alprenolol	7.8 (0.3)	34	7.9 (0.6)	40	4.1 (0.8)	46	2.2 (2.0)	33	1.2 (0.1)	51
Metoprolol	5.2 (0.3)	13	10.0 (0.8)	10	4.7 (0.5)	15	0.3 (0.4)	14	0.05 (0.09)	22
Ranitidine	0.12 (0.01)	3	0.3 (0.1)	8	0.08 (0.05)	2	0.02 (0.03)	0	(nd)	0
Amiloride	4.5 (0.5)	10	6.9 (0.1)	12	(nd)	1	(nd)	0	(nd)	0

176

TABLE 7.8 (Continued)

Sample	+0.6% DA (Model 5.0) P_e(SD)	%R	+1.1% DA (Model 6.0) P_e(SD)	%R	+1.1% PA (Model 8.0) P_e(SD)	%R	+0.6% PG (Model 9.0) P_e(SD)	%R	+1.1% PG (Model 10.0) P_e(SD)	%R
Ibuprofen	5.0 (5.4)	8	19.1 (3.3)	14	7.6 (2.3)	2	0.5 (0.8)	0	—	—
Naproxen	2.0 (0.2)	2	2.8 (0.2)	6	0.3 (0.2)	16	0.4 (0.2)	0	—	—
Theophylline	(nd)	4	0.8 (0.3)	7	(nd)	3	(nd)	0	(nd)	0
Ketoprofen	0.75 (0.09)	1	0.93 (0.08)	5	0.7 (0.1)	1	0.3 (0.1)	0	0.3 (0.1)	0
Hydrochlorothiazide	0.02 (0.02)	2	0.04 (0.04)	4	0.008 (0.007)	1	0.03 (0.01)	0	(nd)	1
Furosemide	0.04 (0.03)	2	0.05 (0.02)	7	0.01 (0.02)	1	0.03 (0.02)	0	0.03 (0.01)	0
Piroxicam	3.3 (0.1)	3	3.3 (0.1)	8	2.5 (0.1)	2	1.8 (0.1)	0	1.5 (0.1)	0
Sulpiride	0.3 (0.5)	3	0.7 (0.2)	7	0.06 (0.08)	2	0.1 (0.1)	0	0.2 (0.1)	0
Terbutaline	(nd)	4	0.1 (0.2)	10	0.02 (0.04)	1	0.06 (0.06)	0	0.02 (0.04)	1
Progesterone	2.3 (0.5)	57	3.0 (0.3)	64	3.5 (0.3)	61	3.9 (0.8)	66	3.5 (1.0)	66
Griseofulvin	8.5 (0.1)	20	9.0 (0.2)	18	7.2 (1.6)	29	6.9 (0.6)	20	8.1 (0.3)	24
Carbamazepine	11.7 (0.8)	9	13.5 (1.4)	11	8.0 (0.4)	10	5.9 (0.1)	4	5.6 (0.4)	6
Antipyrine	1.0 (0.1)	2	1.3 (0.1)	7	1.1 (0.2)	6	1.3 (0.2)	1	1.0 (0.1)	3
Caffeine	1.7 (0.1)	2	1.9 (0.1)	8	1.7 (0.2)	4	1.9 (0.1)	0	1.6 (0.2)	0

[a]All P_e and SD(P_e) are in units of 10^{-6} cm/s; (nd) = compound not detected in the acceptor compartment.

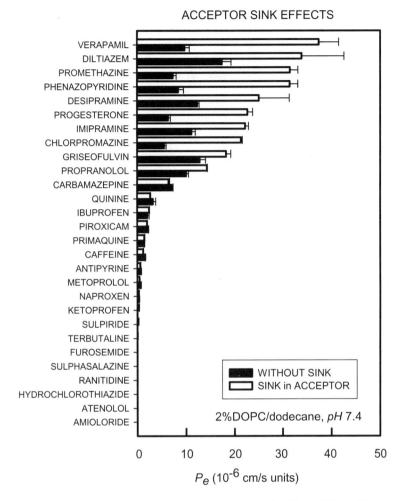

Figure 7.26   Permeabilities with and without sink, 2% DOPC model.

7.7.3.1 DOPC under Sink Conditions

Figure 7.26 shows the effect of the sink condition on the effective permeabilities in the 2% DOPC system (model 1.1). Just about all of the lipophilic bases showed a two- to three-fold increase in P_e. The simplest interpretation of this is that when lipophilic molecules reach the acceptor wells, they are bound to the surfactant, and the concentration of the unbound (membrane-permeating) form of the drug greatly diminishes. Hence, the reverse flux, based on the unbound portion of the concentration $C_A(t)$ in Eq. (7.1), is nil. Thus, half of the UWL resistance effectively disappears, leading to a doubling of P_e for the diffusion-limited molecules. The topic of the UWL is discussed in greater detail in Section 7.7.6. The binding of the positively charged lipophilic molecules by the negatively charged micelles is

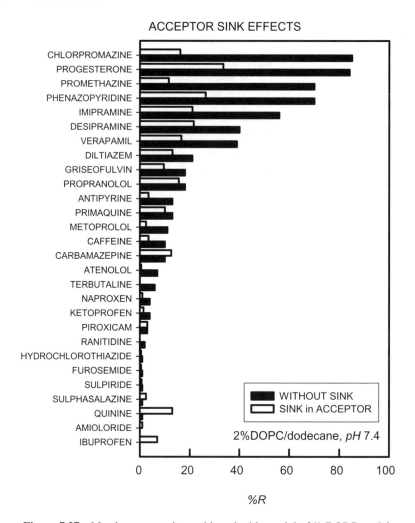

Figure 7.27   Membrane retentions with and without sink, 2% DOPC model.

expected to have a strong electrostatic component, as well as a hydrophobic component.

Furthermore, the membrane retentions of the lipophilic probe molecules are dramatically decreased in the presence of the sink condition in the acceptor wells, as shown in Fig. 7.27. All molecules show $R < 35\%$, with progesterone and phenazopyridine showing the highest values, 34% and 26%, respectively.

The combination of increased P_e and decreased %R allowed the permeation time to be lowered to 4 h, in comparison to the originally specified time of 15 h [547,550], a considerable improvement for high-throughput applications. The quality of the measurements of the low-permeability molecules did not substantially improve with sink conditions or the reduced assay times.

7.7.3.2 DOPC with Dodecylcarboxylic Acid under Sink Conditions

The free fatty acid model 5.1 shows dramatic differences in permeabilities over the neutral-charge model 1.1. For example, quinine, metoprolol, and primaquine are 10, 14, and 16 times more permeable, respectively, in the charged (0.6% wt/vol in dodecane) lipid system. The most remarkable enhancement is that of amiloride. In the DOPC system, no detectable amount of amiloride permeates; however, P_e is 1.6×10^{-6}cm/s when 0.6% DA is added to the dodecane. It is thought that a very strong ion-pair complex forms between the positively-charged amiloride (Fig. 7.21) and the negative-charge dodecylcarboxylate group, through strong electrostatic and hydrogen bonding, perhaps forming an eight-membered ring $-(-C=N^+-H \cdots {}^-O-C=O \cdots H-N-)-$. Uncharged carboxylic acids are known to form dimeric units of a similar sort when dissolved in oil [538].

The increase of negative charge from 0.6% to 1.1% wt/vol in dodecane (modeling the expected increase between BBM and BBB lipid compositions; see Table 7.1) shows further increases to the permeabilities of the dramatically affected molecules, especially amiloride, which becomes effectively more permeable than piroxicam.

Most of the weak-acid probe molecules (ibuprofen, naproxen, ketoprofen, piroxicam) show significant increase in permeabilities with models 5.1 and 6.1, compared to model 1.1. This is surprising, considering that most of the weak-acid probes are negatively charged themselves, and would be expected to be less permeable, due to electrostatic repulsions. Apparently, the increased membrane–water partitioning of weak acids in the two-component lipid models overcomes the expected negative charge repulsions between the ionized acids and the charged membrane components, and leads to increased permeability. Also, membrane surface negative charge is expected to lower the surface pH, thus increasing interfacial f_u (Table 7.4), leading to higher permeabilities of ionizable acids [457].

7.7.3.3 DOPC with Phosphatidic Acid under Sink Conditions

The PA systems (models 6.1 and 7.1) seem to show some of the general patterns of changes seen above, but to a lesser extent. Amiloride permeates in its usual way (poorly). The weak-acid probes are more permeable in the PA models, compared to neutral DOPC, but to a lesser extent than in DA. As a predictor of GIT absorption, the phosphatidic acid system appears to be the best. (The rankings of all the investigated lipid systems are discussed in Section 7.8.3.) Figure 7.28a shows the effect of PA on the permeabilities of the weak-base probe molecules. Dramatic and systematic increases are seen in all the membrane-limited permeabilities (left side of the bar graph). When the permeabilities reach the UWL limit of model 1.1, then no substantial effects due to increasing amounts of PA are seen (right side of the bar graph). So, most of the charged bases are elevated to be nearly diffusion-limited in their permeabilities, when PA is part of the membrane constituents.

Figure 7.28b shows that membrane retention is very systematically increased for almost all of the weak bases. This is a general pattern for bases with any of the negatively charged membrane models, and is probably best explained by the increased electrostatic attractions between the drugs and the membranes. Still, all retentions are below 50%, due to the offsetting sink condition created in the acceptor wells.

(a) Permeability

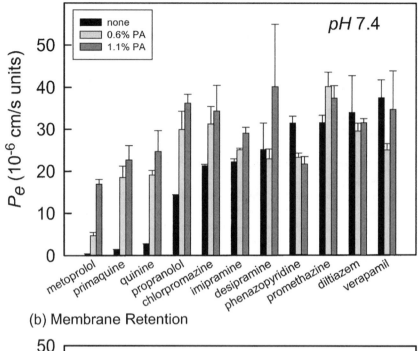

(b) Membrane Retention

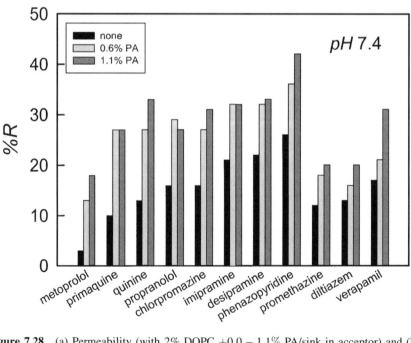

Figure 7.28   (a) Permeability (with 2% DOPC +0.0 − 1.1% PA/sink in acceptor) and (b) membrane retentions as a function of phosphatidic acid in 2% DOPC/dodecane lipid membranes at pH 7.4 for a series of weak bases.

7.7.3.4 DOPC with Phosphatidylglycerol under Sink Conditions

The PG models 9.1 and 10.1 show similar trends as indicated by PA, but the effects are somewhat muted. The increase in PG from 0.6% to 1.1% causes the permeabilities of weak bases to decrease and membrane retentions to increase, with many bases showing $R > 60\%$. Many molecules were not detected in the acceptor compartments by UV spectrophotometry after 4 h permeation times (Table 7.7). These properties of the PG system make it less attractive for high-throughput applications than the other two-component models.

7.7.3.5 DOPC with Negative Lipids without Sink

The two-component lipid models were also characterized in the absence of sink conditions (Table 7.8). Comparisons between models 7.0 (Table 7.7) and 1.0 (Table 7.5) suggest that negative charge in the absence of sink causes the permeabilities of many of the bases to decrease. Exceptions are quinine, prazosin, primaquine, ranitidine, and especially metoprolol. The inclusion of 0.6% PA causes P_e of metoprolol to increase nearly 10-fold, to a value twice that of propranolol, a more lipophilic molecule than metoprolol (based on the octanol-water scale). Naproxen and ketoprofen become notably more permeable in the two-component system. Surprisingly, the neutral progesterone becomes significantly less permeable in this system.

With the noted exceptions above, the other negative-lipid combinations (Table 7.8) show consistently lower permeabilities compared to neutral DOPC. Surprisingly, the retentions are not concomitantly higher than in the neutral DOPC lipid.

7.7.4 Five-Component Anionic Lipid Model (Chugai Model)

The interesting five-component BBM model (11.0 in Table 7.3) proposed by Sugano et al. [561,562] was tested by us (Table 7.9). A small modification was made to the original composition: 1,7-octadiene was replaced by dodecane, due to safety concerns over the use of the octadiene in an unprotected laboratory setting [561]. The permeabilities in the dodecane-modified Chugai model were considerably lower than those shown in *p*ION model 1.0 (and those reported by Sugano's group). This may be due to the lessened "fluidity" of the membrane mixture when the octadiene is replaced by dodecane. Retention is quite considerable in the modified Chugai model, with chlorpromazine and progesterone showing R 95% and 87%, respectively. As discussed later, the Sugano model actually has a good GIT absorption prediction property, about as good as that of model 7.1 (which contains only two lipid components).

The Chugai model was unstable in the presence of a sink-forming surfactant in the acceptor wells, and no further efforts were devoted to the untenable model 11.1. The 1% wt/vol cholesterol in dodecane may have interacted with the sink-forming micelles.

TABLE 7.9 Five-Component Anionic Lipid PAMPA Model (without Sink), pH 7.4[a]

Sample	2%DOPC (Model 1.0) P_e(SD)	%R	2%DOPC +0.5% Cho (Model 1A.0) P_e(SD)	%R	Sugano[b] (Model 11.0) P_e(SD)	%R
Chlorpromazine	5.5 (0.4)	85	6.4 (1.0)	93	6.3 (0.2)	95
Phenazopyridine	8.4 (1.1)	70	7.9 (0.1)	71	6.9 (0.1)	79
Verapamil	9.7 (1.0)	39	7.6 (0.1)	31	6.3 (1.2)	46
Promethazine	7.3 (0.7)	70	6.8 (0.1)	70	6.7 (0.3)	76
Quinine	3.1 (0.6)	1	5.1 (0.1)	10	6.4 (0.1)	22
Imipramine	11.1 (0.8)	56	7.4 (0.1)	53	7.4 (0.2)	64
Diltiazem	17.4 (1.8)	21	7.6 (0.3)	17	7.4 (0.6)	31
Prazosin	0.4 (0.1)	15	3.6 (0.1)	9	5.4 (0.4)	33
Propranolol	10.0 (0.5)	18	6.9 (0.1)	18	7.2 (0.1)	34
Desipramine	12.3 (0.4)	40	7.5 (0.1)	39	7.1 (0.8)	55
Primaquine	1.4 (0.1)	70	5.0 (0.2)	18	6.5 (0.1)	28
Alprenolol	11.8 (0.3)	16	—	—	—	—
Metoprolol	0.69 (0.04)	11	2.0 (0.7)	7	3.8 (0.1)	17
Ranitidine	0.009 (0.004)	2	0.04 (0.01)	0	(nd)	2
Amiloride	0.002 (0.005)	0	(nd)	0	(nd)	0
Ibuprofen	2.7 (0.5)	38	4.8 (1.6)	27	9.8 (2.3)	43
Acetaminophen	0.001 (0.005)	1	—	—	—	—
Naproxen	0.33 (0.03)	4	0.7 (0.1)	2	0.85 (0.01)	2
Sulfasalazine	0.007 (0.004)	1	(nd)	3	(nd)	0
Theophylline	0.04 (0.01)	1	0.18 (0.05)	2	0.28 (0.02)	1
Ketoprofen	0.05 (0.01)	4	0.16 (0.04)	3	0.19 (0.02)	1
Hydrochlorothiazide	0.01 (0.01)	1	(nd)	1	(nd)	0
Furosemide	0.02 (0.01)	1	0.002 (0.005)	0	(nd)	1
Salicyclic acid	0.006 (0.004)	1	—	—	—	—
Piroxicam	2.2 (0.1)	3	2.5 (0.1)	2	2.8 (0.1)	3
Sulpiride	0.01 (0.01)	1	0.01 (0.01)	2	(nd)	1
Terbutaline	0.04 (0.01)	6	0.02 (0.03)	2	(nd)	5
Progesterone	6.3 (0.5)	84	6.2 (0.2)	87	5.5 (0.3)	87
Griseofulvin	12.8 (1.2)	18	7.5 (0.1)	20	7.6 (0.1)	16
Carbamazepine	7.1 (0.3)	10	6.4 (0.1)	6	6.8 (0.3)	7
Antipyrine	0.73 (0.05)	13	0.8 (0.1)	5	0.91 (0.08)	7
Caffeine	1.6 (0.1)	2	1.6 (0.1)	2	1.6 (0.1)	3

[a]All P_e and SD(P_e) are in units of 10^{-6} cm/s; (nd) = compound not detected in the acceptor compartment.
[b]Five-lipid formula as reported by Sugano, except 1,7-octadiene was substituted with dodecane.

DOPC (model 1.0) and DOPC + 0.5% cholesterol (model 1A.0) results are listed in Table 7.9 for comparison with the Chugai model. It is quite surprising that the complex mixture of components in the Chugai model is very closely approximated by the cholesterol-DOPC system (model 1A.0), as shown in Fig. 7.29.

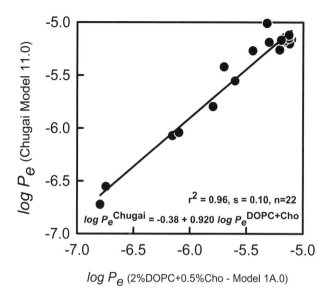

Figure 7.29 Modified Chugai model compared to 2% DOPC + 0.5% cholesterol model.

7.7.5 Lipid Models Based on Lecithin Extracts from Egg and Soy

Hydrogen bonding and electrostatic interactions between the sample molecules and the phospholipid bilayer membranes are thought to play a key role in the transport of such solute molecules. When dilute 2% phospholipid in alkane is used in the artificial membrane [25,556], the effect of hydrogen bonding and electrostatic effects may be underestimated. We thus explored the effects of higher phospholipid content in alkane solutions. Egg and soy lecithins were selected for this purpose, since multicomponent mixtures such as model 11.0 are very costly, even at levels of 2% wt/vol in dodecane. The costs of components in 74% wt/vol (see below) levels would have been prohibitive.

7.7.5.1 Egg Lecithin from Different Sources

Egg lecithins from two sources were considered: Avanti Polar Lipids (Alabaster, AL) and Sigma-Aldrich (St. Louis, MO). The "60% lecithin total extract" grade from Avanti and the "60% lecithin" grade from Sigma-Aldrich were tested. Apparently, different procedures are used to extract the lipids from egg yolk, since the permeability properties of the lecithins from the two sources are significantly different. The Avanti catalog identifies their procedure as a chloroform–methanol extraction. The extract is partitioned against deionized water, and the chloroform phase is concentrated. This extraction procedure is expected to remove proteins and polar (e.g., phenolic) substituents. The Avanti principal lipid components are listed in Table 7.1. The Sigma-Aldrich composition was not available.

TABLE 7.10 Egg Lecithin 10% wt/vol in Dodecane PAMPA Models, pH 7.4[a]

Sample	Avanti (Model 12.0), No Sink P_e(SD)	%R	Avanti (Model 12.1), Sink P_e(SD)	%R	Avanti +0.5%Cho (Model 13.0) No Sink P_e(SD)	%R
Chlorpromazine	—	—	—	—	1.5 (0.1)	83
Phenazopyridine	6.1 (0.4)	91	20.3 (2.9)	44	3.0 (0.5)	84
Verapamil	10.7 (3.0)	73	23.4 (1.1)	20	5.8 (0.5)	58
Promethazine	2.5 (0.5)	85	31.2 (1.2)	17	1.9 (0.8)	82
Quinine	9.2 (2.5)	61	9.9 (4.5)	31	7.3 (0.6)	48
Imipramine	7.0 (1.5)	83	31.8 (4.7)	23	4.0 (0.7)	76
Diltiazem	11.1 (1.5)	50	27.6 (2.5)	12	9.7 (0.8)	46
Prazosin	8.8 (3.2)	28	3.8 (0.5)	20	—	—
Propranolol	5.7 (1.1)	73	16.1 (3.5)	24	5.2 (0.4)	64
Desipramine	5.5 (0.8)	89	21.8 (2.1)	30	7.8 (0.9)	61
Primaquine	—	—	—	—	5.7 (1.8)	62
Alprenolol	12.5 (6.3)	65	23.1 (3.7)	27	—	—
Metoprolol	17.8 (9.7)	71	23.4 (4.9)	19	6.6 (0.5)	18
Ranitidine	0.2 (0.1)	8	0.2 (0.2)	7	0.3 (0.1)	8
Amiloride	0.006 (0.005)	15	(nd)	7	0.03 (0.03)	11
Ibuprofen	7.8	58	10.3 (2.4)	16	4.9 (0.2)	14
Acetaminophen	0.9 (0.3)	0	(nd)	3	—	—
Naproxen	1.4 (0.1)	12	0.9 (0.1)	4	1.6 (0.1)	2
Sulfasalazine	0.002 (0.003)	2	0.01 (0.02)	3	0.003 (0.005)	4
Theophylline	0.3 (0.1)	6	0.4 (0.1)	5	—	—
Ketoprofen	0.5 (0.1)	12	0.6 (0.1)	1	0.5 (0.1)	5
Hydrochlorothiazide	0.01 (0.01)	24	0.1 (0.1)	4	0.005 (0.005)	4
Furosemide	0.01 (0.01)	19	0.06 (0.05)	4	0.03 (0.01)	8
Salicyclic acid	0.04 (0.03)	15	0.9 (0.8)	3	—	—
Piroxicam	2.6 (0.2)	15	2.7 (0.2)	6	2.5 (0.1)	7
Sulpiride	0.2 (0.1)	5	0.04 (0.07)	5	0.17 (0.03)	2
Terbutaline	0.2 (0.1)	11	0.1 (0.2)	6	(nd)	3
Progesterone	2.8 (0.8)	93	29.8 (2.8)	22	2.6 (0.5)	82
Griseofulvin	10.5 (0.5)	42	19.0 (0.5)	15	11.4 (0.6)	18
Carbamazepine	8.6 (0.2)	19	10.7 (0.5)	19	10.8 (1.6)	16
Antipyrine	1.3 (0.1)	27	1.9 (0.5)	3	1.4 (0.1)	5
Caffeine	1.9 (0.1)	6	2.2 (0.3)	5	2.1 (0.4)	8

[a] All P_e and SD(P_e) are in units of 10^{-6} cm/s; (nd) = compound not detected in the acceptor compartment.

Kansy et al. [547,550] used 10% wt/vol egg lecithin in dodecane. Cholesterol was added as well. We also chose to use 10% egg lecithin ("60% grade") in our laboratory. Tables 7.10 and 7.11 list the results of the various 10% egg lecithin models tested at *p*ION. Some of the models were used in conjunction with a sink

TABLE 7.11   Egg Lecithin 10% wt/vol in Dodecane PAMPA Models, pH 7.4[a]

Sample	Sigma (Model 14.0), No Sink P_e(SD)	%R	Sigma (Model 14.1), Sink P_e(SD)	%R	Sigma +0.5%Cho (Model 15.0), No Sink P_e(SD)	%R	Sigma +0.5%Cho (Model 15.1), Sink P_e(SD)	%R
Chlorpromazine	1.2	84	31.9 (6.1)	41	0.7 (0.3)	86	32.1 (8.6)	31
Phenazopyridine	2.7 (0.1)	84	17.4 (1.5)	55	3.1 (0.3)	86	18.8 (1.7)	50
Verapamil	3.1 (0.5)	69	25.4 (5.8)	33	1.8 (1.2)	83	28.4 (3.1)	23
Promethazine	2.2 (0.3)	84	35.3 (0.5)	35	1.3 (0.4)	89	36.4 (3.5)	22
Quinine	4.6 (0.7)	52	9.6 (0.5)	48	4.0 (0.7)	59	12.8 (0.6)	39
Imipramine	2.5	74	34.3 (1.0)	40	3.8 (0.3)	75	35.3 (6.3)	34
Diltiazem	7.1 (3.0)	50	31.3 (4.9)	18	3.8 (0.2)	64	33.2 (3.8)	8
Prazosin	5.3 (0.5)	34	11.8 (0.3)	21	4.4 (0.4)	38	16.9 (1.1)	16
Propranolol	4.1 (1.6)	65	21.2 (0.8)	43	3.5 (0.3)	70	22.3 (0.5)	34
Desipramine	3.9 (0.7)	78	24.3 (7.5)	49	2.7 (0.3)	80	29.2 (8.6)	30
Primaquine	4.4 (0.7)	65	22.8 (1.2)	36	4.4 (0.8)	81	30.0 (0.9)	26
Alprenolol	—	—	—	—	5.5 (0.2)	65	—	—
Metoprolol	4.0	26	4.3 (0.4)	22	3.7 (0.1)	26	8.0 (0.9)	12
Ranitidine	0.3	2	(nd)	9	0.1 (0.1)	7	(nd)	11
Amiloride	(nd)	5	(nd)	4	0.02 (0.03)	3	(nd)	3
Ibuprofen	(nd)	47	(nd)	—	6.9 (3.9)	31	(nd)	—
Acetaminophen	—	—	—	—	—	—	—	—
Naproxen	1.3	6	(nd)	6	1.0 (0.1)	6	1.3 (0.6)	3
Sulfasalazine	0.05	4	(nd)	4	—	—	0.04 (0.06)	2
Theophylline	0.2	11	(nd)	6	0.3 (0.1)	4	0.2 (0.2)	7
Ketoprofen	0.3 (0.1)	8	0.1 (0.1)	—	0.3 (0.1)	5	0.4 (0.1)	2
Hydrochlorothiazide	(nd)	5	(nd)	1	0.006 (0.005)	4	(nd)	3
Furosemide	(nd)	5	(nd)	4	0.01 (0.01)	4	0.09 (0.04)	2
Salicyclic acid	—	—	—	—	—	—	—	—
Piroxicam	2.1 (0.1)	8	2.2 (0.1)	6	2.0 (0.1)	6	2.2 (0.1)	4
Sulpiride	(nd)	9	(nd)	3	0.1 (0.1)	5	(nd)	3
Terbutaline	(nd)	5	(nd)	3	0.06 (0.01)	0	(nd)	2
Progesterone	5.2 (0.6)	80	42.3 (2.7)	31	4.0 (0.7)	88	37.9 (3.2)	33
Griseofulvin	9.7 (2.1)	46	21.4 (1.4)	25	5.1 (0.6)	41	21.7 (0.3)	21
Carbamazepine	9.1 (1.4)	20	13.8 (12.1)	20	5.1 (0.2)	23	15.7 (2.2)	19
Antipyrine	1.4 (0.1)	7	0.9 (0.2)	5	1.1 (0.2)	4	1.4 (0.3)	3
Caffeine	2.3 (0.4)	9	2.0 (0.1)	7	2.3 (0.1)	7	2.0 (0.2)	4

[a]All P_e and SD(P_e) are in units of 10^{-6} cm/s; (nd) = compound not detected in the acceptor compartment.

condition in the acceptor wells. Figure 7.30 shows permeability and membrane retention results for weak-base probes, using the Sigma-Aldrich source of lecithin, with and without sink and 0.5% wt/vol cholesterol. The presence of a sink dramatically increases permeabilities, as indicated in Figure 7.30a. In some cases, further significant increases in permeability were realized by the use of cholesterol, even though its amount was only 0.5%. Only in the diffusion-limited cases (right side of Fig. 7.30a) was there only minimal enhancement due to cholesterol.

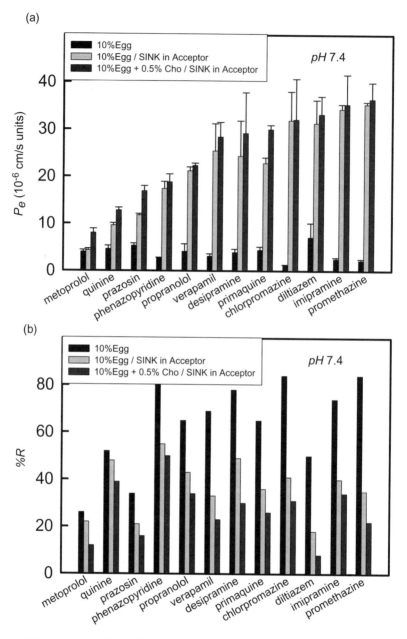

Figure 7.30   (a) Permeabilities [for egg lecithin (Sigma) in dodecane] and (b) membrane retentions for a series of weak bases in various egg lecithin PAMPA models.

Without an artificial sink, the membrane retentions are very high, with many basic probe molecules showing $R > 80\%$. With the imposed sink, many of the retentions dropped by as much as 50%. Furthermore, just 0.5% wt/vol cholesterol in dodecane (in addition to the sink) caused increased retention to drop by at least a further 10–30%. It was not possible to form stable cholesterol-containing lipid models under sink conditions with Avanti's egg lecithin; acceptor buffer solutions turned significantly turbid in the untenable model 13.1.

The peculiar depression of metoprolol and quinine permeabilities in 2% DOPC (model 1.0) was not seen in the egg lecithin models. Metoprolol and quinine were significantly more permeable in the lecithins, in line with expectations based on relative octanol–water lipophilicities and relative in vivo absorptions of β-blockers [593].

7.7.5.2 Soy Lecithin and the Effects of Phospholipid Concentrations

We explored the use of Avanti's "20% lecithin soy lipid extract," dissolved at various concentrations in dodecane. This is not a highly purified grade, and contains 37% unspecified neutral lipids, most likely asymmetric triglycerides. We chose this grade because it contained negatively charged phospholipids, having a charged : zwitterionic lipid ratio about half way between that of BBM and BBB compositions (Table 7.1). Soy-based PAMPA lipid models have been prepared with ("20% extract" grade) soy lecithin, 10–74% wt/vol in dodecane. These newly formulated lipids have net negative charge at pH 7.4, which further increases above pH 8, as the phosphatidic groups ionize (cf. ionization constants in Fig. 7.4). The inositol (predominant negatively charged lipid) content is 4 times higher in soy than in egg lecithin. However, when ≤74% phospholipid fractions are used, severe experimental problems arise. With lipophilic sample molecules, the use of concentrated phospholipid artificial membranes leads to two unwanted effects: (1) nearly complete membrane retention (90–100%) and (2) highly diminished permeability (indeterminate in some cases), both effects presumably due to excessive drug-membrane binding. These adverse effects are nearly eliminated by using an ionic surfactant to create a very strong sink condition in the acceptor compartment of the permeation cell. The negative charge on the micelles formed from the surfactant added to the acceptor compartment appears to play a stabilizing role.

Tables 7.12–7.14 list the pH 7.4 permeability and retention values of the probe series of drug substances, grouped as bases, acids, and neutral molecules. Figures 7.31a–c are graphs of the effective permeabilities with and without sink as a function of increasing soy content, beginning with 2% DOPC for a benchmark. Figures 7.32a–c are plots of the corresponding membrane retentions.

Most of the permeabilities of the bases decrease steadily as the phospholipid fraction increases. There are some significant exceptions. Metoprolol, which is only moderately permeable in the DOPC lipid, becomes appreciably permeable in 10% soy lecithin. But at the 68% soy level, this molecule also shows reduced transport.

The permeabilities of the acid examples rise with increasing phospholipid content, up to 20% lipid, with rank ordering preserved. Naproxen and ketoprofen

TABLE 7.12 Soy Lecithin in Dodecane PAMPA Models (No Sink), pH 7.4[a]

Sample	10% Soy (Model 16.0) P_e(SD)	%R	20% Soy (Model 17.0) P_e(SD)	%R	20% Soy +0.5% Cho (Model 18.0) P_e(SD)	%R	35% Soy (Model 19.0) P_e(SD)	%R	68% Soy (Model 21.0) P_e(SD)	%R
Phenazopyridine	5.8 0.4	95	1.0 (0.3)	94	7.9 (2.4)	96	4.1 (1.2)	98	5.3 (0.5)	99
Verapamil	1.4 (1.3)	94	1.1 (0.1)	94	0.1 (0.2)	96	1.6 (1.5)	95	0.2 (0.3)	94
Promethazine	0.9 (0.8)	94	0.9 (1.0)	97	0.8 (0.7)	93	(nd)	96	0.1 (0.1)	97
Quinine	4.0 (0.1)	94	3.7 (2.6)	95	4.5 (1.4)	96	0.8 (1.4)	98	(nd)	98
Imipramine	0.001 (0.005)	98	0.2 (0.3)	93	1.9 (1.5)	91	3.2 (1.2)	95	1.8 (1.7)	96
Diltiazem	4.6 (1.2)	87	7.3 (0.8)	92	3.2 (2.5)	95	6.9 (2.0)	97	2.0 (1.7)	90
Prazosin	6.7 (0.1)	57	3.5 (0.1)	63	9.7 (0.8)	72	5.5 (0.6)	79	2.2 (0.1)	83
Propranolol	2.4 (1.3)	93	1.8 (0.5)	95	1.3 (0.9)	92	2.8 (1.4)	96	2.5 (0.8)	95
Desipramine	1.2 (1.2)	97	0.6 (0.7)	93	1.1 (1.6)	95	2.7 (4.6)	96	3.2 (3.3)	91
Alprenolol	2.5 (0.9)	92	0.01 (0.03)	90	4.8 (1.0)	93	5.6 (3.8)	95	4.8 (3.1)	95
Metoprolol	6.0 (0.7)	44	8.2 (1.2)	42	8.8 (2.1)	62	7.1 (1.7)	70	3.2 (0.1)	73
Ranitidine	0.41 (0.03)	8	0.36 (0.01)	13	—	—	0.43 (0.02)	22	0.13 (0.05)	16
Amiloride	0.004 (0.005)	14	0.003 (0.006)	13	(nd)	18	0.003 (0.005)	19	(nd)	0

	P_e (SD)	%	P_e (SD)	%	P_e (SD)	%	P_e (SD)	%	P_e (SD)	%
Ibuprofen	4.0 (1.4)	63	5.0 (0.7)	21	5.7 (1.1)	18	6.3 (1.8)	18	1.6 (1.2)	30
Acetaminophen	0.7 (0.1)	4	1.1 (0.2)	14	—	—	0.8 (0.1)	15	(nd)	0
Naproxen	2.0 (0.2)	6	3.4 (0.1)	9	2.2 (0.1)	17	2.3 (0.1)	13	0.5 (0.1)	2
Sulfasalazine	0.001 (0.005)	1	0.01 (0.02)	8	—	—	0.002 (0.005)	6	0.002 (0.005)	3
Theophylline	0.65 (0.04)	6	0.79 (0.08)	8	0.85 (0.09)	12	0.60 (0.04)	3	0.10 (0.02)	7
Ketoprofen	1.0 (0.1)	4	1.5 (0.1)	9	1.1 (0.1)	16	1.0 (0.1)	16	0.2 (0.1)	10
Hydrochlorothiazide	0.02 (0.01)	6	0.01 (0.01)	9	(nd)	10	0.03 (0.01)	20	0.02 (0.04)	3
Furosemide	0.02 (0.01)	4	0.04 (0.02)	11	0.03 (0.02)	13	0.05 (0.03)	18	0.01 (0.01)	17
Salicyclic acid	0.13 (0.01)	2	0.24 (0.03)	10	—	—	0.26 (0.08)	14	0.03 (0.03)	9
Piroxicam	2.6 (0.1)	8	3.6 (0.1)	2	2.8 (0.1)	13	2.7 (0.2)	17	1.6 (0.1)	8
Sulpiride	0.19 (0.03)	12	0.25 (0.03)	14	0.20 (0.04)	14	0.18 (0.02)	21	0.10 (0.06)	16
Terbutaline	0.05 (0.09)	10	0.20 (0.14)	2	—	—	0.01 (0.01)	22	(nd)	12
Progesterone	5.8 (0.4)	91	4.6 (0.9)	90	—	—	1.6 (0.3)	92	2.6 (0.8)	93
Griseofulvin	7.2 (0.5)	44	6.6 (0.2)	54	13.8 (0.9)	56	6.4 (0.5)	62	5.4 (0.2)	73
Carbamazepine	6.1 (0.5)	29	10.8 (0.3)	37	12.7 (3.1)	39	6.4 (0.1)	44	5.3 (0.4)	55
Antipyrine	1.2 (0.1)	7	1.5 (0.1)	9	1.6 (0.2)	15	1.3 (0.5)	25	0.5 (0.1)	8
Caffeine	1.8 (0.1)	6	2.1 (0.1)	10	2.3 (0.1)	18	2.2 (0.1)	17	1.6 (0.1)	13

[a] All P_e and SD(P_e) are in units of 10^{-6} cm/s; (nd) = compound not detected in the acceptor compartment.

TABLE 7.13   Soy Lecithin in Dodecane PAMPA Models (with Sink), pH 7.4[a]

Sample	10% Soy (Model 16.1) P_e(SD)	%R	20% Soy (Model 17.1) P_e(SD)	%R	20% Soy +0.5% Cho (Model 18.1) P_e(SD)	%R
Chlorpromazine	—	—	—	—	30.9 (5.2)	40
Phenazopyridine	15.8 (1.4)	47	20.7 (2.0)	60	18.3 (1.7)	63
Verapamil	25.6 (1.5)	31	31.6 (2.8)	31	32.4 (1.4)	31
Promethazine	26.7 (3.2)	25	27.6 (0.9)	32	37.0 (0.9)	34
Quinine	24.6 (4.1)	44	17.6 (1.0)	49	20.5 (1.3)	46
Imipramine	30.1 (0.9)	38	22.9 (0.6)	40	28.5 (3.4)	37
Diltiazem	35.8 (1.3)	22	35.1 (1.9)	17	37.4 (3.8)	20
Prazosin	28.6 (1.3)	16	19.2 (0.3)	22	36.4 (3.7)	15
Propranolol	27.1 (3.4)	39	25.1 (1.7)	36	26.5 (2.0)	40
Desipramine	33.2 (2.8)	33	29.8 (0.2)	39	28.5 (3.2)	50
Primaquine	—	—	—	—	36.9 (2.6)	34
Alprenolol	30.6 (3.8)	30	26.3 (3.5)	40	—	—
Metoprolol	26.4 ((0.1)	27	26.5 (1.1)	23	29.0 (1.6)	29
Ranitidine	0.34 (0.01)	8	0.31 (0.03)	14	0.51 (0.13)	15
Amiloride	0.01 (0.02)	9	0.007 (0.005)	15	0.1 (0.1)	15
Ibuprofen	3.6 (1.4)	32	7.4 (1.1)	34	16.3 (2.3)	39
Acetaminophen	1.2 (0.2)	8	0.4 (0.1)	7	—	—
Naproxen	1.8 (0.1)	10	2.9 (0.1)	13	3.9 (0.5)	13
Sulphasalazine	0.001 (0.005)	2	0.002 (0.005)	10	0.7 (0.4)	11
Theophylline	0.5 (0.1)	7	0.8 (0.1)	8	1.2 (0.2)	16
Ketoprofen	0.8 (0.1)	9	1.2 (0.1)	12	1.5 (0.2)	19
Hydrochlorothiazide	0.004 (0.010)	11	0.004 (0.004)	12	(nd)	17
Furosemide	0.04 (0.02)	14	0.02 (0.01)	11	0.09 (0.08)	17
Salicyclic acid	0.2 (0.2)	13	0.1 (0.1)	7	—	—
Piroxicam	2.3 (0.1)	6	3.2 (0.2)	17	3.6 (0.1)	14
Sulpiride	0.2 (0.1)	6	0.1 (0.1)	14	(nd)	17
Terbutaline	0.2 (0.3)	14	0.1 (0.2)	13	(nd)	20
Progesterone	37.6 (1.3)	40	27.6 (1.1)	37	33.2 (3.2)	34
Griseofulvin	31.8 (1.5)	25	24.4 (1.2)	23	27.0 (3.3)	25
Carbamazepine	16.5 (1.7)	23	15.2 (0.7)	26	21.2 (0.8)	30
Antipyrine	1.6 (0.1)	6	1.6 (0.1)	13	2.5 (0.3)	19
Caffeine	1.5 (0.1)	8	2.0 (0.1)	14	3.0 (0.1)	19

[a]All P_e and SD(P_e) are in units of 10^{-6} cm/s; (nd) = compound not detected in the acceptor compartment.

TABLE 7.14 Soy Lecithin in Dodecane PAMPA Models (with Sink), pH 7.4[a]

Sample	35% Soy (Model 19.1) P_e(SD)	%R	50% Soy (Model 20.1) P_e(SD)	%R	74% Soy (Model 22.1) P_e(SD)	%R
Phenazopyridine	21.1 (3.5)	66	2.4 (0.2)	58	3.9 (1.1)	75
Verapamil	42.9 (4.0)	43	17.5 (0.1)	34	1.8 (2.2)	71
Promethazine	31.3 (3.0)	36	25.7 (2.9)	45	3.7 (0.3)	61
Quinine	27.6 (1.4)	55	9.6 (2.7)	54	2.6 (1.0)	67
Imipramine	42.9 (6.1)	46	5.2 (4.8)	63	5.0 (1.8)	63
Diltiazem	40.4 (7.1)	32	20.8 (1.0)	36	3.5 (5.3)	61
Prazosin	30.9 (2.4)	25	12.7 (0.7)	38	0.4 (0.4)	49
Propranolol	27.6 (4.0)	54	15.9 (5.0)	47	(nd)	62
Desipramine	37.1 (9.4)	48	18.4 (3.0)	39	1.7 (0.7)	59
Alprenolol	42.3 (5.2)	51	7.8 (2.2)	52	2.6 (2.6)	71
Metoprolol	31.4 (0.8)	42	11.6 (0.9)	43	4.0 (6.9)	52
Ranitidine	0.2 (0.1)	13	0.3 (0.4)	8	(nd)	3
Amiloride	0.02 (0.05)	11	0.05 (0.07)	0	(nd)	5
Ibuprofen	8.1 (4.2)	22	16.5 (3.6)	13	2.0 (3.4)	33
Acetaminophen	1.3 (1.1)	15	0.4 (0.5)	16	(nd)	0
Naproxen	2.5 (0.5)	9	1.4 (0.3)	11	0.2 (0.3)	1
Sulfasalazine	0.04 (0.02)	7	(nd)	—	(nd)	2
Theophylline	0.7 (0.1)	8	0.4 (0.3)	0	0.02 (0.03)	6
Ketoprofen	1.3 (0.6)	33	1.6 (1.4)	30	(nd)	4
Hydrochlorothiazide	0.03 (0.04)	10	0.09 (0.11)	1	0.01 (0.01)	5
Furosemide	0.01 (0.02)	16	(nd)	1	0.001 (0.005)	10
Salicyclic acid	1.1 (0.5)	11	0.3 (0.5)	5	0.2 (0.3)	0
Piroxicam	2.9 (0.2)	18	1.6 (0.1)	13	1.0 (0.2)	6
Sulpiride	0.5 (0.2)	17	0.3 (0.5)	2	0.1 (0.2)	3
Terbutaline	0.1 (0.1)	20	1.3 (1.8)	22	(nd)	1
Progesterone	36.2 (0.5)	36	23.2 (0.5)	65	31.8 (7.2)	39
Griseofulvin	22.1 (2.9)	27	14.6 (1.0)	37	13.4 (4.5)	44
Carbamazepine	15.3 (2.0)	27	9.9 (0.4)	36	2.1 (0.4)	38
Antipyrine	1.8 (1.0)	18	2.5 (1.4)	14	1.0 (0.3)	1
Caffeine	2.0 (0.1)	18	(nd)	9	1.2 (0.3)	8

[a]All P_e and SD(P_e) are in units of 10^{-6} cm/s; (nd) = compound not detected in the acceptor compartment.

show the most dramatic increases in going from 2% DOPC to 10% soy lipid membranes, somewhat higher in soy than in egg. Piroxicam shows less sensitivity to lipid changes. For higher phospholipid concentrations, all the acid permeabilities decrease.

The nonionizable molecules respond to the changes in the phospholipid content. Griseofulvin has the highest permeability in the lowest phospholipid-containing membranes. The most remarkable change of properties in going from 2% to 10%

phospholipid occurs with the membrane retention of the bases. Most of the bases are retained above 90% in all of the soy lecithin cases ($\leq$68% in dodecane). This is thought to be largely due to the added electrostatic attractions between positively charged sample molecules and the negatively-charged membrane constituents.

Acids show small, steady increases in membrane retention with increasing phospholipid content. Even though the acids are negatively charged at pH 7.4, as are a portion of the membrane constituents, the increasing phospholipid content draws the sample molecules in, due to increased hydrogen-bonding and any other lipophilic forces arising from the phospholipids (increased membrane-water partition coefficient). Decreased surface pH due to the membrane negative surface charge [457] may also play a role in increasing permeability of weak acids.

Neutral molecules show a range of retention properties between those of acids and bases. Progesterone membrane retention is very high in all cases. Griseofulvin and carbamazepine retention steeply increase with phospholipid content. The patterns of retention follow the lipophilicity properties of the molecules, as indicated by octanol–water apparent partition coefficients (Table 7.4).

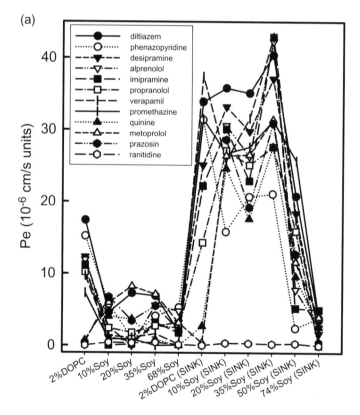

Figure 7.31 Soy lecithin permeabilities at various concentrations in dodecane, with and without sink: (a) bases; (b) acids; (c) neutrals.

(b)

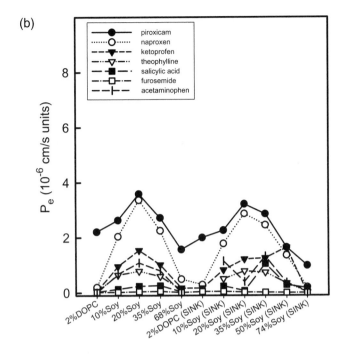

(c)

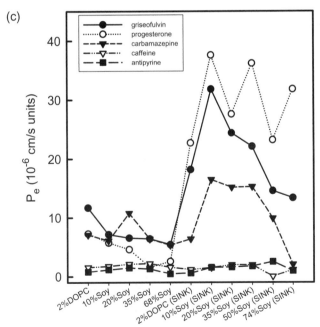

Figure 7.31 *(Continued)*

7.7.5.3 Lipophilicity and Decrease in Permeability with Increased Phospholipid Content in Dodecane

Figures 7.31a–c clearly show that after some critical soy content in dodecane, P_e values decrease with increasing soy, for both sink and sinkless conditions. [This is not due to a neglect of membrane retention, as partly may be the case in Fig. 7.23; permeabilities here have been calculated with Eq. (7.21).] Section 7.6 discusses the Kubinyi bilinear model (Fig. 7.19d) in terms of a three-compartment system: water, oil of moderate lipophilicity, and oil of high lipophilicity. Since liposome(phospholipid)–water partition coefficients (Chapter 5) are generally higher than alkane–water partition coefficients (Chapter 4) for drug-like molecules, soy lecithin may be assumed to be more lipophilic than dodecane. It appears that the increase in soy concentration in dodecane can be treated by the Kubinyi analysis. In the original analysis [23], two different lipid phases are selected at a fixed ratio (e.g., Fig. 7.20), and different molecules are picked over a range of lipophilicities.

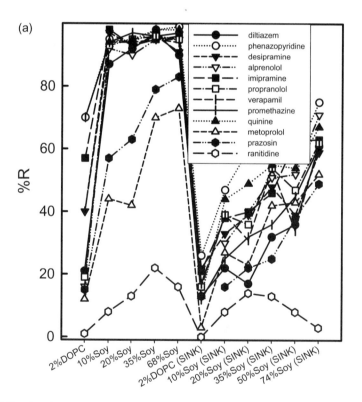

Figure 7.32 Soy lecithin membrane retentions at various concentrations in dodecane, with and without sink: (a) bases; (b) acids; (c) neutrals.

(b)

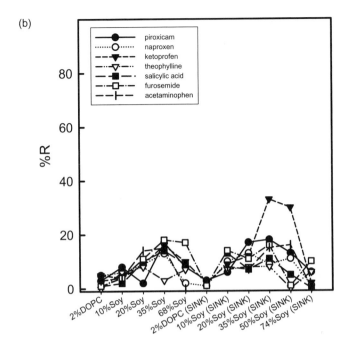

(c)

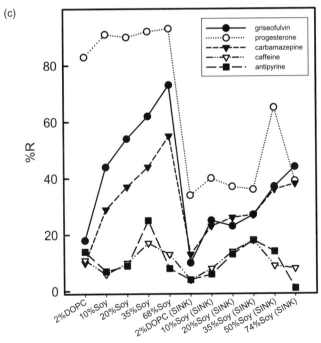

Figure 7.32   (*Continued*)

The more lipophilic molecules preferentially concentrate in the more lipophilic phase, leading to decreased permeabilities, according to the effect of the negative term in Eq. (7.44), as the concentration of solute in the lower-lipophilicity phase decreases. In the soy lecithin models, the lipid phases are systematically varied, with reference to a molecule of a particular lipophilicity. The plots in Figs. 7.31a–c are *orthogonally* equivalent to the Kubinyi model type plots (Fig. 7.19d), with each curve representing a particular molecule and the horizontal axis corresponding to varied lipid ratios. Eq. 7.44 applies and Figs. 7.31a–c may be interpreted as bilinear curves, for both sink and sinkless domains. For example, the maximum permeability for most molecules occurs at about 20% wt/vol lecithin in dodecane. For higher lecithin content, the negative term in Eq. (7.44) dominates, causing the P_e values to decrease.

7.7.5.4 Sink Condition to Offset the Attenuation of Permeability

The preceding section treats the decrease in permeabilities with increasing lecithin content in dodecane in terms of shifting concentration distributions between a weak lipophilic domain (dodecane) and a stronger lipophilic domain (lecithin). Another view of this may be that at the molecular level, as the amount of phospholipid increases, the effects of electrostatic and H-bonding play a more prominent role in the transport process. Generally, %R of the lipophilic molecules increases with increasing lecithin content, most dramatically in the case of lipophilic bases. Such losses of compound to the membrane pose a challenge to the analysis of concentrations, which can be significantly diminished (to undetectable levels at times) in the aqueous compartments. At the same time, the permeability drops to near vanishing values in 68% soy lecithin–dodecane membranes. Under these conditions, the permeabilities of the lipophilic bases and acids converge to similar low values, significantly departing from the expected values based on the octanol–water lipophilicity scale (Table 7.4) and the pH partition hypothesis. This excessive drug–membrane binding would not be expected under in vivo conditions in the small intestine, due to the naturally occurring sink state. There would be competing lipid environments in the receiving compartment (serum proteins, other membrane barriers, etc.), and the solute-binding membrane would release a portion of the retained lipophilic molecules, resulting in a concomitant higher effective permeability.

The transport properties of the molecules in concentrated soy lecithin, Tables 7.12–7.14, do not adequately model the in vivo permeabilities reported by Winiwarter et al. [56] (Table 7.4). The strategy to overcome this shortcoming of the model involves creating a model sink condition. However, the use of BSA or other serum proteins, although easily effected, is not practical in high-throughput screening, since the UV absorption due to the proteins would render determination of the compound concentrations in the acceptor compartments by direct UV spectrophotometry nearly impossible in most cases. Without knowledge of the concentration of sample in the acceptor compartment, the determination of %R would not be practical. Some PAMPA practitioners, using BSA to create sink conditions, make the simplifying assumption that membrane retention is zero. It is neither reasonable nor warranted to expect that membrane retention is eliminated in the presence of serum proteins or other practical substitutes in the acceptor compartment. Figures 7.32a–c clearly show that retention under sink can be substantial.

Since lipophilic molecules have affinity for *both* the membrane lipid and the serum proteins, membrane retention is expected to decrease, by the extent of the relative lipophilicities of the drug molecules in membrane lipid versus serum proteins, and by the relative amounts of the two competitive-binding phases [see Eqs. (7.41)–(7.43)]. Generally, the serum proteins cannot extract all of the sample molecules from the phospholipid membrane phase at equilibrium. Thus, to measure permeability under sink conditions, it is still necessary to characterize the extent of membrane retention. Generally, this has been sidestepped in the reported literature.

We found that the negatively charged surfactant, sodium laurel sulfate, can be successfully substituted for the serum proteins used previously. In low ionic strength solutions, the cmc of the surfactant is 8.1 mM [577]. We explored the use of both sub-CMC (data not shown) and micelle-level concentrations. Saturated micelle solutions are most often used at *p*ION.

The addition of surfactant to the acceptor solution allows for the re-distribution of lipophilic permeants between the PAMPA membrane phase and the surfactant phase in the acceptor compartment, in the manner of Kubini's [23] analysis (Sec. 7.6), according to the relative lipophilicities of the two oil phases. This redistribution can be approximated. Garrone et al. [600] derived a Collander relationship for a series of substituted benzoic acids, relating their lipophilicities in 30–100 mM sodium laurel sulfate to the octanol-water system. The Collander equation comparing the drug partitioning in liposome–water to octanol–water systems (Fig. 5.6) can be combined with that of the above micellar relationship to get the approximate equation: $\log K_{p,\mathrm{mic}} = 1.4 \log K_{p,\mathrm{liposome}} - 1.6$. If it is assumed that the PAMPA membrane lipophilicity can be approximated by that of liposomes, then the strength of the surfactant-created acceptor sink can be compared to that of the PAMPA membrane, according to the latter expression. The most lipophilic molecules will favor the micellar phase when their liposome partition coefficients, $\log K_{p,\mathrm{liposome}}$, are greater than 4. (The micellar and PAMPA lipid volumes are nearly the same.) Positively charged drug molecules will favor additional binding to the negatively charged micelles, unless the PAMPA membrane lipid composition also has negative charge.

The effect of the surfactant is most dramatic for the bases and neutral molecules studied, as shown in Tables 7.13 and 7.14. Permeabilities increased by up to fourfold for the lipophilic bases and neutral molecules, and membrane retentions were decreased by 50% in most cases of bases and neutral compounds (Figs. 7.31 and 7.32).

The transport properties of the acids did not respond significantly to the presence of the sink. This may be because at pH 7.4 the acids are negatively charged, as are the phospholipid membranes and also the surfactant micelles; electrostatic repulsions balanced out the attractive forces due to increased membrane lipophilicity. Lowered surface pH may also play a balancing role [457].

7.7.5.5 *Comparing Egg and Soy Lecithin Models*

The negative-charge lipid content in the egg lecithins is not as high as that found in BBM and especially BBB lipids (Table 7.1). Furthermore, the negative-charge content in the egg lecithin is about one-fourth that in the soy lecithin. This is clearly evident in the membrane retention parameters for the bases at the 10% lecithin levels (models 12.0 or 14.0 in Table 7.8 vs. model 16.0 in Table 7.12), as they are ~20–30% lower for the lipophilic bases in egg, compared to soy.

For acids, the membrane retention actually *increases* in the case of egg lecithin, compared to soy lecithin. This may be due to decreased repulsions between the negatively charged sample and negatively charged phospholipid, allowing H-bonding and hydrophobic forces to more fully realize in the less negatively charged egg lecithin membranes. The neutral molecules display about the same transport properties in soy and egg lecithin, in line with the absence of direct electrostatic effects. These differences between egg and soy lecithins make soy lecithin the preferred basis for further model development.

7.7.5.6 *Titrating a Suspension of Soy Lecithin*

Since soy lecithin ("20% extract" from Avanti) was selected as a basis for absorption modeling, and since 37% of its content is unspecified, it is important to at least establish that there are no titratable substituents near physiological pH. Asymmetric triglycerides, the suspected unspecified components, are not expected to ionize. Suspensions of multilamellar vesicles of soy lecithin were prepared and titrated across the physiological pH range, in both directions. The versatile Bjerrum plots (Chapter 3) were used to display the titration data in Fig. 7.33. (Please note the extremely expanded scale for $\bar{n}_H$.) It is clear that there are no ionizable groups

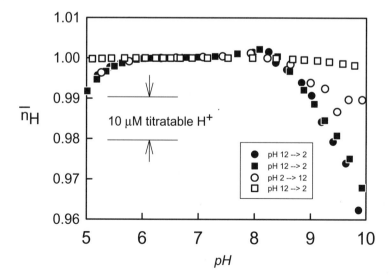

Figure 7.33 Bjerrum plot for titration of a suspension of 1 mM soy lecithin.

between pH 5.5 and 8.0 with concentrations in excess of 1 μM. For pH > 8, phosphatidic and possibly free fatty acids start to ionize, approximately to the extent of 1% of the total soy content by pH 9.

7.7.6 Intrinsic Permeability, Permeability–pH Profiles, Unstirred Water Layers (UWL), and the pH Partition Hypothesis

Up to now we have focused on measurement of permeability and membrane retention at pH 7.4. Since the GIT covers a range of pH values, with pH 5–8 characterizing the small intestine, it is necessary to address the pH dependence of the transport of drug molecules. Even nonionizable molecules may be affected by pH dependence, since several biological membrane components themselves are ionizable (pK_a values listed in Fig. 7.4). For example, with PS, PA, and DA (free fatty acid) undergoing changes in charge state in the pH 5–8 interval. In this section, we examine the consequences of pH dependence.

7.7.6.1 Unstirred Water Layer Effect (Transport across Barriers in Series and in Parallel)

Passive transport across a membrane barrier is a combination of diffusion through the membrane and also diffusion through the stagnant water layers at the two sides of the membrane. Stirring the bulk aqueous solution does not appreciably disturb the water layers in contact with the membrane. When the solute is introduced into the bulk aqueous phase, convective mixing resulting from applied stirring, quickly positions the drug molecule next to the so-called unstirred water layer (UWL). At that point, the passage through the UWL is governed by the laws of diffusion, and is independent of stirring. In simple hydrodynamic models [534–538] the UWL is postulated to have a distinct boundary with the rest of the bulk water. The UWL can be made thinner with more vigorous stirring, but it cannot be made to vanish. Extensions of the simple UWL models have been discussed in the literature [539,540], but such models are not often used in practice.

The actual thickness of the unstirred water layer depends somewhat on the transport model system. The in vivo UWL is significantly different from the in vitro assay measuring cell UWL. Because of the efficient mixing near the surface of the epithelium, the in vivo UWL is estimated to be 30–100 μm thick. The UWL in the endothelial microcapillaries of the brain is nil, given that the diameter of the capillaries is ~7 μm and the efficiency of the mixing due to the passage of erythrocytes [612]. However, in unstirred in vitro permeation cells, the UWL values can be 1500–2500 μm thick, depending on cell geometry and dimensions.

It may be assumed that the total resistance to passive transport across the trilamellar (UWL–membrane–UWL) barrier is the sum of the resistances of the membrane and the UWL on each side of it. Resistance is the inverse of permeability. So

$$\frac{1}{P_e} = \frac{1}{P_m} + \frac{1}{P_u} \tag{7.46}$$

where P_e refers to the measured *effective* permeability, P_u refers to the total UWL permeability, P_m is the permeability of the *membrane* (which would be measured if the UWL were made vanishingly thin). If it is possible to separate the donor and acceptor contributions to the UWL, then the total P_u can be allocated between its parts according to $1/P_u = 1/P_u^{(D)} + 1/P_u^{(A)}$. In Caco-2 literature, equations like Eq. (7.46) often have a fitter, P_f, component, to account for resistance of the water-filled pores of the fitter. In PAMPA, all pores are filled with lipid, and no consideration of filter contributions are needed.

The UWL permeability is nearly the same for drugs of comparable size, and is characterized by the water diffusivity (D_{aq}) of the drug divided by twice the thickness of the layer (h_{aq}), $P_u = D_{aq} / (2\,h_{aq})$, in a symmetric permeation cell [40]. The unstirred water layer permeability can be determined experimentally in a number of ways: based on pH dependency of effective permeability [25,509,535–538], stirring rate dependence [511–514,552,578], and transport across lipid-free microfilters [25,546].

7.7.6.2 Determination of UWL Permeability using pH Dependence (pK_a^{flux}) Method

The membrane permeabilities P_m may be converted to *intrinsic* permeabilities P_0, when the pK_a is taken into consideration. An ionizable molecule exhibits its intrinsic permeability when it is in its uncharged form and there is no water layer resistance. The relationship between P_m and P_0 is like that between the pH-dependent apparent partition coefficient (log K_d) and the true partition coefficient (log K_p), respectively. This relationship can be rationalized by the mass balance. Take, for example, the case of a monoprotic acid, HA. The total substance concentration is

$$C_{HA} = [HA] + [A^-] \qquad (7.47)$$

Using the ionization quotient expression [Eq. (3.1)], $[A^-]$ may be expressed in terms of [HA]:

$$
\begin{aligned}
C_{HA} &= [HA] + \frac{[HA]K_a}{[H^+]} \\
&= [HA]\left(1 + \frac{K_a}{[H^+]}\right) \\
&= [HA](1 + 10^{-pK_a + pH})
\end{aligned}
\qquad (7.48)
$$

In the UWL, HA and A^- diffuse in parallel; the total UWL flux, J_u, is the sum of the two individual flux components. If it is assumed that the transport is under steady state and that the aqueous diffusivities of HA and A^- are the same, the UWL flux becomes

$$
\begin{aligned}
J_u &= J_u^{\text{HA}} + J_u^{\text{A}} \\
&= P_u^{\text{HA}} \Delta[\text{HA}] + P_u^{\text{A}} \Delta[\text{A}^-] \\
&= P_u \Delta C_{\text{HA}}
\end{aligned}
\tag{7.49}
$$

where ΔC_{HA} represents the drop in total concentration across the entire trilamellar barrier. If the pH partition hypothesis holds, then the flux in the membrane is related to the concentration gradient of the uncharged solute

$$
J_m = P_0 \Delta[\text{HA}]
\tag{7.50}
$$

where $\Delta[\text{HA}]$ represents the drop in concentration of the uncharged species in the membrane. Since the membrane and the UWL are in series, the total flux J may be expressed as

$$
\begin{aligned}
\frac{1}{J} &= \frac{1}{J_u} + \frac{1}{J_m} \\
&= \frac{1}{P_e \Delta C_{\text{HA}}}
\end{aligned}
\tag{7.51}
$$

Multiplying this expression by the total sample concentration change, we obtain

$$
\begin{aligned}
\frac{1}{P_e} &= \frac{1}{P_u} + \frac{\Delta C_{\text{HA}}}{\Delta[\text{HA}]P_0} \\
&= \frac{1}{P_u} + \frac{1 + K_a/[\text{H}^+]}{P_0} \\
&= \frac{1}{P_u} + \frac{(1 + 10^{-\text{p}K_a+\text{pH}})}{P_0}
\end{aligned}
\tag{7.52}
$$

Equating Eqs. (7.52) and (7.46) reveals the relationship between intrinsic and membrane permeabilities, Eq. (7.53), for the case of weak acids. Similar steps lead to expressions for weak bases and ampholytes, Eqs. (7.54) and (7.55):

$$P_0 = P_m(1 + 10^{-\text{p}K_a+\text{pH}}) \qquad \text{(weak acid)} \tag{7.53}$$

$$P_0 = P_m(1 + 10^{\text{p}K_a-\text{pH}} \qquad \text{(weak base)} \tag{7.54}$$

$$P_0 = P_m(1 + 10^{\text{p}K_{a1}-\text{pH}} + 10^{-\text{p}K_{a2}+\text{pH}}) \qquad \text{(ampholyte)} \tag{7.55}$$

For ionizable molecules, the intrinsic P_0 and the UWL P_u can be deduced from the pH dependence of P_e, as shown by Gutknecht and co-workers [535–537].

As can be seen from the second line of Eq. (7.52), a plot of $1/P_e$ versus $1/[H^+]$ is expected to be linear (for a weak acid), with the intercept: $1/P_u + 1/P_0$ and the slope K_a/P_0. When the pK_a of the molecule is known, then both P_0 and P_u can be determined. If P_u can be independently determined, then, in principle, the ionization constant may be determined from the pH dependence of the effective permeability.

Figure 7.34 shows the pH dependence of the effective permeability of ketoprofen (measured using pION's PAMPA system with 2% DOPC in dodecane membrane lipid) [558], a weak acid with pK_a 4.12 (0.01 M ionic strength, 25°C). Figure 7.34a shows that the log P_e curve has a flat region for pH $< pK_a$ and a region with a slope of -1 for pH $> pK_a$. At pH 7.4, ketoprofen has a very low permeability, since it is almost entirely in a charged form. The molecule shows increasing permeabilities with decreasing pH, approaching 18×10^{-6} cm/s (thick curve, Fig. 7.34b inset). This is close to the value of the UWL permeability, 21×10^{-6} cm/s (log P_e − 4.68). The small difference vanishes for very lipophilic molecules, such as imipramine. For lipophilic acids, when pH $< pK_a$, the transport is said to be "diffusion-limited." For pH $> pK_a$, the P_e curve coincides with the P_m curve, where transport is "membrane-limited." In general, highly permeable molecules all show nearly the same maximum effective permeability when measured in the same apparatus. In order to deduce the uncharged molecule membrane permeability (top of the dashed curve in Fig. 7.34a), it is necessary to analyze the P_e–pH curve by the Gutknecht method [535–537]; thus, Eq. (7.52) is solved for P_u and P_0, when pK_a is known. Such analysis produces the dashed curve in Figs. 7.34a,b.

The P_m curve (dashed line) is not shifted to the right of the "fraction neutral substance" curve f_u, (see inset in Fig. 7.34b). It just looks that way when unmatched scaling is used [554]. The two curves are exactly superimposed when the vertical coordinates of the P_m and f_u are normalized to a common value. The P_e curve, in contrast, is shifted to the right for weak acids and to the left for weak bases. In the log–log plot (log P_e vs. pH), the pH value at the intersection of the slope 0 and slope -1 curve segments indicates an *apparent* pK_a (Fig. 7.34a).

We have seen many instances of slope–(0, ±1) log–log plots (e.g., Figs. 2.2, 4.2–4.4, 4.6, 5.7, 5.11, 6.1–6.4, 6.12). Behind each tetrad equilibrium (e.g., Figs. 4.1, 5.1, 6.5) there is such a log–log plot, and associated with each such log–log plot is an apparent pK_a. We have called these pK_a^{oct}, pK_a^{mem}, pK_a^{gibbs}. In permeability, there is yet another one: pK_a^{flux} (Fig. 7.34a). If we take the difference between pK_a and pK_a^{flux}, we can deduce the difference between log P_0 and log P_u:

$$\log P_0 = \log P_u + |pK_a - pK_a^{flux}| \tag{7.56}$$

The shapes of permeability–pH profiles mirror those of solubility–pH (see, Figs. 6.1a, 6.2a, and 6.3a), with slopes of *opposite* signs. In solutions *saturated* with an insoluble compound, the product of solubility and permeability ("flux," as described in Chapter 2) is pH-independent! This is indicated in Fig. 2.2 as the maximum flux portions of the curves.

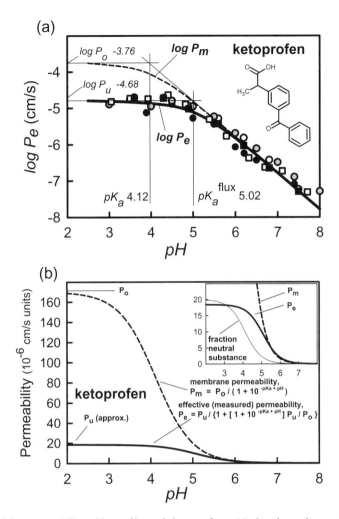

Figure 7.34 Permeability–pH profiles of ketoprofen: (a) log–log plot; solid curve represents effective permeability, and the dashed curve is the membrane permeability, calculated by Eq. (7.53). The latter curve levels off at the intrinsic permeability, P_0. The effective curve levels off to approximately the unstirred water layer permeability, P_u. (b) Direct plot; the inset curve for the fraction neutral substance levels of at 100% (scale not shown). [Avdeef, A., *Curr. Topics Med. Chem.*, **1**, 277–351 (2001). Reproduced with permission from Bentham Science Publishers, Ltd.]

Figure 7.35 shows the characteristic log P_e–pH curve for a weak base, phenazopyridine (pK_a 5.15). With bases, the maximum permeability is realized at high pH values. As in Fig. 7.34, the PAMPA assays were performed under iso-pH conditions (same pH in donor and acceptor wells), using the 2% DOPC in dodecane lipid system.

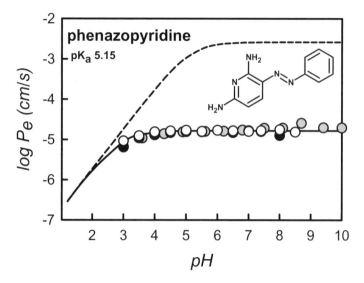

Figure 7.35 Permeability–pH profile of phenazopyridine under iso-pH conditions. [Based on data in Ref. 558.]

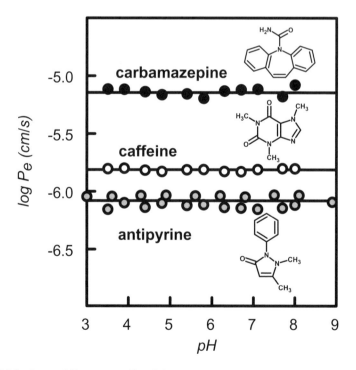

Figure 7.36 Permeability–pH profile of three neutral molecules under iso-pH conditions.

Figure 7.36 shows the log P_e–pH plots for three nonionizable molecules: carbamazepine, caffeine, and antipyrine. As is expected, there is no pH dependence shown; neither the molecules, nor the 2% DOPC/dodecane membrane show charge-state changes in the pH interval from pH 3 to 9.

Table 7.15 lists the intrinsic permeabilities and the unstirred water permeabilities of 16 drugs, determined by the Gutknecht method. The average unstirred water layer permeability is 16×10^{-6} cm/s. Since the aqueous diffusivity (D_{aq}) of most of the drugs in Table 7.15 is near 8×10^{-6} cm^2/s, the average thickness of the unstirred water layer on each side of the membrane is $\sim$2500 μm in the unagitated 96-well microtiter plates, used by pION's PAMPA system. The permeation cell dimensions in typical Caco-2 assays indicate UWL of about 1500 μm (when the plates are unstirred) [554]. The thickness of the unstirred water layer can be driven down to values as low as 300–500 μm if the plate is vigorously stirred during permeation [546,554,556].

The intrinsic membrane permeabilities in Table 7.15 span about eight orders of magnitude, whereas the effective (measured in the in vitro assay) permeabilities are confined to a much narrower range, limited by the UWL. Since the in vivo UWL in the gut is estimated to be about 50 μm [541], it is more appropriate to use P_m than P_e values in oral absorption prediction strategies.

7.7.6.3 Determination of UWL Permeabilities using Stirring Speed Dependence

Caco-2 assay permeabilities corrected for the UWL usually include P_u determined as a function of the stirring speed (since the cells are not stable over a wide pH range), as in Eq. (7.57) [511–514,552,578]

$$P_u = \kappa v^x \qquad (7.57)$$

where κ is a constant descriptive of the diffusivity of the solute and v is the stirring speed (rpm). If the thickness of the UWL is different on the two sides of the membrane, then there are two different values of κ [514]. Equation (7.57) may be substituted into Eq. (7.46) to obtain

$$\frac{1}{P_e} = \frac{1}{P_m} + \frac{1}{\kappa v^x} \qquad (7.58)$$

Measurements of P_e in fixed-pH solutions but at various different stirring speeds need to be made. The double-reciprocal analysis, $1/P_e$ versus $1/v^x$, for Caco-2 permeability measurements in the Transwell (Corning Costar) system produced a linear plot for $x = 0.8$ [514]. The intercept yields the membrane permeability for the particular pH value in the study; the slope determines the κ constant. From the analysis of testosterone transport, for the stirring speed of 25 rpm (planar rotating shaker), the thickness of each UWL (assuming symmetric geometry) was calculated to be 465 μm; at 150 rpm, $h_{aq} = 110$ μm [514]. Karlsson and Artursson [512] found $x = 1.0$ to best represent their stirring-based analysis of the UWL permeability.

TABLE 7.15 Intrinsic Permeabilities and Unstirred Water Layer Permeabilities Determined from Iso-pH Dependence of Effective Permeabilities: 2% DOPC in Dodecane

Compound	P_0 (cm/s) (SD)[a]	$\log P_0$ (SD)	$P_u 10^{-6}$ cm/s (SD)[b]	$\log P_u$ (SD)	MW	pK_a ($I = 0.01$ M, 25°C)	pK_a^{FLUX}	pH Range[c]	n^d	GOF[e]
Imipramine	3.8 (0.3)	+0.58 (0.04)	17.0 (0.8)	-4.77 (0.02)	280.4	9.51	4.2	3.0–7.6	11	0.8
Verapamil	1.5 (0.1) × 10⁻¹	-0.82 (0.03)	13.2 (0.3)	-4.88 (0.01)	454.6	9.07	5.6	4.7–8.1	36	0.6
Propranolol	5.7 (0.3) × 10⁻³	-2.24 (0.03)	13.3 (0.7)	-4.88 (0.02)	259.3	9.53	7.1	5.5–8.1	95	1.3
Ibuprofen	3.0 (0.3) × 10⁻³	-2.52 (0.04)	12.3 (0.6)	-4.91(0.02)	206.3	4.59	6.9	4.0–9.0	9	0.6
Phenazopyridine	2.6 (0.5) × 10⁻³	-2.58 (0.08)	16.7 (0.5)	-4.78 (0.01)	213.2	5.15	3.4	2.9–10.1	35	1.1
Piroxicam	1.1 (0.1) × 10⁻³	-2.96 (0.04)	16.9 (0.8)	-4.77 (0.02)	331.4	2.3, 5.22	6.8	2.9–9.8	33	1.4
Naproxen	4.9 (0.4) × 10⁻⁴	-3.31 (0.04)	18.2 (1.2)	-4.74 (0.03)	230.3	4.32	5.7	3.5–9.8	43	1.5
Ketoprofen	1.7 (0.1) × 10⁻⁴	-3.76 (0.04)	20.7 (1.5)	-4.68 (0.03)	254.3	4.12	5.0	2.9–8.8	54	1.5
Metoprolol	7.5 (0.4) × 10⁻⁵	-5.13 (0.02)	6.3 (0.3)	-5.20 (0.05)	267.4	9.56	8.4	6.8–10.1	56	0.7
Quinine	6.0 (0.4) × 10⁻⁵	-4.22 (0.03)	12.8 (0.7)	-4.89 (0.02)	324.4	4.09, 8.55	7.8, 4.0	6.6–10.1	24	0.6
Salicylic acid	2.9 (0.1) × 10⁻⁵	-4.53 (0.02)	33.0 (6.0)	-4.48 (0.01)	138.1	3.02	3.3	2.9–6.1	12	0.4
Carbamazepine	1.6 (0.3) × 10⁻⁵	-4.8 (0.1)	—	—	236.3	—[f]	—	3.5–8.0	11	0.5
Caffeine	1.8 (0.1) × 10⁻⁶	-5.7 (0.2)	—	—	194.2	—[f]	—	3.5–8.0	11	0.2
Antipyrine	8.7 (0.3) × 10⁻⁷	-6.1 (0.1)	—	—	188.2	—[f]	—	3.0–9.0	21	0.5
Acetaminophen	5 (1) × 10⁻⁸	-7.3 (0.1)	—	—	151.2	9.78	—	3.5–8.5	8	1.3
Hydrochlorothiazide	7 (3) × 10⁻⁹	-8.2(0.1)	—	—	297.7	8.76, 9.95	—	3.5–8.5	10	0.7

Source: Based on data in Ref. 558.

[a] P_0 = intrinsic permeability, SD = estimated standard deviation.
[b] P_u = unstirred water permeability.
[c] Data range actually used in the regression analysis.
[d] Number of P_e measurements.
[e] GOF = goodness of fit in the weighted nonlinear regression analysis.
[f] Carbamazepine, caffeine, and antipyrine are neutral molecules. Their effective permeabilities were corrected for the unstirred water layer using the average unstirred water layer permeability of 1.6×10^{-5} cm/s, determined by the other molecules.

Similar analysis can be applied to side-by-side diffusion cell systems, where stirring is effected by bubbling an O_2/CO_2 gas mixture. For a bubbling rate of 40 mL gas/min, each UWL was estimated to be 282 μm [515].

7.7.6.4 Determination of UWL Permeabilities from Transport across Lipid-Free Microfilters

An infrequently used method (in pharmaceutical research) for determining the UWL permeability involves measuring transport of molecules across a high-porosity microfilter that is not coated by a lipid. The molecules are able to diffuse freely in the water channels of the microfilter. The filter barrier prevents convective mixing between the donor and acceptor sides, and an UWL forms on each sides of the microfilter. Camenisch et al. [546] measured the effective permeabilities of a series of drug molecules in 96-well microtiter plate–filterplate (Millipore GVHP mixed cellulose ester, 0.22 μm pore) "sandwich" where the filters were not coated by a lipid. The permeabilities were nearly the same for all the molecules, as shown in Fig. 7.8a. Our analysis of their data, Fig. 7.8b, indicates $h_{aq} = 460$ μm (sandwich stirred at 150 rpm). We have been able to confirm similar results in our laboratory with different microfilters, using the lipid-free method.

7.7.6.5 Estimation of UWL Thickness from pH Measurements Near the Membrane Surface

Antonenko and Bulychev [84] measured local pH changes near BLM surfaces using a variably positioned 10 μm antimony-tip pH microelectrode. Shifts in pH near the membrane surface were induced by the addition of $(NH_4)_2SO_4$. As the neutral NH_3 permeated, the surface on the donor side of the BLM accumulated excess H^+ and the surface on the acceptor side of the membrane was depleted of H^+ as the permeated NH_3 reacted with water. These effects took place in the UWL. From measurement of the pH profile as a function of distance from the membrane surface, it was possible to estimate h_{aq} as 290 μm in the stirred solution.

7.7.6.6 Prediction of Aqueous Diffusivities D_{aq}

The method preferred in our laboratory for determining the UWL permeability is based on the pH dependence of effective permeabilities of ionizable molecules [Eq. (7.52)]. Nonionizable molecules cannot be directly analyzed this way. However, an approximate method may be devised, based on the assumption that the UWL depends on the aqueous diffusivity of the molecule, and furthermore, that the diffusivity depends on the molecular weight of the molecule. The thickness of the unstirred water layer can be determined from ionizable molecules, and applied to nonionizable substances, using the (symmetric) relationship $P_u = D_{aq}/2h_{aq}$. Fortunately, empirical methods for estimating values of D_{aq} exist. From the Stokes–Einstein equation, applied to spherical molecules, diffusivity is expected to depend on the inverse square root of the molecular weight. A plot of log D_{aq} versus log MW should be linear, with a slope of -0.5. Figure 7.37 shows such a log–log plot for 55 molecules, with measured diffusivities taken from several

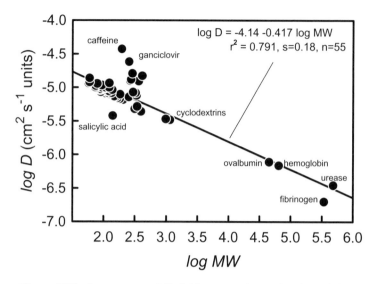

Figure 7.37   Log aqueous diffusivities versus log molecular weights.

sources [40,553,594]. Molecular weights spanned from ~100 to 500,000 Da. The linear regression equation from the analysis is

$$\log D_{\text{aq}} = -4.14 - 0.417 \ \log \text{MW} \tag{7.59}$$

with $r^2 = 0.79$, $s = 0.2$, $n = 55$. The slope is close to the theoretically expected value of -0.5.

The P_u values in Table 7.15 can be combined with Eq. (7.59) to determine approximate h_{aq}. The plot of $\log P_u$ versus $\log$ MW for 11 molecules is shown in Fig. 7.38. The solid line in the plot was determined from the equation (based on $P_u = D_{\text{aq}}/h$)

$$\begin{aligned}\log P_u &= \log D_{\text{aq}} - \log h \\ &= -4.14 - 0.417 \ \log \text{MW} - \log h \end{aligned} \tag{7.60}$$

where h is the sum UWL thickness. The best-fit value of h was determined by regression to be 4.5 mm. Thus each UWL thickness is ~2300 μm. Note that this represents approximately the thickness of the water layer in the unagitated microtiter plate sandwich configuration of the *pION* system. The two highest deviation points in Fig. 7.38 correspond to metoprolol and salicylic acid. These deviations are due mainly to the weak UV spectra of these molecules in the acceptor wells in the PAMPA iso-pH assay.

7.7.6.7 *Intrinsic Permeability–log K_p Octanol–Water Relationship*
Once the 2% DOPC/dodecane permeability data have been corrected for pH and UWL effects, the resulting intrinsic permeabilities P_0 should be linearly related

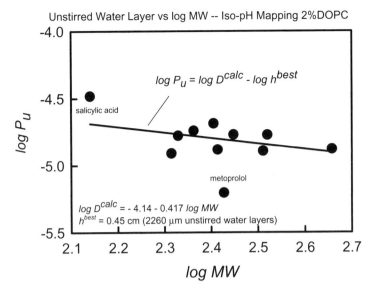

Figure 7.38 Log unstirred water permeabilities versus log molecular weights, based on analysis of iso-pH data.

to the partition coefficients, octanol–water K_p, provided the octanol–water system is a suitable model for the phospholipid system. Ideally, a plot of $\log P_0$ versus $\log K_p$ would represent case (a) in Fig. 7.19. For permeability data uncorrected for UWL effects, a case (b) relationship would be expected. The case (c) pattern in Fig. 7.19 would be expected if water pore transport were an available mechanism in PAMPA. Figure 7.39, showing $\log P_0$ (Table 7.10) versus $\log K_p$ (Table 7.4), indicates that the relationship is approximately linear (r^2 0.79) over eight orders of magnitude of permeability, suggesting the absence of water pores.

7.7.6.8 Iso-pH Permeability Measurements using Soy Lecithin–Dodecane–Impregnated Filters

The above iso-pH measurements are based on the 2% DOPC/dodecane system (model 1.0 over pH 3–10 range). Another membrane model was also explored by us. Table 7.16 lists iso-pH effective permeability measurements using the soy lecithin (20% wt/vol in dodecane) membrane PAMPA (models 17.1, 24.1, and 25.1) The negative membrane charge, the multicomponent phospholipid mixture, and the acceptor sink condition (Table 7.1) result in different intrinsic permeabilities for the probe molecules. Figure 7.40 shows the relationship between the 2% DOPC and the 20% soy iso-pH PAMPA systems for ketoprofen. Since the intrinsic permeability of ketoprofen in the soy lecithin membrane is about 20 times greater than in DOPC membrane, the flat diffusion-limited transport region of the $\log P_e$

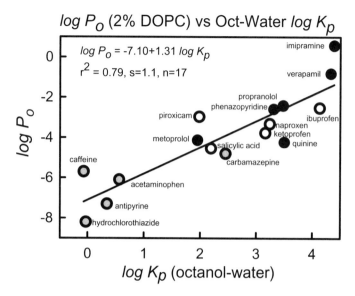

Figure 7.39   Intrinsic permeabilities (iso-pH data analysis) versus octanol–water partition coefficients.

curve is extended to higher pH values. Thus less evidence of membrane-limited transport is visible in the physiological pH range when the soy lecithin system is used. For this reason, correction for the UWL effect is all the more important when devising oral absorption prediction models, which reflect the pH gradient found in the small intestine.

TABLE 7.16   Permeability (10^{-6} cm/s units) and Retention in 20% wt/vol Soy Lecithin, at Iso-pH 5.0, 6.5, 7.4 with Sink in Acceptor Wells

Sample	pH 5.0	%R	pH 6.5	%R	pH 7.4	%R
Desipramine	10.4	35	19.4	35	29.7	39
Propranolol	37.4	31	26.0	37	25.8	40
Verapamil	9.1	30	20.7	20	31.6	31
Metoprolol	2.9	17	16.1	25	28.6	26
Ranitidine	0.00	4	0.03	2	0.31	14
Piroxicam	10.2	24	8.9	12	3.2	17
Naproxen	11.8	50	6.6	12	2.3	13
Ketoprofen	9.5	37	6.5	12	1.2	12
Furosemide	0.8	25	0.0	2	0.0	11
Carbamazepine	19.5	27	17.9	18	15.3	26
Antipyrine	0.9	17	3.0	11	1.7	14

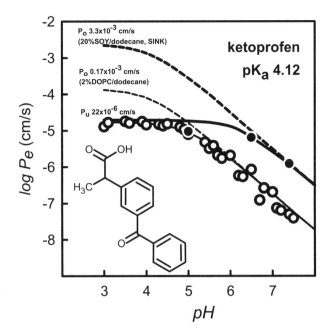

Figure 7.40 Permeability–pH profiles for ketoprofen under iso-pH conditions for two different PAMPA models: unfilled circles = 2% DOPC/dodecane, filled circles = 20% soy lecithin/dodecane. [Reprinted from Avdeef, A., in van de Waterbeemd, H.; Lennernäs, H.; Artursson, P. (Eds.). *Drug Bioavailability. Estimation of Solubility, Permeability, Absorption and Bioavailability.* Wiley-VCH: Weinheim, 2003 (in press), with permission from Wiley-VCH Verlag GmbH.]

7.7.6.9 Gradient pH Effects

The gradient pH soy lecithin, acceptor sink systems (models 26.1–30.1 in Table 7.3) were explored in the search for the best GIT PAMPA model [559]. Figures 41a–c show examples of three bases: verapamil, propranolol, and metoprolol (in order of decreasing lipophilicities; see Table 7.4). In each case, the acceptor pH was 7.4, but the donor pH values ranged from 3 to 10. Figures 42a–c show examples of three acids: naproxen, ketoprofen, and piroxicam (decreasing lipophilicity order). In all of the examples above, the diffusion-controlled zone spans a much wider pH range, compared to the DOPC system (Figs. 7.34 and 7.35). This is the consequence of increased intrinsic permeabilities in the soy-based system. Figure 7.43 shows examples of two neutral molecules: carbamazepine and antipyrine. It was possible to approximate the membrane permeability curve for carbamazepine (dashed line in Fig. 7.43), based on the analysis of the UWL permeabilities of the ionizable molecules. Antipyrine is hydrophilic and has equivalent P_m and P_e curves.

Table 7.17 summarizes the analysis of the gradient pH experiments. The range of intrinsic permeabilities spans 11 orders of magnitude! The UWL permeabilities ranged from 16 to 52×10^{-6} cm/s. Those molecules that appeared to bind strongly to the sink-forming acceptor surfactant showed UWL P_u values that were about twice those calculated from the iso-pH nonsink assays (Table 7.15). The strong binding between the solute and the surfactant in the acceptor wells drives the unbound fraction of the solute molecules to near zero. According to the pH partition hypothesis, it is the unbound neutral species which crosses the membrane. Since its

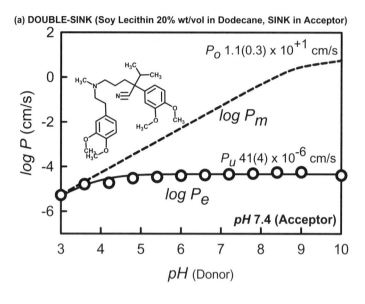

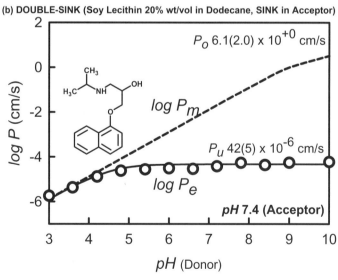

Figure 7.41 Gradient pH profiles for three weak bases with double-sink conditions, 20% wt/vol soy lecithin in dodecane: (a) verapamil (pK_a 9.07); (b) propranolol (pK_a 9.53); (c) metoprolol (pK_a 9.56).

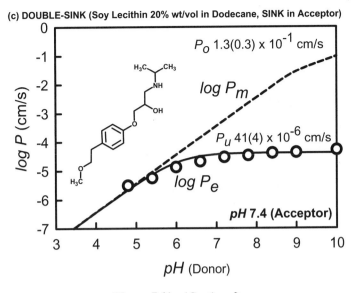

Figure 7.41   (*Continued*)

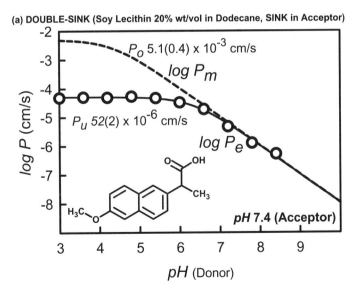

Figure 7.42 Gradient pH profiles for three weak acids with double-sink conditions, 20% wt/vol soy lecithin in dodecane: (a) naproxen (pK_a 4.32); (b) ketoprofen (pK_a 4.12); (c) piroxicam (pK_a 5.22, 2.3).

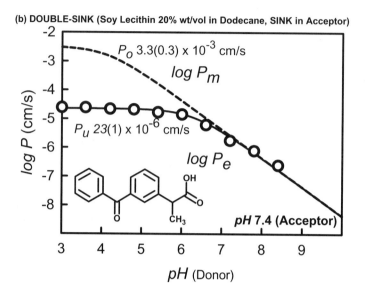

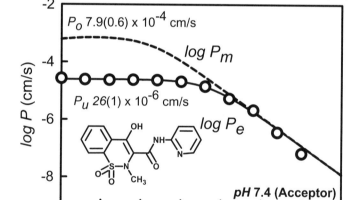

Figure 7.42   *(Continued)*

concentration is near zero, the acceptor-to-donor backflux is nil [see Eq. (7.26)]. So the UWL resistance on the acceptor side is of little consequence in the transport process. When strong binding takes place under the simulated sink condition, only the UWL on the donor side directly contributes to the overall resistance. Hence, P_u values are calculated to be about twice as large as in the case of no-sink iso-pH (Table 7.15).

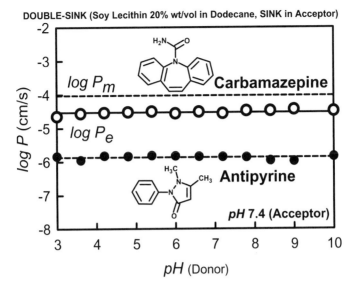

Figure 7.43 Gradient pH profiles for two nonionizable molecules: double-sink conditions, 20% wt/vol soy lecithin in dodecane.

Table 7.18 lists the interpolated apparent and membrane permeabilities, along with the membrane retentions, of the probe molecules used in the gradient pH study, at pH values 5.0, 5.5, 6.0, 6.5, and 7.4.

7.7.6.10   Collander Relationship between 2% DOPC and 20% Soy Intrinsic Permeabilities

The 20% soy lecithin (Table 7.17) and the 2% DOPC (Table 7.15) intrinsic permeabilities may be compared in a Collander equation, as shown in Fig. 7.44. The slope of the regression line, soy versus DOPC, is greater than unity. This indicates that the soy membrane is more lipophilic than the DOPC membrane. Intrinsic permeabilities are generally higher in the soy system. Three molecules were significant outliers in the regression: metoprolol, quinine, and piroxicam. Metoprolol and quinine are less permeable in the DOPC system than expected, based on their apparent relative lipophilicities and in vivo absorptions [593]. In contrast, piroxicam is more permeable in DOPC than expected based on its relative lipophilicity. With these outliers removed from the regression calculation, the statistics were impressive at r^2 0.97.

7.7.7   Evidence of Transport of Charged Species

In Section 4.8 the topic of charged-species absorption ("fact or fiction") was first considered. The partitioning properties of some lipophilic charged molecules in the

TABLE 7.17 Intrinsic Permeabilities and the Unstirred Water Layer Permeabilities Determined from Gradient–pH Dependence of Effective Permeabilities: 20% Soy Lecithin in Dodecane, Sink in Acceptor

Compound	P_0 (cm/s) (SD)[a]	$\log P_0$ (SD)	P_u (10^{-6} cm/s) (SD)[b]	$\log P_u$ (SD)	MW	pK_a ($I = 0.01$ M, 25° C)	pK_a^{FLUX}	pH Range[c]	n[d]	GOF[e]
Desipramine	$1.9\ (0.8) \times 10^{+2}$	+2.27 (0.19)	42.1 (4.1)	−4.38 (0.04)	302.8	10.16	3.7	3.0–10.0	12	2.4
Verapamil	$1.1\ (0.3) \times 10^{+1}$	+1.03 (0.14)	41.2 (3.5)	−4.39 (0.04)	454.6	9.07	3.9	3.0–9.0	11	2.0
Propranolol	6 (2)	+0.79 (0.15)	41.8 (4.7)	−4.38 (0.05)	259.3	9.53	4.5	3.0–10.0	12	2.6
Metoprolol	$1.3\ (0.3) \times 10^{-1}$	−0.88 (0.10)	41.2 (3.5)	−4.39 (0.04)	267.4	9.56	6.0	4.8–10.0	9	1.7
Quinine	$5.0(0.7) \times 10^{-2}$	−1.30 (0.06)	15.6 (0.6)	−4.81 (0.02)	324.4	4.09, 8.55	5.1	3.6–10.0	11	0.7
Naproxen	$5.1\ (0.4) \times 10^{-3}$	−2.30 (0.04)	51.8 (1.6)	−4.29 (0.01)	230.3	4.32	6.2	3.0–8.4	8	0.7
Ketoprofen	$3.3\ (0.3) \times 10^{-3}$	−2.49 (0.05)	22.8 (1.0)	−4.64 (0.02)	254.3	4.12	6.2	3.0–8.4	8	0.7
Piroxicam	$7.9\ (0.6) \times 10^{-4}$	−3.11 (0.03)	26.3 (0.8)	−4.58 (0.01)	331.4	2.33, 5.22	6.7	3.0–9.0	11	0.5
Furosemide	$1.3\ (0.1) \times 10^{-4}$	−3.90 (0.03)	17.0 (1.0)	−4.77(0.03)	330.8	3.67, 10.93	4.6	3.0–6.6	7	0.4
Carbamazepine	$9.4\ (0.6) \times 10^{-5}$	−4.03 (0.03)	44[f]	−4.36	236.3	—	—	3.0–10.0	12	1.0
Ranitidine	$5.6\ (0.6) \times 10^{-6}$	−5.25 (0.04)	19.8 (10.2)	−4.70 (0.22)	350.9	8.31	8.2	6.0–10.0	7	0.4
Antipyrine	$1.4\ (0.2) \times 10^{-6}$	−5.86 (0.06)	17[f]	−4.77	188.2	—	—	3.0–10.0	12	0.5
Terbutaline	$1.2\ (0.2) \times 10^{-7}$	−6.91 (0.08)	17[f]	−4.77	274.3	8.67, 10.12	—	3.0–10.0	12	1.0
Hydrochlorothiazide	$5\ (6) \times 10^{-9}$	−8.30 (0.49)	17[f]	−4.77	297.7	9.81, 10.25	—	3.0–10.0	5	0.8

[a] P_0 = intrinsic permeability, SD = estimated standard deviation.
[b] P_u = unstirred water permeability.
[c] Data range actually used in the regression analysis.
[d] Number of P_e measurements.
[e] GOF = goodness of fit in the weighted nonlinear regression analysis.
[f] Carbamazepine, antipyrine, terbutaline, and hydrochlorothiazide were treated as neutral molecules. Their effective permeabilities were corrected for the unstirred water layer using estimated unstirred water layer permeabilities, determined by the other molecules of similar lipophilicities and size.

TABLE 7.18 Interpolated Apparent and Membrane Permeabilities Determined from Double-Sink Conditions: 20% Soy Lecithin in Dodecane

Compound	pH 5.0			pH 5.5			pH 6.0			pH 6.5			pH 7.4		
	P_a (10^{-6} cm/s)	P_m (cm/s)	%R	P_a (10^{-6} cm/s)	P_m (cm/s)	%R	P_a (10^{-6} cm/s)	P_m (cm/s)	%R	P_a (10^{-6} cm/s)	P_m (cm/s)	%R	P_a (10^{-6} cm/s)	P_m (cm/s)	%R
Verapamil	39	9.2×10^{-4}	28	41	2.9×10^{-3}	30	41	9.2×10^{-3}	29	41	0.029	32	41	0.23	34
Quinine	7	1.4×10^{-5}	55	12	4.5×10^{-5}	56	14	1.4×10^{-4}	58	15	4.4×10^{-4}	58	16	3.3×10^{-3}	59
Propranolol	34	1.8×10^{-4}	35	39	5.7×10^{-4}	38	41	1.8×10^{-3}	39	42	5.7×10^{-3}	39	42	0.045	39
Desipramine	41	1.3×10^{-3}	34	42	4.0×10^{-3}	35	42	0.013	36	42	0.040	37	42	0.32	39
Metoprolol	3	3.6×10^{-6}	19	9	1.1×10^{-5}	20	19	3.6×10^{-5}	24	30	1.1×10^{-4}	24	39	9.0×10^{-4}	25
Ranitidine	0.003	2.7×10^{-9}	16	0.009	8.7×10^{-9}	16	0.03	2.7×10^{-8}	17	0.09	8.6×10^{-8}	17	0.6	6.1×10^{-7}	19
Naproxen	49	2.7×10^{-4}	22	44	3.1×10^{-4}	22	35	1.0×10^{-4}	21	20	3.3×10^{-5}	20	4	4.2×10^{-6}	13
Ketoprofen	22	3.8×10^{-4}	21	19	1.3×10^{-4}	21	15	4.2×10^{-5}	20	9	1.3×10^{-5}	19	2	1.7×10^{-6}	16
Hydrochlorothiazide	0.005	5.0×10^{-9}	16	0.005	5.0×10^{-9}	16	0.005	5.0×10^{-9}	16	0.005	5.0×10^{-9}	14	0.005	5.0×10^{-9}	14
Furosemide	4.2	5.6×10^{-6}	42	1.7	1.8×10^{-6}	34	0.6	5.9×10^{-7}	32	0.2	1.9×10^{-7}	24	0.02	2.3×10^{-8}	19
Piroxicam	25	4.9×10^{-4}	28	24	2.7×10^{-4}	25	21	1.1×10^{-4}	23	16	3.9×10^{-5}	17	4	5.1×10^{-6}	16
Terbutaline	0.1	1.2×10^{-7}	18	0.1	1.2×10^{-7}	18	0.1	1.2×10^{-7}	18	0.1	1.2×10^{-7}	19	0.1	1.2×10^{-7}	19
Carbamazepine	30	9.4×10^{-5}	29	30	9.4×10^{-5}	29	30	9.4×10^{-5}	29	30	9.4×10^{-5}	29	30	9.4×10^{-5}	29
Antipyrine	1.3	1.4×10^{-6}	14	1.3	1.4×10^{-6}	14	1.3	1.4×10^{-6}	14	1.3	1.4×10^{-6}	14	1.3	1.4×10^{-6}	14

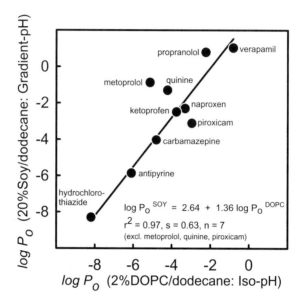

Figure 7.44 Collander relationship between intrinsic permeabilities of 20% soy lecithin versus 2% DOPC PAMPA models.

octanol–water system might suggest that, given a background solution of a lipophilic counterion, ion pair transmembrane transport takes place (e.g., Section 4.5). Such hypotheses can be tested in a direct way in the PAMPA assay. If the charged species, especially quaternary ammonium drugs, appear in the acceptor compartment, the case for charged species transport could be further advanced. It is very difficult to make the case for charged-species transmembrane transport using the in vitro cultured cell model, because of the simultaneous presence of several possible transport mechanisms [1].

7.7.7.1 The Case for Charged-Species Transport from Cellular and Liposomal Models

Trimethylaminodiphenylhexatriene chloride (TMADPH; Fig. 7.45) is a fluorescent quaternary ammonium molecule that appears to permeate cell membranes [595]. TMADPH fluoresces only when it is in the bilayer, and not when it is dissolved in water. Therefore, its location in cells can be readily followed with an imaging fluorescence microscope. One second after TMADPH is added to the extracellular solution bathing HeLa cell types, the charged molecule fully equilibrates between the external buffer and the extracellular (outer) leaflet bilayer. Washing the cells for one minute removes >95% of the TMADPH from the outer leaflet. If the cells are equilibrated with TMADPH for 10 min at 37°C, followed by a one-minute wash that removed the TMADPH from the outer leaflet, the fluorescent molecule is

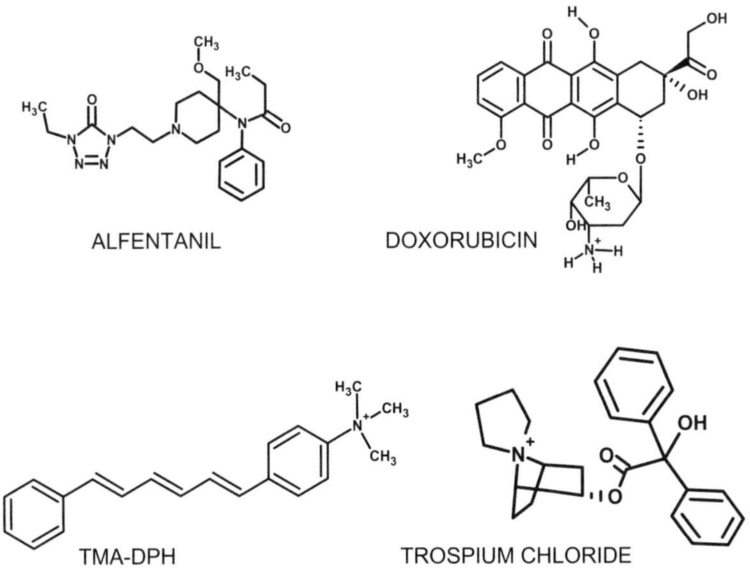

Figure 7.45   Molecules that may violate the pH partition hypothesis.

seen concentrated in the perinuclear and the mitochondrial membranes inside the cytoplasm. This indicates that the charged molecule somehow crossed the cell wall. Endocytosis is not likely to be the influx mechanism, because the charged molecule would not have been able to interact with the perinuclear and mitochondrial membranes. P-gp transfected HeLa cells showed decreased intracellular fluorescence, but the concentration of the fluorescent molecule in the outer leaflet was not affected by P-gp presence. When cyclosporin A, a known P-gp inhibitor, was added, TMADPH intracellular accumulation was reestablished. Since P-gp is postulated to interact with its substrates brought to the active site at the inner leaflet position of the bilayer [596], TMADPH must be somehow crossing the bilayer to get into the inner leaflet. These observations led Chen et al. [595] to propose a flip-flop mechanism, since active transporters for TMADPH were not seen. However, the possibility of a surface protein assisted transport could not be ruled out. Since several transport mechanisms are possible, the unequivocal route is not established with certainty. An ideal follow-up experiment would have utilized "ghost" vesicles formed from protein-free reconstituted HeLa cell lipids. Such an experiment has not been reported.

Regev and Eytan [597] studied the transport properties of doxorubicin (Fig. 7.45) across bilayers, using model liposomes formed from anionic phosphatidylserine and "ghost" erythrocytes. Doxorubicin, unlike TMADPH, can undergo charge-state changes. At neutral pH, the amine on the daunosamine moiety is expected to be positively charged ($pK_a \sim 8.6$). The phenolic protons are expected to have

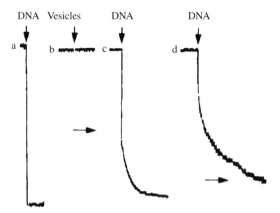

Figure 7.46 Fluorescence quenching of doxorubicin by DNA [597]: (a) doxorubicin in aqueous solution, quenched immediately on addition of DNA; (b) doxorubicin fluorescence not affected by vesicles; (c) Doxorubicin preequilibrated with vesicles, and then subjected to DNA. The fraction bound to the outer membrane leaflet is immediately quenched by the DNA. (d) Same as (c), but multilamellar vesicles used. The left arrow represents a 5-min interval and applies to the first three cases; the right arrow represents 30-min interval and applies to (d) only. [Reprinted from Ronit Regev and Gera D. Eylan, *Biochemical Pharmacology*, vol. 54, 1997, pp. 1151–1158. With permission from Elsevier Science.]

$pK_a > 11$, due to the likely formation of six-membered ring intramolecular H bonds. Doxorubicin is mildly lipophilic, with an octanol–water $\log K_p$ 0.65 (slightly less than morphine) and $\log K_d$ of -0.33. It is not very permeable across 2% DOPC/dodecane PAMPA membranes $(P_e \sim 4 \times 10^{-9}\,\text{cm/s})$. About 90% of doxorubicin is surface-bound in PS liposomes [597]. Doxorubicin is fluorescent in water. Its fluorescence is quickly quenched by interactions with DNA; an aqueous solution of doxorubicin is immediately quenched by the addition of DNA, as shown in curve (a) of Fig. 7.46, where the left arrow represents 5 min and applies to curves (a)–(c) in Fig. 7.46. Vesicles don't affect the fluorescence [Fig. 7.46, curve (b)]. However, a solution equilibrated with doxorubicin and unilamellar liposomes, is 50% quenched instantly, and 100% quenched after about 5 min (1.1–1.3 min half-life at 23°C), as shown in curve (c) of Fig. 7.46 [597]. This indicates that the outer leaflet doxorubicin (50% of the total) is immediately quenched, and the intravesicular doxorubicin takes ~1 min to permeate out, by crossing the bilayer, presumably as a charged species at neutral pH. Curve (d) of Fig. 7.46 represents a multilamellar liposome extraction quenching, where the right arrow is ~30 min long. About 20% of the doxorubicin is quickly quenched, but the rest of the drug takes about 2 h to quench, since many bilayers need to be crossed by the positively charged molecule. Still, these observations do not prove that the actual permeating molecule is charged. The molecule (charged in the aqueous phase) may be permeating as the neutral species (in the membrane phase). The only clue that perhaps some degree of charged species permeation is taking place comes from the observation that at pH 9.7, the transcellular transport is increased only twofold. If the pH partition hypothesis were valid, and the pK_a were 8.6, then changing pH from 7.4 to 9.7 should have increased transport by much more than a factor of 2. It would have been interesting to perform the experiments of Regev and Eytan using TMADPH, to unequivocally demonstrate the violation of the pH partition hypothesis.

Trospium chloride, a quaternary ammonium drug (Fig. 7.45), appears to be a substrate of P-gp, and it can be taken up by cells quicky [597]. The evidence for transmembrane diffusion appears substantial. The molecule is very soluble in water (>50 mg/mL), but not in lipids (9.2 µg/mL in mineral oil); the octanol-water log K_p is −1.22 [598]. The human intestinal absorption (HIA) is 11%; the molecule is not metabolized. In cell intestinal patch uptake studies, trospium is absorbed from a 7.5 mM solution at the rate of 7 µg/h, after a slow 60-min buildup to an approximate steady state flux. At donor concentration of 0.5 mM, rat Caco-2 P_e is 8×10^{-7} cm s^{-1}. At the higher concentration of 45 mM, the permeability increases to 2.2×10^{-6} cm s. This suggests that an efflux transporter is saturable. At 5 mM trospium concentration, the apical-to-basolateral permeability is 7 times lower than the basolateral-to-apical permeability. Verapamil (P-gp inhibitor) equalizes the above two permeabilities. Since the mechanism of P-gp efflux involves the interaction of the substrate from the inner leaflet of the bilayer [596], trospium is somehow crossing lipid bilayers. But since cells were used, it is difficult to rule out a carrier-mediated transport. More light could be shed with simpler models, perhaps using "ghost" erythrocytes or PAMPA.

Palm et al. [578] studied the Caco-2 permeabilities of two molecules, alfentanil (Fig. 7.45) and cimetidine, whose pK_a values were near neutral (6.5 and 6.8, respectively), but whose octanol–water partition coefficients, log K_p, were more than an order of magnitude different (2.2 and 0.4, respectively). The group studied the permeabilities over a range of pH values, from 4 to 8, something that is very rarely done in Caco-2 assays. The viability of the cells was demonstrated for pH 4.8–8.0. The analysis of the pH-dependent permeability data indicated that the positive-charge form of alfentanyl had a permeability coefficient $(1.5 \times 10^{-6} \text{ cm/s})$ that was substantially greater than that of cimetidine $(5 \times 10^{-8} \text{ cm/s})$. Since alfentanyl has a molecular weight of 416 (cimetidine has 252), it is not expected to transport by the paracellular route. The authors proposed that the charged form of the drug can permeate membrane by passive transcellular diffusion.

7.7.7.2 PAMPA Evidence for the Transport of Charged Drugs

It is difficult to prove that quaternary ammonium compounds can cross lipid bilayers using cell uptake experiments, since several mechanisms may be operative, and separating contributions from each may be very difficult [1]. It may be an advantage to use PAMPA to investigate transport properties of permanently ionized molecules. Of all the molecules whose permeabilities were measured under iso-pH conditions in 2% DOPC/dodecane, verapamil, propranolol, and especially quinine seem to partially violate the pH partition hypothesis, as shown in Figs. 7.47a–c. In Fig. 7.47c, the solid line with slope of +1 indicates the expected effective permeability if the pH partition hypothesis were strictly adhered to. As can be seen at pH 4 in the figure, quinine is about 100 times more permeable than predicted from the pH partition hypothesis. Instances of acids violating the pH partition hypothesis have not been reported.

When negatively charged membranes are used, the weak bases no longer appear to violate the pH partition hypothesis, as indicated in Fig. 7.48 for quinine. It appears that the negative membrane surface charge and the positive drug charge leads to electrostatic interactions that inhibit the passage of charged drugs through the membranes. These observations will be further explored in our laboratory.

7.7.8 Δ log P_e–Hydrogen Bonding and Ionic Equilibrium Effects

Most drug-like molecules dissolved in water form hydrogen bonds with the solvent. When such a molecule transfers from water into a phospholipid bilayer, the solute–water hydrogen bonds are broken (desolvation), as new solute–lipid H bonds form in the lipid phase. The free-energy difference between the two states of solvation has direct impact on the ability of the molecules to cross biological barriers.

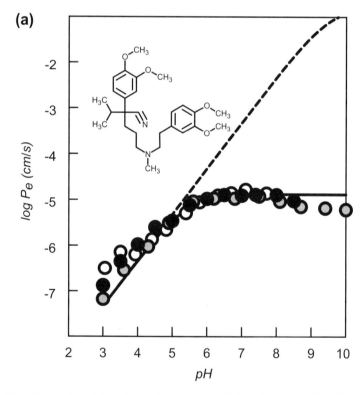

Figure 7.47 Examples of three bases that appear to violate the pH partition hypothesis in 2% DOPC/dodecane PAMPA models: (a) verapamil (pK_a 9.07); (b) propranolol; (c) guinine (pK_a 8.55, 4.24).

(b)

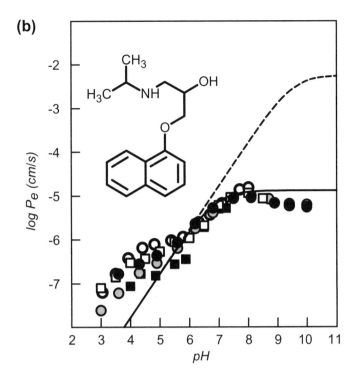

(c)

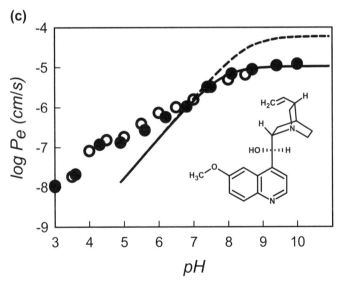

Figure 7.47 (*Continued*)

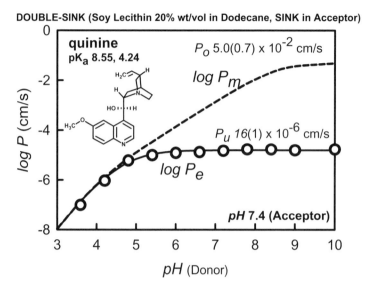

Figure 7.48 Negatively charged PAMPA models showing no evidence for violation of the pH partition hypothesis.

Seiler [250] proposed a way of estimating the extent of hydrogen bonding in solute partitioning between water and a lipid phase by measuring the so-called $\Delta \log P$ parameter. The latter parameter is usually defined as the difference between the partition coefficient of a solute measured in the octanol–water system and that measured in an inert alkane–water suspension: $\Delta \log P = \log K_{p,\,oct} - \log K_{p,\,alk}$.

Young et al. [599] demonstrated the usefulness of the $\Delta \log P$ parameter in the prediction of brain penetration of a series of H_2-receptor histamine antagonists. Neither $\log K_{p,\,oct}$ nor $\log K_{p,\,alk}$ was found to correlate with brain penetration, $\log BB$ (where BB is defined as the ratio of the compound concentration in the brain and the compound concentration in plasma). However, the difference between the two partition coefficients correlated well, as shown in Fig. 7.49. When the difference is large, so is the H bonding expressed by the solute, and less brain penetration is expected. It was suggested that the $\Delta \log P$ parameter accounts for H-bonding ability and reflects two distinct processes—alkane encodes the partitioning into nonpolar regions of the brain and octanol encodes protein binding in the peripheral blood. El Tayar et al. [255] elaborated that the parameter contains information on the capacity of a solute to donate H bonds; the rate-limiting step in brain penetration was proposed to be the donation of H bonds of solute to hydrophilic parts of lipids in the blood–brain barrier (BBB). Van de Waterbeemd and Kansy [251] reexamined Young et al. [599] data with solvatochromic equations for identifying physicochemical properties governing solubility and partitioning. They suggested that the combination of calculated molar volumes and just the $\log \log K_{p,\text{alk}}$ could

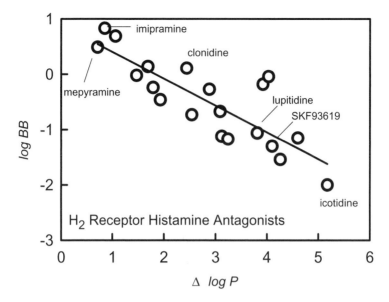

Figure 7.49   Brain penetration and $\Delta \log P$ (from Young et al. [599]).

substitute for two-lipid partition measurements, thus reducing the amount of measurement needed. Also, they introduced the use of polar surface areas as an interesting alternative to the use of $\Delta \log P$. Abraham et al. [257] analyzed the $\Delta \log P$ parameter in terms of the Abraham descriptors to broaden the understanding of the concept. Von Geldern et al. [252] used the $\Delta \log P$ parameter to optimize structural modifications to a series of endothelin A-receptor antagonists to improve gut absorption. A urea fragment in their series of molecules had NH residues systematically replaced with NCH_3, O, and CH_2, and correlations between $\Delta \log P$ and antagonist selectivity effectively guided the optimization procedure.

Avdeef et al. [556] measured the PAMPA permeabilities of a series of drug molecules and natural products using both dodecane- and (dodecane + 2%DOPC)-coated filters. It was proposed that a new H-bonding scale could be explored, based not on partition coefficients but on permeabilities.

$$\Delta \log P_e = \log P_e^{2\%DOPC} - \log P_e^{dodecane} \qquad (7.61)$$

Figure 7.50 shows $\Delta \log P_e$ (difference permeability) versus $\log P_e$ (dodecane-treated filters) for a series of common drugs and research compounds at pH 7.4. Some of the differences are positive, and some are negative. For example, phenazopyridine is attenuated by the presence of DOPC in the dodecane, but diltiazem is accelerated by the DOPC [556]. The effects are most pronounced where the permeability in pure dodecane is less than about $3 \times 10^{-6} \, cm/s$. That is, molecules that are very permeable in dodecane are unaffected by the presence of DOPC, as

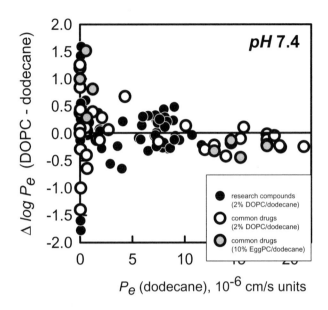

Figure 7.50 Hydrogen bonding/electrostatics scale based on phospholipid–alkane permeability differences: $\Delta \log P_e$ versus P_e (dodecane).

indicated in Fig. 7.50. Both H-bonding and ionic interactions may be encoded in the $\Delta \log P_e$ parameter. This topic is the subject of further investigation at pION, with the aim of developing BBB PAMPA models.

7.7.9 Effects of Cosolvent in Donor Wells

Many research compounds are poorly soluble in water. When very lipophilic molecules precipitate in the donor wells, it is possible to filter the donor solution before the PAMPA sandwich is prepared. On occasion, the filtered donor solution contains such small amounts of the compound that determination of concentrations by UV spectrophotometry becomes impractical. One strategy to overcome the precipitation of the sample molecules in the donor wells is to add a cosolvent to the solutions (Section 7.4.4). It is a strategy of compromise and practicality. Although the cosolvent may solubilize the lipophilic solute molecule, the effect on transport may be subtle and not easy to predict. At least three mechanisms may cause P_e and membrane retention ($\%R$) values to alter as a result of the cosolvent addition. To a varying extent, all three mechanisms may simultaneously contribute to the observed transport:

1. The cosolvent will lower the dielectric constant of the mixed solvent, independent of the properties of the solute molecule. The ionization constant of acids will increase and that of bases will decrease (see Sections 3.3.3 and 3.3.4), the result of which is to increase the fraction of uncharged substance in

solution (f_u in Table 7.4). With an increased concentration of the uncharged species in the donor solution, both P_e and %R are expected to *increase*. Generally, this effect is minimal for cosolvent amounts less than 10% vol/vol [119,172].

2. The cosolvent may increase the aqueous solubility of the sparingly soluble molecules, which would lower the membrane–donor solution partition coefficient. According to Eq. (2.3), P_e will *decrease*. Since %R is related to lipophilicity (Section 7.7.2), the retention is also expected to *decrease*.

3. Sparingly soluble surface-active molecules, such as chlorpromazine, may form water-soluble high-molecular-weight (HMW) aggregates (Sections 6.2, 6.5.2, 6.5.3). Their diffusion in the unstirred water layer will decrease according to Eq. (7.60). Cosolvent may break up these aggregates, resulting in *increased* P_e, and to a lesser extent, an *increased* %R.

Table 7.19 summarizes the PAMPA (2% DOPC in dodecane) transport properties of several molecules, with and without 10% 1-propanol in the donor wells. This particular cosolvent was selected for its low UV absorbance and low volatility.

The most dramatic effects are with the bases. The first seven bases in Table 7.19 are the most lipophilic. Cosolvent causes their %R to decrease, consistent with effect (2) listed above. For the three least-lipophilic bases, %R increases with cosolvent, consistent with effect (1). Chlorpromazine and verapamil experience

TABLE 7.19   Effect of 10% 1-Propanol, pH 7.4[a]

Sample	2% DOPC (Model 1.0) P_e(SD)	%R	2% DOPC (Cosolvent) P_e(SD)	%R
Chlorpromazine	5.5 (0.4)	85	18.0 (8.9)	71
Phenazopyridine	8.4 (1.1)	70	6.5 (0.3)	50
Verapamil	9.7 (1.0)	39	19.4 (3.1)	25
Promethazine	7.3 (0.7)	70	3.1 (0.2)	34
Propranolol	10.0 (0.5)	18	8.3 (1.7)	12
Desipramine	12.3 (0.4)	40	5.3 (0.4)	22
Primaquine	1.4 (0.1)	70	2.6 (0.4)	26
Alprenolol	11.8 (0.3)	16	10.2 (2.5)	28
Metoprolol	0.69 (0.04)	11	1.5 (0.1)	27
Amiloride	0.002 (0.005)	0	0.03 (0.04)	19
Naproxen	0.33 (0.03)	4	1.6 (0.2)	25
Sulfasalazine	0.007 (0.004)	1	0.04 (0.01)	26
Theophylline	0.04 (0.01)	1	0.33 (0.05)	15
Salicyclic acid	0.006 (0.004)	1	0.13 (0.05)	19
Sulpiride	0.01 (0.01)	1	(nd)	10
Terbutaline	0.04 (0.01)	6	(nd)	12

[a]All P_e and SD(P_e) are in units of 10^{-6} cm/s; (nd) = compound not detected in the acceptor compartment.

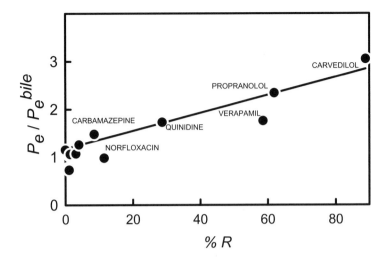

Figure 7.51   Effect of bile salt on permeability in donor well at pH 6.5.

significantly elevated P_e, consistent with effects (1) and (3). The four acids in the Table 7.19 behave according to effect (1) listed above; both P_e and %R are elevated. The two ampholytes may also be affected this way, judging by the increased %R.

7.7.10   Effects of Bile Salts in Donor Wells

An alternative method to overcome the solubility problem mentioned in the last section is to use bile salts to solubilize lipophilic molecules in the donor wells. Figure 7.51 shows a plot of relative permeability (P_e without bile/P_e with bile) versus membrane retention, which is related to lipophilicity (Section 7.7.2). As the plot shows, the most lipophilic molecules (carvedilol, propranolol, and verapamil) have attenuated permeabilities (by a factor of 3 in the case of carvedilol). The effective partition coefficient between the PAMPA membrane phase and the aqueous phase containing bile salt micelles [577] is expected to be lower for lipophilic molecules, which should result in lower P_e values. This is evident in the figure.

7.7.11   Effects of Cyclodextrin in Acceptor Wells

The method for creating acceptor sink condition discussed so far is based on the use of a surfactant solution. In such solutions, anionic micelles act to accelerate the transport of lipophilic molecules. We also explored the use of other sink-forming reagents, including serum proteins and uncharged cyclodextrins. Table 7.20 compares the sink effect of 100 mM β-cyclodextrin added to the pH 7.4 buffer in the acceptor wells to that of the anionic surfactant. Cyclodextrin creates a weaker sink for the cationic bases, compared to the anionic surfactant. The electrostatic binding force between charged lipophilic bases and the anionic surfactant micelles

TABLE 7.20 Effect of 100 mM β-Cyclodextrin in Acceptor Wells, pH 7.4[a]

Sample	20% Soy (Model 17.0) P_e(SD)	%R	20% Soy (Cyclodextrin) P_e(SD)	%R	20% Soy (Model 17.1) P_e(SD)	%R
Verapamil	1.1 (0.1)	94	3.5 (0.4)	53	31.6 (2.8)	31
Propranolol	1.8 (0.5)	95	3.4 (0.4)	61	25.1 (1.7)	36
Desipramine	0.6 (0.7)	93	9.3 (5.7)	56	29.8 (0.2)	39
Metoprolol	8.2 (1.2)	42	21.0 (0.7)	26	26.5 (1.1)	23
Ranitidine	0.36 (0.01)	13	0.20 (0.04)	8	0.31 (0.03)	14
Naproxen	3.4 (0.1)	9	3.6 (0.2)	9	2.9 (0.1)	13
Ketoprofen	1.5 (0.1)	9	1.2 (0.1)	11	1.2 (0.1)	12
Hydrochlorothiazide	0.01 (0.01)	9	0.01 (0.03)	9	0.004 (0.004)	12
Furosemide	0.04 (0.02)	11	0.05 (0.01)	0	0.02 (0.01)	11
Piroxicam	3.6 (0.1)	2	3.2 (0.2)	9	3.2 (0.2)	17
Terbutaline	0.20 (0.14)	2	(nd)	7	0.1 (0.2)	13
Carbamazepine	10.8 (0.3)	37	22.9 (0.9)	25	15.2 (0.7)	26
Antipyrine	1.5 (0.1)	9	2.1 (0.3)	9	1.6 (0.1)	13

[a]All P_e and SD(P_e) are in units of 10^{-6} cm/s; (nd) = compound not detected in the acceptor compartment.

is missing in the cyclodextrin system. Some molecules (e.g., metoprolol, carbamazepine) may have the suitable shape to take advantage of strong cyclodextrin binding, and thus indicate substantially increased permeabilities.

7.7.12 Effects of Buffer

Gutknecht and Tosteson [535] considered the effect of buffer on the transport of salicylic acid across a single bilayer (BLM). Buffers affect the magnitude of the pH gradient formed in the unstirred water layer as the result of the diffusion of ionizable permeants. (This is in addition to bulk solution pH gradient conditions formed by the added buffers.) In turn, the pH at the membrane–water interface affects the concentration of the uncharged (membrane-permeant) species, and thus contributes to the magnitude of the permeant concentration gradient in the membrane phase. The gradient pH permeation cell considered in the abovementioned study [535] (unbuffered in Fig. 7.52a or buffered in Fig. 7.52b) consisted of a pH 3.9 donor solution, a membrane, and a phosphate buffered acceptor solution. The flux (10^{-8} mol cm^{-2} s^{-1} units) was measured to be 0.09 in the unbuffered solution and 3.9 in the buffered solution. The buffer attenuates the pH gradient in the donor-side unstirred water layer and causes the pH at the donor-side surface of the membrane to be 4.81, (Fig. 7.52a) compared to pH 7.44 (Fig. 7.52b) in the unbuffered donor solution. With the lower pH, the fraction of uncharged salicylic acid at the membrane–water interface is higher, and so transport is increased (43 times), over the condition of the unbuffered solution.

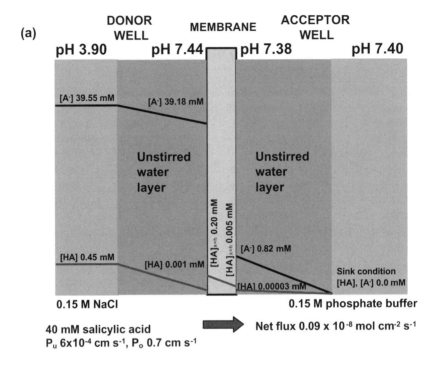

(a)

DONOR WELL   MEMBRANE   ACCEPTOR WELL
pH 3.90   pH 7.44   pH 7.38   pH 7.40

[A⁻] 39.55 mM   [A⁻] 39.18 mM

Unstirred water layer   Unstirred water layer

$[HA]_{x=h}$ 0.20 mM
$[HA]_{x=h}$ 0.005 mM

[A⁻] 0.82 mM

[HA] 0.45 mM   [HA] 0.001 mM

Sink condition
[HA], [A⁻] 0.0 mM

[HA] 0.00003 mM

0.15 M NaCl   0.15 M phosphate buffer

40 mM salicylic acid   Net flux 0.09 × 10⁻⁸ mol cm⁻² s⁻¹
P_u 6×10⁻⁴ cm s⁻¹, P_o 0.7 cm s⁻¹

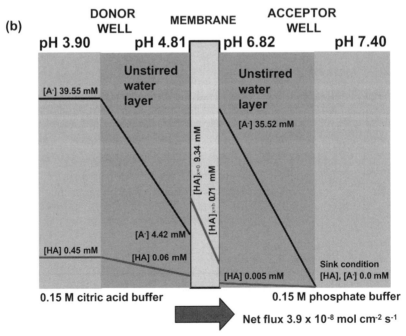

(b)

DONOR WELL   MEMBRANE   ACCEPTOR WELL
pH 3.90   pH 4.81   pH 6.82   pH 7.40

[A⁻] 39.55 mM

Unstirred water layer   Unstirred water layer

[A⁻] 35.52 mM

$[HA]_{x=0}$ 9.34 mM
$[HA]_{x=h}$ 0.71 mM

[A⁻] 4.42 mM
[HA] 0.45 mM   [HA] 0.06 mM

Sink condition
[HA], [A⁻] 0.0 mM

[HA] 0.005 mM

0.15 M citric acid buffer   0.15 M phosphate buffer

Net flux 3.9 × 10⁻⁸ mol cm⁻² s⁻¹

Figure 7.52 Effect of buffer on flux.

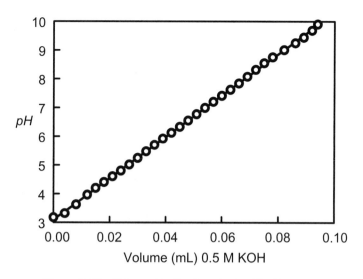

Figure 7.53 Universal buffer for robotic pH adjustment.

Antonenko et al. [540] considered pH gradients forming in the UWL under bulk solution iso-pH conditions. They elegantly expanded on the buffer effect model and made it more general by considering multicomponent buffer mixtures. Direct measurements of the pH gradients (using wire-coated micro-pH electrodes) near the membrane-water interface were described.

A 10 mM ionic strength universal buffer mixture, consisting of Good zwitterionic buffers, [174] and other components (but free of phosphate and boric acid), is used in the pION apparatus [116,556]. The 5-pK_a mixture produces a linear response to the addition of base titrant in the pH 3–10 interval, as indicated in Fig. 7.53. The robotic system uses the universal buffer solution for all applications, automatically adjusting the pH with the addition of a standardized KOH solution. The robotic system uses a built-in titrator to standardize the *pH mapping* operation.

7.7.13 Effects of Stirring

Stirring the permeation cell solution increases the effective permeability, by decreasing the thickness of the UWL (Section 7.7.6.3). Since the PAMPA sandwich (Fig. 7.9) has no airgaps in the bottom wells, and since the solution volumes are small (200–400 µL), the use of rotary-motion platforms to stir the plate is not very effective. Avdeef et al. [556] described the effects of stirring up to speeds of 600 rpm, and noted that the stirring efficiency is about 4 times greater along the periphery of the plate compared to the center locations. This is demonstrated in Fig. 7.54 by 96-replicate verapamil permeability measurements in a plate stirred at 500 rpm. The use of individual magnetic stir bars in each bottom well is a more effective way to stir the solutions. This is currently being developed at pION.

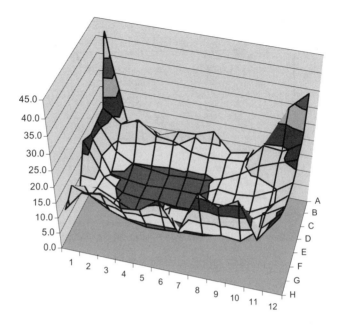

Figure 7.54 Effect of stirring: verapamil permeabilities (in units of 10^{-6} cm/s) in 96 replicates, orbital shaker at 500 rpm.

7.7.14   Errors in PAMPA: Intraplate and Interplate Reproducibility

Figure 7.55 shows a plot of over 2000 2%DOPC/dodecane P_e measurements (10^{-6} cm/s units), each representing at least three intra-plate replicates, vs. the estimated standard deviations, $\sigma(P_e)$. Over 200 different drug-like compounds were measured. The %CV (coefficient of variation $100 \times \sigma(P_e)/P_e$) is about 10% near $P_e 10 \times 10^{-6}$ cm/s, and slightly increases for higher values of permeability, but rapidly increases for $P_e < 0.1 \times 10^{-6}$ cm/s, as shown in Table 7.21. These statistics accurately reflect the errors that should be expected in general. For some molecules, such as caffeine and metoprolol, %CV has been typically about 3–6%.

The errors mentioned above represent the reproducibility obtained on the same microtiter plate when the sample molecule is assayed in several different wells. When the reproducibility of P_e measurement is assessed on the basis of assays performed at different times over a long period of time, more systematic sources of errors show up, and the reproducibility can be about 2–3 times worse. Figure 7.56 shows reproducibility of standard compounds taken over a period of about 12 months. Carbamazepine show a long-term reproducibility error of ~15%. The other compounds show somewhat higher errors.

Considering that PAMPA is a high-throughput screening method, the errors are low enough to encourage the use of the method to study mechanistic properties, as our group at *p*ION has done since 1997.

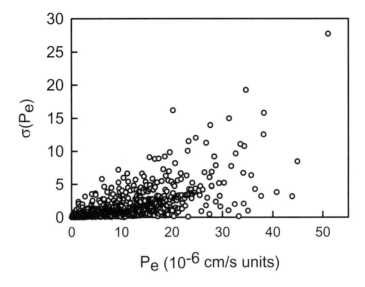

Figure 7.55 Intraplate errors in PAMPA measurement in 2% DOPC model.

7.7.15 UV Spectral Data

The use of direct UV spectrophotometry to measure sample concentrations in pharmaceutical research is uncommon, presumably because of the prevalence and attractiveness of HPLC and LC/MS methods. Consequently, most researchers are unfamiliar with how useful direct UV can be. The UV method is much faster than the other methods, and this is very important in high-throughput screening.

If samples are highly impure or decompose readily, the UV method is inappropriate to use. LC/MS has been demonstrated to be a suitable detection system

TABLE 7.21   Approximate Intraplate Errors in PAMPA Measurement[a]

$P_e(10^{-6}\,\text{cm/s})$	%CV
<0.01	>100%
0.1	60%
0.5	25%
1	15%
5	10%
10	10%
20	15%
30	20%
50	25%

[a]Based on $\sim$6000 measurements of >200 different compounds using the 2% DOPC/dodecane (model 1.0) PAMPA system.

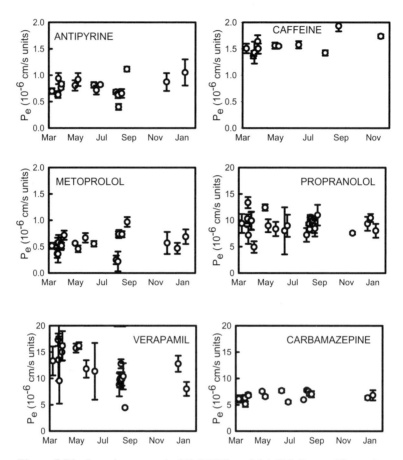

Figure 7.56   Interplate errors in 2% DOPC model (pH 7.4) over 12 months.

under those conditions [556]. When used carefully, LC/MS produces excellent results. However, when LC/MS data-taking is driven very rapidly (e.g., 20 min/plate), disappointing results have been noted in collaborative studies [data not shown].

Figures 7.57a–c show the acceptor, donor, and reference spectra of 48 µM propranolol at the end of 15 h PAMPA assay using 20% wt/vol soy lecithin in dodecane. The sum of the donor (3 µM) and the acceptor (<1 µM) well concentrations indicates that 45 µM is lost to the membrane. In the absence of sink-creating surfactant, only a trace of propranolol reached the acceptor wells at the end of 15 h, with 94% of the compound trapped in the membrane, compared to 19% in the 2% wt/vol DOPC case (Table 7.5). The effective permeability in 20% soy dropped to 1.8×10^{-6} cm/s, compared to the DOPC value of 10.2×10^{-6} cm/s.

With surfactant-created sink condition in the acceptor compartment, the amount of propranolol reaching the acceptor wells is dramatically increased (Fig. 7.57d),

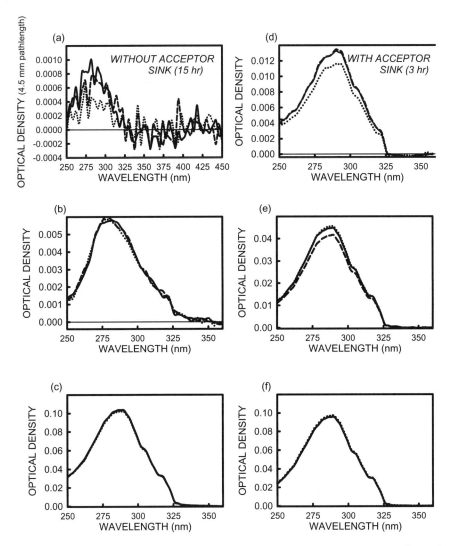

Figure 7.57 UV spectra of propranolol: (a,d) acceptor wells; (b,e) donor wells; (c,f) reference wells (pH 7.4, 47 µm).

with the concomitant decrease in membrane retention from 94% to 41%. Furthermore, the effective permeability rises to 25.1×10^{-6} cm/s, more than a 10-fold increase, presumably due to the desorption effect of the surfactant. Only 3 h permeation time was used in the case (Figs. 7.57d–f). With such a sink at work, one can lower the permeation time to less than 2 h and still obtain very useful UV spectra. This is good for high-throughput requirements.

Figure 7.57a shows that reproducible absorbances can be measured with optical density (OD) values as low as 0.0008, based on a spectrophotometric pathlength of 0.45 cm. The baseline noise (OD in the range 350–500 nm in Fig. 7.57a) is estimated to be about ± 0.0002 OD units.

7.8 THE OPTIMIZED PAMPA MODEL FOR THE GUT

7.8.1 Components of the Ideal GIT Model

The examination of over 50 PAMPA lipid models has led to an optimized model for gastrointestinal tract (GIT) absorption. Table 7.22 shows six properties of the GIT, which distinguish it from the blood–brain barrier (BBB) environment.

1. The in vitro measurements of permeability by the cultured-cell or PAMPA model underestimate true membrane permeability, because of the UWL, which ranges in thickness from 1500 to 2500 μm. The corresponding in vivo value is 30–100 μm in the GIT and nil in the BBB (Table 7.22). The consequence of this is that highly permeable molecules are (aqueous) diffusion limited in the in vitro assays, whereas the membrane-limited permeation is operative in the in vivo case. Correcting the in vitro data for the UWL effect is important for both GIT and BBB absorption modeling.

2. The in vivo environment of the GIT is characterized by a pH gradient; the pH value is constant at 7.4 in the receiving compartment (blood), and varying in the donor compartment (lumen) from ~ 5 to ~ 8 from the start to the end of the small intestine. In contrast, the BBB has a constant iso-pH 7.4. Modeling the two environments requires proper pH adjustment in the in vitro model, as indicated in Table 7.22.

3. The receiver compartment in the GIT has a strong sink condition, effected by serum proteins. In contrast, the BBB does not have a strong sink condition. In the GIT, lipophilic molecules are swept away from the site of absorption; in

TABLE 7.22 In Vitro Double-Sink PAMPA Models for GIT and BBB Targets

	In Vivo GIT	In Vivo BBB	In Vitro Double-Sink GIT Model (20% Soy Lecithin)	In Vitro Double-Sink BBB Model (20% Soy Lecithin)
Unstirred water layer (μm)	30–100	0	2300→30 (corr.)	2300→0 (corr.)
pH donor/receiver	5–8 /7.4	7.4/7.4	5.0/7.4, 6.2/7.4, 7.4/7.4	7.4/7.4
Receiver sink	Yes	No	Yes	No
Mixed micelles in lumen[a]	Yes	No	Yes	No
Negative-charge lipids (% wt/wt)	13	27	16	16
Cholesterol + triglycerides + cholesterol ester (% wt/wt)	37	27	37	37

[a]Proposed simulated intestinal fluid containing fasted-state mixed micelle, 3 mM sodium taurocholate + 0.75 mM lecithin, or fed-state mixed micelle, 15 mM sodium taurocholate +3.75 mM lecithin [61].

the brain, lipophilic molecules accumulate in the endothelial cells. Consequently, the in vitro GIT model calls for a sink condition; the BBB model does not.

4. Highly insoluble molecules are in part transported in the GIT by partitioning into the mixed micelles injected into the lumen from the biliary duct in the duodenum (Fig. 2.3). Mixed micelles consist of a 4 : 1 mixture of bile salts and phospholipids (Fig. 7.13). In contrast, at the point of absorption in the BBB, highly insoluble molecules are transported by serum proteins. This distinction is expected to be important in in vitro assay modeling. The use of simulated intestinal fluids is appealing.

5. The GIT has about 13% wt/wt negatively charged lipid-to-zwitterionic phospholipid ratio. It is about twice as large in the BBB. Factoring this into the in vitro model is expected to be important.

6. The white fat content of the GIT is higher than that of the BBB. The use of triglycerides and cholesterol in in vitro modeling seems important.

The strategy for the development of the oral absorption model at pION is illustrated in Fig. 7.58. The human jejunal permeabilities reported by Winiwarter et al. [56] were selected as the *in vivo* target to simulate by the in vitro model. In particular, three acids, three bases and two nonionized molecules studied by the University of Uppsala group were selected as probes, as shown in Fig. 7.58. They are listed in the descending order of permeabilities in Fig. 7.58. Most peculiar in the ordering is that naproxen, ketoprofen, and piroxicam are at the top of the list, yet these three acids are ionized under *in vivo* pH conditions and have lipophilicity $(\log K_d)$ values near or below zero. The most lipophilic molecules tested, verapamil and carbamazepine

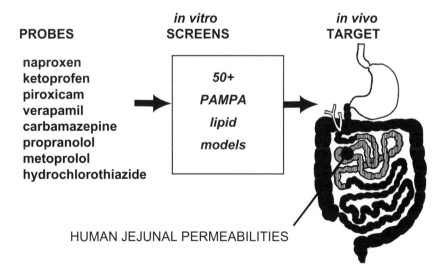

Figure 7.58   Strategy for oral absorption model (from Winiwarter et al. [56]).

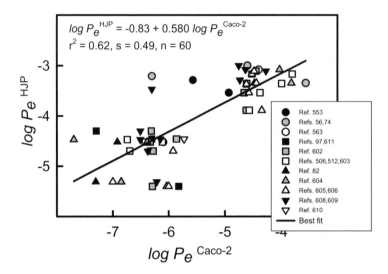

Figure 7.59 Human jejunal permeabilities compared to Caco-2 permeabilities from several groups.

($\log K_d \sim 2.5$; cf. Table 7.4), are in second rank ordering. We took it as a challenge to explain these anomalies in our optimized in vitro GIT model.

As Fig. 7.58 indicates, our task was to explain the ordering of the eight probe molecules in the human in vivo target, but subjecting the eight probe molecules to each of the 50 PAMPA lipid models. For each PAMPA model, the regression correlation coefficient, r^2, was used to assess the appropriateness of the model.

7.8.2 How Well Do Caco-2 Permeability Measurements Predict Human Jejunal Permeabilities?

Since the widely accepted in vitro permeability model in the pharmaceutical industry is based on the use of cultured cells, such as Caco-2 or MDCK, it was appropriate to analyze the regression correlation coefficients based on the comparisons of Caco-2 $\log P_e$ and the $\log P_e$ values based on the human jejunal measurements [56].

Figure 7.59 shows a plot of $\log P_e^{HJP}$ (human jejunal permeabilities) vs. $\log P_e^{Caco-2}$ taken from the literature, based on the work of more than 11 laboratories. The r^2 for the correlation is 0.62. It is clear from the plot that some laboratories better predicted the HJP than other laboratories. Figure 7.60 shows the plot of the results published by Artursson's group [506,512,603], where r^2 was calculated as 0.95, the most impressive value of all the comparisons. It is noteworthy that naproxen, ketoprofen, and piroxicam were not available for the comparison in the Fig. 7.60 plot.

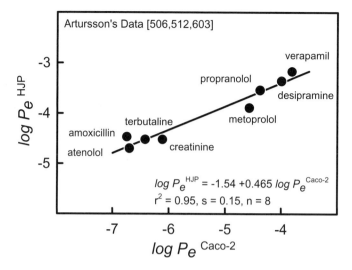

Figure 7.60 Human jejunal permeabilities compared to Caco-2 permeabilities from Artursson's group.

7.8.3 How Well Do PAMPA Measurements Predict the Human Jejunal Permeabilities?

Table 7.23 shows the results for 47 specific PAMPA models tested at pION, according the the scheme in Fig. 7.58. The two columns on the right are the r^2 values in the comparisons. The neutral-lipid models (1.0, 1A.0, 2.0, 3.0, and 4.0) at pH 7.4 do not explain the permeability trend indicated in the human jejunal permeabilities [56]. Octanol was least effective, with r^2 0.01. This should not be too surprising, since we did note that the appearance of naproxen, ketoprofen, and piroxicam at the top of the HJP ordering was unexpected. Our "expectations" were based on the octanol–water lipophilicity scale, which clearly does not correlate with the HJP trend. Adding a sink condition to the 2% DOPC model (model 1.1) improves correlation (r^2 increases from 0.33 to 0.53). The addition of cholesterol to the 2% DOPC/dodecane system made the model unstable to the surfactant-created sink condition.

Introducing negative-charge phospholipids to the 2% DOPC at pH 7.4 improved the correlations significantly (models 5.0, 6.0, 7.0, 8.0, 9.0, 10.0). Sink conditions only marginally improved the correlations for the dodecylcarboxylic acid (1.1% DA) and phosphatidic acid (0.6% PA) models (models 6.1 and 7.1). The phosphatidylglycerol (PG) models (models 9.1 and 10.1) did not correlate well under sink conditions. The modified Chugai model at pH 7.4 performed well (r^2 0.60), but was unstable under sink conditions.

Several egg lecithin models were tested at pH 7.4. The Avanti egg lecithin behaved differently from the Sigma-Aldrich egg lecithin, and was unstable under sink conditions when cholesterol was added. The correlation coefficients were

TABLE 7.23   Correlation (r^2) between Human Jejunal and PAMPA Permeabilities

No.	Type	Composition	pH_{DON}/pH_{ACC}	No Sink	With Sink
1	Neutral	2% DOPC	7.4	0.33	0.53
1A		2% DOPC + 0.5% Cho	7.4	0.61	(Turbid)
2		100% olive oil	7.4	0.36	—
3		100% octanol	7.4	0.01	—
4		100% dodecane	7.4	0.32	—
5	2-Component anionic	2% DOPC + 0.6% DA	7.4	0.58	0.53
6		2% DOPC + 1.1% DA	7.4	0.53	0.61
7		2% DOPC + 0.6% PA	7.4	0.60	0.61
8		2% DOPC + 1.1% PA	7.4	0.52	0.33
9		2% DOPC + 0.6% PG	7.4	0.55	0.10
10		2% DOPC + 1.1% PG	7.4	0.79	0.25
11	5-Component anionic	0.8% PC + 0.8% PE + 0.2% PS + 0.2% PI + 1% Cho	7.4	0.60	(Turbid)
12	Lecithin extracts anionic	10% eggPC (Avanti)	7.4	0.47	0.22
13		10% eggPC (Avanti) + 0.5% Cho	7.4	0.60	(Turbid)
14		10% eggPC (Sigma)	7.4	0.65	0.17
15		10% eggPC (Sigma) + 0.5% Cho	7.4	0.58	0.57
16		10% soyPC	7.4	0.62	0.48
17		20% soyPC	7.4	0.65	0.55
18		20% soyPC + 0.5% Cho	7.4	0.56	0.63
19		35% soyPC	7.4	0.58	0.42
20		50% soyPC	7.4	—	0.36
21		68% soyPC	7.4	0.29	—
22		74% soyPC	7.4	—	0.04
24	Iso-pH	20% soyPC	6.5 / 6.5	—	0.77
25		20% soyPC	5.0 / 5.0	—	0.86
26	Gradient–pH	20% soyPC	6.5 / 7.4	—	0.52
27	(Corrected UWL)	20% soyPC	6.0 / 7.4	—	0.72
28		20% soyPC	5.5 / 7.4	—	0.89
29		20% soyPC	5.0 / 7.4	—	0.97
30		20% soyPC	4.5 / 7.4	—	0.95

slightly better with the Sigma-Aldrich source of egg lecithin. In all cases, the sink condition caused the correlation coefficients to be lower at pH 7.4.

The soy lecithin models (Avanti) were tested most thoroughly at pH 7.4. Figure 7.61 shows the plot of r^2 versus the amount of soy lecithin dissolved in dodecane, from 10% to 74% wt/vol, with and without acceptor sink condition. In the plot, the maximum r^2 was achieved at about 20% wt/vol. The sink condition depressed the (dashed) curve by about 0.15 in r^2 at pH 7.4. The 20% soy lecithin formulation was selected for all subsequent testing.

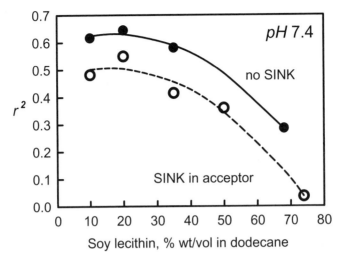

Figure 7.61 Correlation between human jejunal permeabilities and soy lecithin models (in dodecane) at pH 7.4.

Considerable improvements were achieved when iso-pH solutions were tested, at pH 6.5 and 5.0 (Table 7.23). At pH 5, r^2 reached 0.86.

The best correlations were observed under gradient pH and sink conditions ("double-sink" set at the bottom of Table 7.23), with the donor pH 5 and acceptor pH 7.4 producing r^2 0.97. The r^2/donor pH plot is shown in Fig. 7.62. The data represented by the solid line corresponds to P_m values (P_e corrected for the

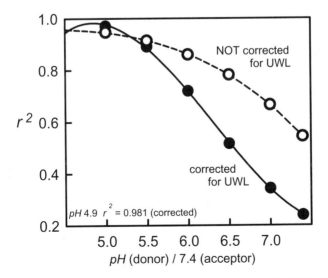

Figure 7.62 Correlation between human jejunal permeabilities [vs. PAMPA (double-sink)] and soy lecithin models under gradient–pH conditions.

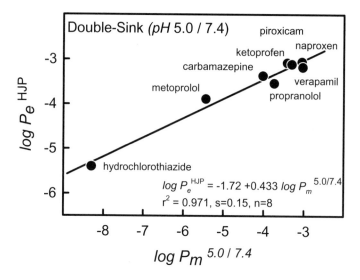

Figure 7.63 Human jejunal permeabilities compared to *p*ION's double-sink sum-P_e PAMPA GIT model.

UWL), and the data represented by the dashed line corresponds to P_e values (uncorrected for the UWL, r^2 data not shown in Table 7.23).

The *p*ION double-sink GIT model, with donor pH 5, predicts the human jejunal permeabilities as well as the best reported Caco-2 model (Artursson's), and a lot better than the rest of the reported Caco-2 models, as shown in Fig. 7.63.

7.8.4 Caco-2 Models for Prediction of Human Intestinal Absorption (HIA)

The strategy of the preceding sections was based on predicting the permeabilities of drug compounds in the human jejunum. The rest of the intestinal tract has higher pH, and this needs to be factored in when considering models to predict *not* human permeabilities, but human *absorption* (see Fig. 2.3 and Table 7.2).

Caco-2 permeabilities have been used to predict human intestinal absorption (HIA) in the literature. Figure 7.64 is a plot of %HIA versus log P_e^{Caco-2}, drawing on the published work of about a dozen laboratories. The plot in Fig. 7.64 resembles "rain," and perhaps very little can be learned from such a plot. This may be an example of what Lipinski [1] pointed out as the consequences of using multimechanistic ADME measurements—the more data points are brought in, the worse the plot looks. Another way of looking at this is that each laboratory has a somewhat differently expressed Caco-2 line, and interlaboratory comparisons can only be done in a rank-order sense. When individual-laboratory data are examined, some groups have better correlations than others. Figure 7.65 shows the results from

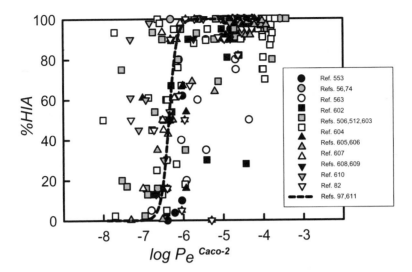

Figure 7.64   Human intestinal absorption compared to Caco-2 permeabilities from several groups.

Yazdanian's group [602], which seem to be marginally better than those of most of the other groups. Griseofulvin is a false positive outlier, which can be rationalized by recognizing that very low solubility of the molecule may be responsible for the low HIA value. The other outlier is nadolol, which has good aqueous

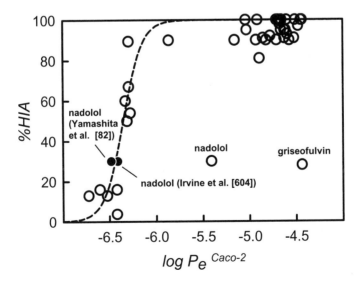

Figure 7.65   Human intestinal absorption compared to Caco-2 permeabilities from Yazdanian's group [602].

solubility. The results from Irvine et al. [604] and Yamashita et al. [82] place nado-lol on the best-fit curve (dashed line). The pharmacokinetic data indicates low HIA for nadolol, possibly due to P-gp efflux attenuating absorption. The Caco-2 result from Yazdanian's group may be high because of the use of high drug concentrations, enough to saturate the P-gp transport in Caco-2. There appears to be no consensus on what sample concentrations to use in Caco-2 assays, and every laboratory appears to have slightly different protocols, when it comes to Caco-2 measurements.

7.8.5 Novartis max-P_e PAMPA Model for Prediction of Human Intestinal Absorption (HIA)

The PAMPA strategy to predict HIA is based on recognizing that gradient pH conditions need to be incorporated into the in vitro models, and that the donor pH values must reflect the properties of the entire GIT (Table 7.2 and Fig. 2.3). Weak acids ought to be better absorbed in the jejunum, where the pH is well below 7.4. However, at the low pH, weak bases may not be well absorbed, since they are positively charged. In the ileum, where the pH may be as high as 8, the absorption of weak bases should be higher than that of weak acids, since the fraction of uncharged form of the bases will be higher at pH 8, compared to pH 5.

In a screening application, where the acid–base properties of discovery molecules may not be certain, it is necessary to screen at least at two pH values, to reflect the conditions of the small intestine. The higher of the two measured permeabilities can then be used to predict HIA. For example, if pH 5 and 7.4 were the two pH values in the PAMPA assay, a weak acid may show very high P_e at pH 5 but a very low P_e at pH 7.4. A single-pH assay at pH 7.4 may have classed the weak acid as a negative, whereas its absorption may have been excellent in the jejunum (pH 5), but this would not have been recognized in the single-pH assay. If a two-pH PAMPA assay is used, then the selection of the maximum P_e of the two measured values would avoid the case of false negatives. This strategy was recognized by Avdeef [26], Faller and Wohnsland [509,554], and Zhu et al. [549]. Figure 7.66 shows the plot of percent absorption versus PAMPA %flux [509]. Figure 7.66a shows the values were taken from just a single pH 6.8; Fig. 7.66b shows the correlation when the max-P_e value is selected from the range pH 5–8. The 50–70% absorption region, shows improvement in the max-P_e model (Fig. 7.66b).

7.8.6 pION Sum-P_e PAMPA Model for Prediction of Human Intestinal Absorption (HIA)

The preceding section can be further generalized, to properly account for absorption of nonionized molecules. The selection of the maximum P_e for HIA prediction implicitly recognized that only a fraction of the small intestine is available for the maximum absorption of acids (with pK_a near 4) and bases (with pK_a near 9). But when this approach is applied to nonionizable molecules, then the absorption may be underestimated, since absorption should be uniform across the whole intestinal tract. The remedy is to sum the two P_e values. This is roughly equivalent to

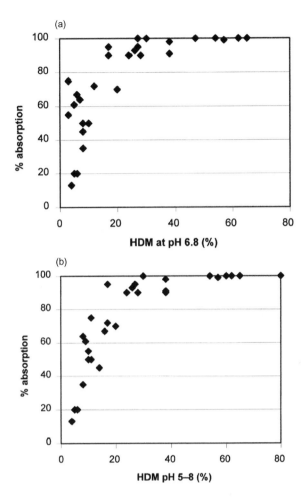

Figure 7.66 Novartis' max-P_e PAMPA model [509]. [Reprinted from Faller, B.; Wohnsland, F., in Testa, B.; van de Waterbeemd, H.; Folkers, G.; Guy, R. (Eds.). *Pharmacokinetic Optimization in Drug Research*, Verlag Helvetica Chimica Acta: Zürich and Wiley-VCH: Weinheim, 2001, pp. 257–274. With permission from Verlag Helvetica Chimica Acta AG.]

integrating a system with parallel absorption taking place in different parts of the intestine. Our preference is to perform PAMPA assay at three gradient pH conditions, with acceptor sink included (double-sink method): donor pH 5, 6.2, and 7.4, with acceptor pH always at 7.4. Figure 7.67a shows such a double-sink sum-P_m (P_e data corrected for the UWL) plot. Figure 7.67b shows the plot of $\log P_e^{\text{HJP}}$ versus %HIA—human permeability data attempting to predict human absorption. As can be seen, the PAMPA data and the HJP data perform equally and tolerably well. One is a lot cheaper to do than the other! Of particular note is that the PAMPA scale covers nearly eight orders of magnitude, compared to

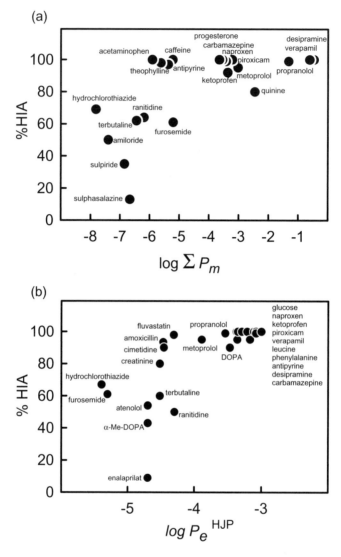

Figure 7.67 Human intestinal absorption compared to (a) *p*ION's double-sink sum-P_e PAMPA GIT model and (b) human jejunal permeabilities [56].

about two and a half orders for the HJP data. Such a spread in the PAMPA data could facilitate the selection of well-absorbed molecules from those poorly absorbed.

In conclusion, the double-sink sum-P_e PAMPA in vitro GIT assay seems to predict human absorption as well as in vivo human permeability measurements (see Figs. 7.66a,b) and in vitro Caco-2 permeability measurements (see Figs. 7.60 and 7.63), but at a lower cost and higher speed.

CHAPTER 8

SUMMARY AND SOME SIMPLE RULES

We began Chapter 2 with a simple Fick's law of diffusion model for absorption, with the key components: permeability, solubility, and charge state (the pH effect). The BCS scheme is more or less constructed along these lines. Closely related to permeability are partitioning in the well-trodden octanol–water and in the lesser-traveled liposome–water systems. We carefully examined the recent literature, with a focus on describing experimental methods which can yield high-quality results, including fast methods. Sometimes forgotten classic works were also revisited. The "it is not just a number" idea was drilled thoroughly with the tetrad–equilibria speciation diagrams for octanol, liposomes, and solubility. The log–log plots having $(0, \pm 1)$ slopes were evoked in several places, to relate the true pK_a to the apparent pK_a and learn something about the "apparency." Out of these efforts emerged the practical concepts underlying pK_a^{oct}, pK_a^{mem}, pK_a^{gibbs}, and pK_a^{flux}.

The charge-state section highlighted the value of Bjerrum plots, with applications to 6- and a 30-pK_a molecules. Water-miscible cosolvents were used to identify acids and bases by the slope in the apparent pK_a/wt% cosolvent plots. It was suggested that extrapolation of the apparent constants to 100% methanol could indicate the pK_a values of amphiphilic molecules embedded in phospholipid bilayers, a way to estimate pK_a^{mem} using the dielectric effect.

Using such dielectric-based predictions, when the methanol-apparent solubility, log S_0^ϵ versus wt% methanol is extrapolated to 0% cosolvent, the aqueous solubility, log S_0, can be estimated; when log S_0^ϵ is extrapolated to 100% cosolvent, the membrane solubility, log S_0^{mem}, can be estimated. The approximate membrane partition

Absorption and Drug Development: Solubility, Permeability, and Charge State. By Alex Avdeef
ISBN 0-471-423653. Copyright © 2003 John Wiley & Sons, Inc.

coefficient can be calculated from the difference between the two solubilities: $\log P_{mem}^N = \log S_0^{mem} - \log S_0$, a concept we called the "solubility-partition unification." Very little of this kind of prediction has been reported elsewhere.

Ion pair partitioning in octanol–water was carefully reviewed. The 'parabola vs. step' shape log D plots of peptides should no longer be subjects of controversy. But "Ion-pairing, fact or fiction?" needs to be further explored. The significance of the partitioning behavior of quaternary ammonium drugs in octanol–water is not entirely resolved. For example, when the permeability of warfarin across *phospholipid*-impregnated filters at high pH is measured, no evidence of warfarin in the acceptor compartment is seen. Given the complex structure of octanol and observations of the sodium dependence of the permeability of warfarin at high pH through *octanol*-impregnated filters suggests that such permeability is more a characteristic of octanol than "real" biological membranes. A postulate that orally administered amphiphilic molecules, even charged ones, can enter the bloodstream by going across the epithelial cell barrier "under the skin of the tight junction," as depicted in scheme 3a → 3b → 3c in Fig. 2.7 is worthy of exploration. Not enough is really known of how such amphiphilic molecules can cross the tight junction.

The study of octanol–water ion pair partitioning has suggested the "*diff* 3–4" rule. With it, ion pair partition coefficients can be predicted from knowledge of just the neutral-species log P. With the liposome–water system, the rule slips to "*diff*$_{mem}$ 1–2." Knowing these rules-of-thumb can prevent ill-guided use of equations to convert single-point log D values to log P values. An analogous "*sdiff* 3–4" rule for solubility was proposed. This may help to predict effects of salts in the background of a physiological concentration of NaCl or KCl.

The study of drug partitioning into liposomes has revealed some puzzling observations, in terms of the δ parameter. Why does acyclovir have such a high liposome–water log P and such a low octanol-water log P? Are such anomalies observed in IAM chromatography? The review of the literature hints that the high liposome–water log P values indicate a surface phenomenon (H-bonding, enthalpy-driven) that attenuates or even prevents membrane transport. Sometimes, high membrane log P or long retention times in IAM chromatography just means that the molecule is stuck on the membrane surface, and does not permeate. This idea needs to be further explored.

The concept of the Δ shift in HTS solubility measurements is quite exciting. It means that DMSO can be used in solubility measurements and the measured values later corrected to DMSO-free conditions. So we can have speed and accuracy at the same time! The pharmaceutical industry needs speed and accuracy, and will need these more in the future. In silico methods are no better than the data used to train them.

Solubility and dissolution are processes that take place in the gastric and the luminal fluids, not on the surface of epithelial cells. Measurement of solubility ideally needs to take place at pH 1.7 (stomach) and pH 5–8 (small intestinal tract). Ideally, the screen media should resemble intestinal fluids and contain bile acid-lecithin mixed micelles. Fast and reliable techniques for assessing solubility in such environments are available. Industrywide consensus on solubility measurement protocols is needed, so that clinically relevant measurements are produced.

Permeability is a property closely tied to the environment of the epithelial cell surface. There is little point in measuring permeability at pH 1.7, if the microclimate barrier has pH ≥ 5 and ≤ 8, averaging ~ 6. An in vitro permeability screen based on donor pH 5.0–7.4 and acceptor pH 7.4 seems about right. It will be useful to correct the data for the unstirred water layer effect, using computational methods.

Weak acids and bases can be better assessed if the shapes of the flux–pH profiles were considered, as far as predicting the outcome of a particular choice of assay pH.

The lengthy permeability chapter (Chapter 7) recounts the study of many different artificial membrane formulations, comparing transport results of each to human jejunal permeabilities. A very promising in vitro screening system was described: the double-sink sum-P_e PAMPA GIT model. It is most applicable to molecules that are classified as "soluble" in the BCS scheme.

When molecules have the "insoluble" BCS classification, the expected absorption profile is exemplified in Fig. 2.2. The upper horizontal line (solid) in Fig. 2.2, representing log P_0, can be determined by the methods described in Chapter 7. The "slope 0,±1" segments curve (dashed), representing log C_0, the concentration of the uncharged form of an ionizable molecule, can be determined by the methods described in the Chapter 6. The summation of log P_0 and log C_0 curves produces the log flux–pH profile. Such plots indicate under what pH conditions the absorption should be at its highest potential.

The Dressman–Amidon–Fleisher absorption potential concept [45], originally based on octanol–water partition coefficients, can be made more predictive, by using PAMPA permeabilities, instead of partition coefficients, for all the reasons discussed in Chapter 7. Such a scheme can be used to minimize false positive predictions of HIA.

Semiquantitative schemes, like the *maximum absorbable dose* (MAD) system described by Curatolo [53], can be made more predictive by applying solubilities measured by clinically-relevant protocols and PAMPA permeabilities.

The BCS scheme can be made more useful by incorporating a further improved basis of physicochemical profiling. For example, the role of pH in permeability measurements could be better defined. The use of simulated intestinal fluids for solubility measurements could be better promoted. The effects of fed/fasted states on absorption could be better address, in methods that have optimum clinical relevance.

In this book, a conceptually rigorous effort was made to describe the state-of-the-art physical methods that underlie the processes related to absorption. The aim was to give conceptual tools to the analytical chemists in pharmaceutical companies who do such measurements, so that they could in turn convey to the medicinal chemists, who make the molecules, how structural modifications can affect those physical properties that make candidate molecules "drugable."

As Taylor suggested in the introductory chapter, "There are great advances and great opportunities in all this, ..."

REFERENCES

1. Lipinski, C. A., Drug-like properties and the causes of poor solubility and poor permeability, *J. Pharmacol. Tox. Meth.* **44**, 235–249 (2000).
2. Martin, E. J.; Blaney, J. M.; Siani, M. A.; Spellmeyer, D. C.; Wong, A. K.; Moos, W. H., Measuring diversity: Experimental design of combinatorial libraries for drug discovery, *J. Med. Chem.* **38**, 1431–1436 (1995).
3. Drews, J., Drug discovery: a historical perspective, *Science* **287**, 1960–1963 (2000).
4. Pickering, L., Developing drugs to counter disease, *Drug Discov. Dev.* 44–47 (Feb 2001).
5. Dennis, C. et al., The human genome, *Nature* **409**, 813–958 (2001).
6. Handen, J. S., High throughput screening challenges for the future, *Drug. Disc. World* 47–50 (Summer 2002).
7. Testa, B.; Caldwell, J., The "ad hoc" approach as a compliment to ligand design, *Med. Chem. Rev.* **16**, 233–241 (1996).
8. Stenberg, P.; Luthman, K.; Artursson, P., Virtual screening of intestinal drug permbeability, *J. Control. Rel.* **65**, 231–243 (2000).
9. Baum, R. M., Combinatorial approaches provide fresh leads for medicinal chemistry, *Chem. Eng. News* 20–26 (Feb. 7 1994).
10. Halliday, R. G.; Drasco, A. L.; Lumley, C. E.; Walker, S. R., The allocation of resources for R&D in the world's leading pharmaceutical companies, *Res. Dev. Manage.* **27**, 63–77 (1997).
11. Gaviraghi, G.; Barnaby, R. J.; Pellegatti, M., Pharmacokinetic challenges in lead optimization, in Testa, B.; van de Waterbeemd, H.; Folkers, G.; Guy, R. (eds.), *Pharmacokinetic Optimization in Drug Research*, Verlag Helvetica Chimica Acta, Zürich and Wiley-VCH, Weinheim, 2001, pp. 3–14.

Absorption and Drug Development: Solubility, Permeability, and Charge State. By Alex Avdeef
ISBN 0-471-423653. Copyright © 2003 John Wiley & Sons, Inc.

12. Lipinski, C. A.; Lombardo, F.; Dominy, B. W.; Feeney, P. J., Experimental and computational approaches to estimate solubility and permeability in drug discovery and development settings, *Adv. Drug Deliv. Rev.* **23**, 3–25 (1997).

13. Pickering, L., ADME/Tox models can speed development, *Drug Discov. Dev.* 34–36 (Jan. 2001).

14. Kaplita, P. V.; Magolda R. L.; Homon, C. A., HTS: Employing high-throughput automation to better characterize early discovery candidates, *PharmaGenomics* 34–39 (2002).

15. Lipinski, C. A., Drug-like properties and the causes of poor solubility and poor permeability, *J. Pharmacol. Tox. Meth.* **44**, 235–249 (2000).

16. Lipinski, C. A., Avoiding investment in doomed drugs—is solubility an industry wide problem? *Curr. Drug Discov.* 17–19 (April 2001).

17. Taylor, P. J., Hydrophobic properties of drugs, in Hansch, C.; Sammes, P. G.; Taylor, J. B., (eds.), *Comprehensive Medicinal Chemistry*, Vol. 4, Pergamon, Oxford, 1990, pp. 241–294.

18. Fick, A., Ueber diffusion, *Ann. Phys.* **94**, 59–86 (1855).

19. Artursson, P., Application of physicochemical properties of molecules to predict intestinal permeability, AAPS Workshop on Permeability Definitions and Regulatory Standards, Arlington, VA, Aug. 17–19, 1998.

20. van de Waterbeemd, H. Intestinal permeability: Prediction from theory, in Dressman, J. B.; Lennernäs, H. (eds.), *Oral Drug Absorption—Prediction and Assessment*, Marcel Dekker, New York, 2000, pp. 31–49.

21. McFarland, J. W., On the parabolic relationship between drug potency and hydrophobicity, *J. Med. Chem.* **13**, 1192–1196 (1970).

22. Dearden, J. C.; Townsend, M. S., Digital computer simulation of the drug transport process, in *Proc. 2nd Symp. Chemical Structure-Biological Activity Relationships: Quantitative Approaches* (Suhl), Akademie-Verlag, Berlin, 1978, pp. 387–393.

23. Kubinyi, H., Lipophilicity and biological activity, *Arzneim.-Forsch./Drug Res.* **29**, 1067–1080 (1979).

24. Kubinyi, H., Strategies and recent technologies in drug discovery, *Pharmazie* **50**, 647–662 (1995).

25. Avdeef, A., Physicochemical profiling (solubility, permeability, and charge state), *Curr. Top. Med. Chem.* **1**, 277-351 (2001).

26. Avdeef, A., High-throughput measurements of solubility profiles, in Testa, B.; van de Waterbeemd, H.; Folkers, G.; Guy, R. (eds.), *Pharmacokinetic Optimization in Drug Research*, Verlag Helvetica Chimica Acta; Zürich and Wiley-VCH, Weinheim, 2001, pp. 305–326.

27. Avdeef, A., High-throughput measurements of permeability profiles, in van de Waterbeemd, H.; Lennernäs, H.; Artursson, P. (eds.), *Drug Bioavailability. Estimation of Solubility, Permeability, Absorption and Bioavailability* Wiley-VCH: Weinheim, 2003 (in press)

28. Avdeef, A.; Testa, B., Physicochemical profiling in drug research: A brief state-of-the-art of experimental techniques, *Cell. Mol. Life Sci.* 2002 (in press).

29. Kerns, E. H., High throughput physicochemical profiling for drug discovery, *J. Pharm. Sci.* **90**, 1838–1858 (2001).

30. Kerns, E. H.; Di, L., Multivariate pharmaceutical profiling for drug discovery, *Curr. Top. Med. Chem.* **2**, 87–98 (2002).

31. Roberts, S. A., High-throughput screening approaches for investigating drug metabolism and pharmacokinetics, *Xenobiotica* **31**, 557–589 (2001).

32. van de Waterbeemd, H.; Smith, D. A.; Jones, B. C., Lipophilicity in PK design: Methyl, ethyl, futile, *J. Comp.-Aided Mol. Design* **15**, 273–286 (2001).

33. van de Waterbeemd, H.; Smith, D. A.; Beaumont, K.; Walker, D. K., Property-based design: optimization of drug absorption and pharmacokinetics, *J. Med. Chem.* **44**, 1313–1333 (2001).

34. Perrin, D. D.; Dempsey, B.; Serjeant, E. P., *pKa Prediction for Organic Acids and Bases*, Chapman & Hall, London, 1981.

35. Albert, A.; Serjeant, E. P., *The Determination of Ionization Constants*, 3rd ed., Chapman & Hall, London, 1984.

36. Sangster, J., *Octanol-Water Partition Coefficients: Fundamentals and Physical Chemistry*, Wiley Series in Solution Chemistry, Vol. 2, Wiley, Chichester, UK, 1997.

37. Grant, D. J. W.; Higuchi, T., *Solutility Behavior of Organic Compounds*, Wiley, New York, 1990.

38. Yalkowsky, S. H.; Banerjee, S., *Aqueous Solubility: Methods of Estimation for Organic Compounds*, Marcel Dekker, New York, 1992.

39. Streng, W. H., *Characterization of Compounds in Solution—Theory and Practice*, Kluwer Academic/Plenum Publishers, New York, 2001.

40. Flynn, G. L.; Yalkowsky, S. H.; Roseman, T. J., Mass transport phenomena and models: Theoretical concepts, *J. Pharm. Sci.* **63**, 479–510 (1974).

41. Weiss, T. F., *Cellular Biophysics*, Vol. I, *Transport*, MIT Press, Cambridge, MA, 1996.

42. Dawson, D. C., Principles of membrane transport, in Field, M.; Frizzell, R. A. (eds.), *Handbook of Physiology*, Sect. 6: The gastrointestinal system, Vol. IV, *Intestinal Absorption and Secretion*, Am. Physiol. Soc., Bethesda, MD, 1991, pp. 1–44.

43. Wells, J. I., *Pharmaceutical Preformulation: The Physicochemical Properties of Drug Substances*, Ellis Horwood, Chichester, UK, 1988.

44. Schanker, L. S.; Tocco, D. J.; Brodie, B. B.; Hogben, C. A. M., Absorption of drugs from the rat small intestine, *J. Am. Chem. Soc.* **123**, 81–88 (1958).

45. Dressman, J. B.; Amidon, G. L.; Fleisher, D., Absorption potential: Estimating the fraction absorbed for orally administered compounds, *J. Pharm. Sci.* **74**, 588–589 (1985).

46. Seydel, J. K.; Coats, E. A.; Cordes, H. P.; Weise, M., Drug membrane interactions and the importance for drug transport, distribution, accumulation, efficacy and resistance, *Arch. Pharm.* (Weinheim) **327**, 601–610 (1994).

47. Chan, O. H.; Stewart, B. H., Physicochemical and drug-delivery considerations for oral drug bioavailability, *Drug Discov. Today* **1**, 461–465 (1996).

48. Borchardt, R. T.; Smith, P. L.; Wilson, G., *Models for Assessing Drug Absorption and Metabolism*, Plenum Press, New York, 1996.

49. Camenisch, G.; Folkers, G.; van de Waterbeemd, H., Review of theoretical passive drug absorption models: historical background, recent developments and limitations, *Pharm. Acta Helv.* **71**, 309–327 (1996).

50. Grass, G. M., Simulation models to predict oral drug absorption from in vitro data, *Adv. Drug. Del. Rev.* **23**, 199–219 (1997).

51. Stewart, B. H.; Chan, O. H.; Jezyk, N.; Fleisher, D., Discrimination between drug candidates using models from evaluation of intestinal absorption, *Adv. Drug Del. Rev.* **23**, 27–45 (1997).

52. Dowty, M. E.; Dietsch, C. R., Improved prediction of in vivo peroral absorption from in vitro intestinal permeability using an internal standard to control for intra- & inter-rat variability, *Pharm. Res.* **14**, 1792–1797 (1997).

53. Curatolo, W. Physical chemical properties of oral drug candidates in the discovery and exploratory settings, *Pharm. Sci. Tech. Today* **1**, 387–393 (1998).

54. Camenisch, G.; Folkers, G.; van de Waterbeemd, H., Shapes of membrane permeability-lipophilicity curves: Extension of theoretical models with an aqueous pore pathway, *Eur. J. Pharm. Sci.* **6**, 321–329 (1998).

55. Ungell, A.-L.; Nylander, S.; Bergstrand, S.; Sjöberg, Å.; Lennernäs, H., Membrane transport of drugs in different regions of the intestinal tract of the rat, *J. Pharm. Sci.* **87**, 360–366 (1998).

56. Winiwarter, S.; Bonham, N. M.; Ax, F.; Hallberg, A.; Lennernäs, H.; Karlen, A., Correlation of human jejunal permeability (in vivo) of drugs with experimentally and theoretically derived parameters. A multivariate data analysis approach, *J. Med. Chem.* **41**, 4939–4949 (1998).

57. Kristl, A.; Tukker, J. J., Negative correlation of n-octanol/water partition coefficient and transport of some guanine derivatives through rat jejunum in vitro, *Pharm. Res.* **15**, 499–501 (1998).

58. Krämer, S. D., Absorption prediction from physicochemical parameters, *Pharm. Sci. Tech. Today* **2**, 373–380 (1999).

59. Barthe, L.; Woodley, J.; Houin, G. Gastrointestinal absorption of drugs: Methods and studies, *Funda. Clin. Pharmacol.* **13**, 154–168 (1999).

60. Raevsky, O. A.; Fetisov, V. I.; Trepalina, E. P.; McFarland, J. W.; Schaper, K.-J., Quantitative estimation of drug absorption in humans for passively transported compounds on the basis of their physico-chemical parameters, *Quant. Struct.-Act. Relat.* **19**, 366–374 (2000).

61. Dressman, J. B.; Lennernäs, H. (eds.), *Oral Drug Absorption—Prediction and Assessment*, Marcel Dekker, New York, 2000.

62. Wilson, J. P., Surface area of the small intestine in man, *Gut* **8**, 618–621 (1967).

63. Madara, J. L., Functional morphology of epithelium of the small intestine, in Field, M.; Frizzell, R. A. (eds.), *Handbook of Physiology*, Sect. 6: The gastrointestinal system, Vol. IV, *Intestinal Absorption and Secretion*, Am. Physiol. Soc., Bethesda, MD, 1991, pp. 83–120.

64. Dressman, J. B.; Bass, P.; Ritschel, W. A.; Friend, D. R.; Rubinstein, A.; Ziv, E., Gastrointestinal parameters that influence oral medications, *J. Pharm. Sci.* **82**, 857–872 (1993).

65. Gray, V. A.; Dressman, J. B., *Pharm. Forum* **22**, 1943–1945 (1996).

66. Kararli, T. T., Comparative models for studying absorption, AAPS Workshop on Permeability Definitions and Regulatory Standards for Bioequivalence, Arlington, Aug. 17–19, 1998.

67. Charman, W. N.; Porter, C. J.; Mithani, S. D.; Dressman, J. B., The effect of food on drug absorption—a physicochemical and predictive rationale for the role of lipids and pH, *J. Pharm. Sci.* **86**, 269–282 (1997).

68. Dressman, J. B.; Amidon, G. L.; Reppas, C.; Shah, V., Dissolution testing as a prognostic tool for oral drug absorption: Immediate release dosage forms, *Pharm. Res.* **15**, 11–22 (1998).

69. Berne, R. M.; Levy, M. N., *Physiology*, 4th ed., Mosby Yearbook, St. Louis, 1998, pp. 654–661.

70. Rechkemmer, G., Transport of weak electrolytes, in Field, M.; Frizzell, R. A. (eds.), *Handbook of Physiology*, Sect. 6: The gastrointestinal system, Vol. IV, *Intestinal Absorption and Secretion*, Am. Physiol. Soc., Bethesda, MD, 1991, pp. 371–388.

71. Dressman, J. B.; Reppas, C., In vitro–in vivo correlations for lipophilic, poorly water-soluble drugs, *Eur. J. Pharm. Sci.* **11** (Suppl. 2), S73–S80 (2000).

72. Larhed, A. W.; Artursson, P.; Gråsjö, J.; Björk, E., Diffusion of drugs in native and purified gastrointestinal mucus. *J. Pharm. Sci.* **86**, 660–665 (1997).

73. Shiau, Y.-F.; Fernandez, P.; Jackson, M. J.; McMonagle, S., Mechanisms maintaining a low-pH microclimate in the intestine, *Am. J. Physiol.* **248**, G608–G617 (1985).

74. Lennernäs, H., Human intestinal permeability, *J. Pharm. Sci.* **87**, 403–410 (1998).

75. Lutz, K. L.; Siahaan, T. J., Molecular structure of the apical junction complex and its contributions to the paracellular barrier, *J. Pharm. Sci.* **86**, 977–984 (1997).

76. Jackson, M. J.; Tai, C.-Y., Morphological correlates of weak electrolyte transport in the small intestine, in Dinno, M. A. (ed.), *Structure and Function in Epithelia and Membrane Biophysics*, Alan R. Liss, New York, 1981, pp. 83–96.

77. Winne, D., Shift of pH-absorption curves, *J. Pharmacokinet. Biopharm.* **5**, 53–94 (1977).

78. Said, H. M.; Blair, J. A.; Lucas, M. L.; Hilburn, M. E., Intestinal surface acid microclimate in vitro and in vivo in the rat, *J. Lab Clin. Med.* **107**, 420–424 (1986).

79. Jackson. M. J., Drug transport across gastrointestinal epithelia, in Johnson, L. R. (ed.), *Physiology of the Gastrointestinal Tract*, 2nd ed., Raven Press, New York, 1987, pp. 1597–1621.

80. Takagi, M.; Taki, Y.; Sakane, T.; Nadai, T.; Sezaki, H.; Oku, N.; Yamashita, S., A new interpretation of salicylic acid transport across the lipid bilayer: Implication of pH-dependence but not carrier-mediated absorption from the gi tract, *J. Pharmacol. Exp. Therapeut.* **285**, 1175–1180 (1998).

81. Kimura, Y.; Hosoda, Y.; Shima, M.; Adachi, S.; Matsuno, R., Physicochemical properties of fatty acids for assessing the threshold concentration to enhance the absorption of a hydrophilic substance, *Biosci. Biotechnol. Biochem.* **62**, 443–447 (1998).

82. Yamashita, S.; Furubayashi, T.; Kataoka, M.; Sakane, T.; Sezaki, H.; Tokuda, H., Optimized conditions for prediction of intestinal drug permeability using Caco-2 cells, *Eur. J. Pharm. Sci.* **10**, 109–204 (2000).

83. Asokan, A.; Cho, M. J., Exploitation of intracellular pH gradients in the cellular delivery of macromolecules, *J. Pharm. Sci.* **91**, 903–913 (2002).

84. Antonenko, Y. N.; Bulychev, A. A., Measurements of local pH changes near bilayer lipid membrane by means of a pH microelectrode and a protonophore-dependent membrane potential. Comparison of the methods, *Biochim. Biophys. Acta* **1070**, 279–282 (1991).

85. Rojanasakul, Y.; Robinson, J. R., The cytoskeleton of the cornea and its role in tight junction permeability, *Int. J. Pharm.* **68**, 135–149 (1991).

86. Schneeberger, E. E.; Lynch, R. D., Structure, function, and regulation of cellular tight junctions, *Am. J. Physiol.* **262**, L647–L661 (1992).

87. Anderberg, E. K.; Lindmark, T.; Artursson, P., Sodium caprate elicits dilations in human intestinal tight junctions and enhances drug absorption by the paracellular route, *Pharm. Res.* **10**, 857–864 (1993).

88. Bhat, M.; Toledo-Velasquez, D.; Wang, L. Y.; Malanga, C. J.; Ma, J. K. H.; Rojanasakul, Y., Regulation of tight junction permeability by calcium mediators and cell cytoskeleton in rabbit tracheal epithelium, *Pharm. Res.* **10**, 991–997 (1993).

89. Noach, A. B. J. Enhancement of paracellular drug transport across epithelia—in vitro and in vivo studies, *Pharm. World Sci.* **17**, 58–60 (1995).

90. Lutz, K. L.; Jois, S. D. S.; Siahaan, T. J., Secondary structure of the HAV peptide which regulates cadherin-cadherin interaction, *J. Biomol. Struct. Dynam.* **13**, 447–455 (1995).

91. Tanaka, Y.; Taki, Y.; Sakane, T.; Nadai, T.; Sezaki, H.; Yamashita, S., Characterization of drug transport through tight-junctional pathway in Caco-2 monolayer: Comparison with isolated rat jejunum and colon, *Pharm. Res.* **12**, 523–528 (1995).

92. Brayden, D. J.; Creed, E.; Meehan, E.; O'Malley, K. E., Passive transepithelial diltiazem absorption across intestinal tissue leading to tight junction openings, *J. Control. Rel.* **38**, 193–203 (1996).

93. Lutz, K. L.; Szabo, L. A.; Thompson, D. L.; Siahaan, T. J., Antibody recognition of peptide sequence from the cell-cell adhesion proteins: N- and E-cadherins, *Peptide Res.* **9**, 233–239 (1996).

94. Borchard, G.; Lueßen, H. L.; de Boer, A. G.; Verhoef, J. C.; Lehr, C.-M.; Junginger, H. E., The potential of mucoadhesive polymers in enhancing intestinal peptide drug absorption. III: Effects of chitosan-glutamate and carbomer on epithelial tight junctions in vitro, *J. Control. Rel.* **39**, 131–138 (1996).

95. Lutz, K. L.; Siahaan, T. J., Molecular structure of the apical junction complex and its contributions to the paracellular barrier, *J. Pharm. Sci.* **86**, 977–984 (1997).

96. Pal, D.; Audus, K. L.; Siahaan, T. J., Modulation of cellular adhesion in bovine brain microvessel endothelial cells by a decapeptide, *Brain Res.* **747**, 103–113 (1997).

97. Gan, L.-S. L.; Yanni, S.; Thakker, D. R., Modulation of the tight junctions of the Caco-2 cell monolayers by H2-antagonists, *Pharm. Res.* **15**, 53–57 (1998).

98. Hansch, C.; Leo, A., *Substituent Constants for Correlation Analysis in Chemistry and Biology*, Wiley-Interscience, New York, 1979.

99. Iwahashi, M.; Hayashi, Y.; Hachiya, N.; Matsuzawa, H.; Kobayashi, H., Self-association of octan-1-ol in the pure liquid state and in decane solutions as observed by viscosity, self-diffusion, nuclear magnetic resonance and near-infrared spectroscopy measurements, *J. Chem. Soc. Faraday Trans.* **89**, 707–712 (1993).

100. Franks, N. P.; Abraham, M. H.; Lieb, W. R., Molecular organization of liquid n-octanol: An X-ray diffraction analysis, *J. Pharm. Sci.* **82**, 466–470 (1993).

101. Amidon, G. L.; Lennernäs, H.; Shah, V. P.; Crison, J. R., A theoretical basis for a biopharmaceutic drug classification: the correlation of in vitro drug product dissolution and in vivo bioavailability, *Pharm. Res.* **12**, 413–420 (1995).

102. Amidon, G. L., The rationale for a biopharmaceutics drug classification, in *Biopharmaceutics Drug Classification and International Drug Regulation*, Capsugel Library, 1995, pp. 179–194.

103. Amidon, G. L.; Walgreen, C. R., Rationale and implementation of a biopharmaceutics classification system (bcs) for new drug regulation, in *Biopharmaceutics Drug Classification and International Drug Regulation*, Capsugel Library, 1998, pp. 13–27.

104. Hussain, A. S., Methods for permeability determination: A regulatory perspective. AAPS Workshop on Permeability Definitions and Regulatory Standards for Bioequivalence, Arlington, VA, Aug. 17–19, 1998.

105. *CPMP Note for Guidance on the Investigation of Bioavailability and Bioequivalence* (CPMP/EWP/QWP/1401/98 Draft), Dec. 1998.

106. *FDA Guidance for Industry Waiver of in vivo Bioavailability and Bioequivalence Studies for Immediate Release Solid Oral Dosage Forms Containing Certain Active Moieties/ Active Ingredients Based on a Biopharmaceutics Classification System*, CDER-GUID\2062dft.wpd Draft, Jan. 1999.

107. Blume, H. H.; Schug, B. S., The biopharmaceitics classification system (BCS): Class III drugs—better candidates for BA/BE waiver? *Eur. J. Pharm. Sci.* **9**, 117–121 (1999).

108. Lentz, K. A.; Hayashi, J.; Lucisano, L. J.; Polli, J. E., Development of a more rapid, reduced serum culture system for Caco-2 monolayers and application to the biopharmaceutics classification system, *Int. J. Pharm.* **200**, 41–51 (2000).

109. Chen, M.-L.; Shah, V.; Patnaik, R.; Adams, W.; Hussain, A.; Conner, D.; Mehta, M.; Malinowski, H.; Lazor, J.; Huang, S.-M.; Hare, D.; Lesko, L.; Sporn, D.; Williams, R., Bioavailability and bioequivalence: an FDA regulatory overview, *Pharm. Res.* **18**, 1645–1650 (2001).

110. Rege, B. D.; Yu, L. X.; Hussain, A. S.; Polli, J. E., Effect of common excipients on Caco-2 transport of low-permeability drugs, *J. Pharm. Sci.* **90**, 1776–1786 (2001).

111. Ritschel, W. A., in Fraske, D. E.; Whitney, H. A. K., Jr. (eds.), *Perspectives in Clinical Pharmacy*, Drug Intelligence Publications, Hamilton, IL, 1972, pp. 325–367.

112. Avdeef, A., *Applications and Theory Guide to pH-Metric pK_a and log P Measurement*, Sirius Analytical Instruments Ltd. Forest Row, UK, 1993.

113. Levy, R. H.; Rowland, M., Dissociation constants of sparingly soluble substances: nonlogarithmic linear titration curves, *J. Pharm. Sci.* **60**, 1155–1159 (1971).

114. Purdie, N.; Thomson, M. B.; Riemann, N., The thermodynamics of ionization of polycarboxylic acids, *J. Solut. Chem.* **1**, 465–476 (1972).

115. Streng, W. H.; Steward, D. L., Jr., Ionization constants of an amino acid as a function of temperature, *Int. J. Pharm.* **61**, 265–266 (1990).

116. Avdeef, A.; Bucher, J. J., Accurate measurements of the concentration of hydrogen ions with a glass electrode: Calibrations using the Prideaux and other universal buffer solutions and a computer-controlled automatic titrator, *Anal. Chem.* **50**, 2137–2142 (1978).

117. Avdeef, A., Weighting scheme for regression analysis using pH data from acid-base titrations, *Anal. Chim. Acta* **148**, 237–244 (1983).

118. Avdeef, A., STBLTY: Methods for construction and refinement of equilibrium models, in Leggett, D. J. (ed.), *Computational Methods for the Determination of Formation Constants*, Plenum Press, New York, 1985, pp. 355–473.

119. Avdeef, A.; Comer, J. E. A.; Thomson, S. J., pH-metric logP. 3. Glass electrode calibration in methanol-water, applied to pK_a determination of water-insoluble substances, *Anal. Chem.* **65**, 42–49 (1993).

120. Lingane, P. J.; Hugus, Z. Z. Jr., Normal equations for the gaussian least-squares refinement of formation constants with simultaneous adjustment of the spectra of the absorbing species, *Inorg. Chem.* **9**, 757–762 (1970).

121. Hugus, Z. Z., Jr.; El-Awady, A. A., The determination of the number of species present in a system: A new matrix rank treatment of spectrophotometric data, *J. Phys. Chem.* **75**, 2954–2957 (1971).

122. Alcock, R. M.; Hartley, F. R.; Rogers, D. E., A damped non-linear least-squares computer program (dalsfek) for the evaluation of equilibrium constants from spectrophotometric and potentiometric data, *J. Chem. Soc. Dalton Trans.* 115–123 (1978).

123. Maeder, M.; Gampp, H., Spectrophotometric data reduction by eigenvector analysis for equilibrium and kinetic studies and a new method of fitting exponentials, *Anal. Chim. Act* **122**, 303–313 (1980).

124. Gampp, H.; Maeder, M.; Zuberbühler, A. D., General non-linear least-squares program for the numerical treatment of spectrophotometric data on a single-precision game computer, *Talanta* **27**, 1037–1045 (1980).

125. Kralj, Z. I; Simeon, V., Estimation of spectra of individual species in a multisolute solution, *Anal. Chim. Acta* **129**, 191–198 (1982).

126. Gampp, H.; Maeder, M.; Meyer, C. J.; Zuberbühler, A. D., Calculation of equilibrium constants from multiwavelength spectroscopic data. I. Mathematical considerations, *Talanta* **32**, 95–101 (1985).

127. Gampp, H.; Maeder, M.; Meyer, C. J.; Zuberbühler, A. D., Calculation of equilibrium constants from multiwavelength spectroscopic data. III. Model-free analysis of spectro-photometric and ESR titrations. *Talanta* **32**, 1133–1139 (1985).

128. Alibrandi, G., Variable-concentration kinetics, *J. Chem. Soc., Chem. Commun.* 2709–2710 (1994).

129. Alibrandi, G.; Coppolino, S.; Micali, N.; Villari, A., Variable pH kinetics: an easy determination of pH-rate profile, *J. Pharm. Sci.* **90**, 270–274 (2001).

130. Saurina, J.; Hernández-Cassou, S.; Izquierdo-Ridorsa, A.; Tauler, R., pH-gradient spectrophotometric data files from flow-injection and continuous flow systems for two- and three-way data analysis, *Chem. Intell. Lab. Syst.* **50**, 263–271 (2000).

131. Allen, R. I.; Box, K. J.; Comer, J. E. A.; Peake, C.; Tam, K. Y., Multiwavelength spectrophotometric determination of acid dissociation constants of ionizable drugs, *J. Pharm. Biomed. Anal.* **17**, 699–712 (1998).

132. Tam, K. Y.; Takács-Novák, K., Multiwavelength spectrophotometric determination of acid dissociation constants. Part II. First derivative vs. target factor analysis, *Pharm. Rese.* **16**, 374–381 (1999).

133. Mitchell, R. C.; Salter, C. J.; Tam, K. Y., Multiwavelength spectrophotometric determination of acid dissociation constants. Part III. Resolution of multi-protic ionization systems, *J. Pharm. Biomed. Anal.* **20**, 289–295 (1999).

134. Tam, K. Y.; Hadley, M.; Patterson, W., Multiwavelength spectrophotometric determination of acid dissociation constants. Part IV. Water-insoluble pyridine derivatives, *Talanta* **49**, 539–546 (1999).

135. Box, K. J.; Comer, J. E. A.; Hosking, P.; Tam, K. Y.; Trowbridge, L.; Hill, A., Rapid physicochemical profiling as an aid to drug candidate selection, in Dixon, G. K.; Major, J. S.; Rice, M. J. (eds.), *High Throughput Screening: The Next Generation*, Bios Scientific Publishers, Oxford, 2000, pp. 67–74.

136. Comer, J. E. A. High-throughput pK_a and log P determination, in van de Waterbeemd, H.; Lennernäs, H.; Artursson, P. (eds.), *Drug Bioavailability. Estimation of Solubility, Permeability, Absorption and Bioavailability*, Wiley-VCH, Weinheim, 2003 (in press).

137. Tam, K. Y.; Takács-Novák, K., Multi-wavelength spectroscopic determination of acid dissociation constants: A validation study, *Anal. Chim. Acta* **434**, 157–167 (2001).

138. Tam, K. Y., Multiwavelength spectrophotometric resolution of the micro-equilibria of a triprotic amphoteric drug: Methacycline, *Mikrochim. Acta* **136**, 91–97 (2001).

139. Bevan, C. D.; Hill, A. P.; Reynolds, D. P., *Patent Cooperation Treaty*, WO 99/13328, March 18, 1999.

140. Takács-Novák, K.; Tam, K. Y., Multiwavelength spectrophotometric determination of acid dissociation constants. Part V. Microconstants and tautomeric ratios of diprotic amphoteric drugs, *J. Pharm. biomed. anal.* **17**, 1171–1182 (2000).

141. Tam, K. Y., Multiwavelength spectrophotometric determination of acid dissociation constants. Part VI. Deconvolution of binary mixtures of ionizable compounds, *Anal. Lett.* **33**, 145–161 (2000).

142. Tam, K. Y.; Quéré, L., Multiwavelength spectrophotometric resolution of the micro-equilibria of cetirizine, *Anal. Sci.* **17**, 1203–1208 (2001).

143. Hendriksen, B. A.; Sanchez-Felix, M. V.; Tam, K. Y., A new multiwavelength spectrophotometric method for the determination of the molar absorption coefficients of ionizable drugs, *Spectrosc. Lett.* **35**, 9–19 (2002).

144. Cleveland, J. A. Jr.; Benko, M. H.; Gluck, S. J.; Walbroehl, Y. M., Automated pK_a determination at low solute concentrations by capillary electrophoresis, *J. Chromatogr. A* **652**, 301–308 (1993).

145. Ishihama, Y.; Oda, Y.; Asakawa, N., Microscale determination of dissociation constants of multivalent pharmaceuticals by capillary electrophoresis, *J. Pharm. Sci.* **83**, 1500–1507 (1994).

146. Jia, Z.; Ramstad, T.; Zhong, M., Medium-throughput pK_a screening of pharmaceuticals by pressure-assisted capillary electrophoresis, *Electrophoresis* **22**, 1112–1118 (2001).

147. Ishihama, Y.; Nakamura, M.; Miwa, T.; Kajima, T.; Asakawa, N., A rapid method for pK_a determination of drugs using pressure-assisted capillary electrophoresis with photodiode array detection in drug discovery, *J. Pharm. Sci.* **91**, 933–942 (2002).

148. Oumada, F. Z.; Ràfols, C.; Rosés, M.; Bosch, E., Chromatographic determination of aqueous dissociation constants of some water-insoluble nonsteroidal antiinflammatory drugs, *J. Pharm. Sci.* **91**, 991–999 (2002).

149. Avdeef, A.; Box, K. J.; Comer, J. E. A.; Hibbert, C.; Tam, K. Y., pH-metric logP. 10. Determination of vesicle membrane–water partition coefficients of ionizable drugs, *Pharm. Res.* **15**, 208–214 (1997).

150. Slater, B.; McCormack, A.; Avdeef, A.; Comer, J. E. A., pH-metric logP. 4. Comparison of partition coefficients determined by shake-flask, hplc and potentiometric methods, *J. Pharm. Sci.* **83**, 1280–1283 (1994).

151. Avdeef, A.; Barrett, D. A.; Shaw, P. N.; Knaggs, R. D.; Davis, S. S., Octanol-, chloroform-, and PGDP-water partitioning of morphine-6-glucuronide and other related opiates, *J. Med. Chem.* **39**, 4377–4381 (1996).

152. Davies, C. W., *Ion Association*, Butterworths, London, 1962, p. 41.

153. Avdeef, A., pH-Metric logP. 2. Refinement of partition coefficients and ionization constants of multiprotic substances, *J. Pharm. Sci.* **82**, 183–190 (1993).

154. Kortnum, G.; Vogel, W.; Andrussow, K., *Dissociation Constants of Organic Acids in Aqueous Solution*, Butterworths, London, 1961.

155. Sillén, L. G.; Martell, A. E., *Stability Constants of Metal-Ion Complexes*, Special Publication 17, Chemical Society, London, 1964.

156. Perrin, D. D., *Dissociation Constants of Organic Bases in Aqueous Solution*, Butterworths, London, 1965.

157. Sillén, L. G.; Martell, A. E., *Stability Constants of Metal-Ion Complexes*, Special Publication 25, Chemical Society, London, 1971.

158. Serjeant, E. P.; Dempsey, B., *Ionization Constants of Organic Acids in Aqueous Solution*, Pergamon, Oxford, 1979.

159. Smith, R. M.; Martell, A. E., *Critical Stability Constants*, Vols. 1–6, Plenum Press, New York, 1974.

160. Smith, R. M.; Martell, A. E.; Motekaitis, R. J., *NIST Critically Selected Stability Constants of Metal Complexes Database*, Version 5, NIST Standards Reference Database 46, U. S. Dept. Commerce, Gaithersburg, MD, 1998.

161. Avdeef, A., *Sirius Technical Application Notes* (*STAN*), Vol. 1, Sirius Analytical Instruments Ltd., Forest Row, UK, 1994.

162. Avdeef, A.; Box, K. J., *Sirius Technical Application Notes* (*STAN*), Vol. 2, Sirius Analytical Instruments Ltd., Forest Row, UK, 1995.

163. Bjerrum, J., *Metal-Ammine Formation in Aqueous Solution*, Haase, Copenhagen, 1941.

164. Irving, H. M.; Rossotti, H. S., The calculation of formation curves of metal complexes from pH titration curves in mixed solvents, *J. Chem. Soc.* 2904–2910 (1954).

165. Avdeef, A.; Kearney, D. L.; Brown, J. A.; Chemotti, A. R. Jr., Bjerrum plots for the determination of systematic concentration errors in ttitration data, *Anal. Chem.* **54**, 2322–2326 (1982).

166. Takács-Novák, K.; Box, K. J.; Avdeef, A., Potentiometric pK_a determination of water-insoluble compounds. Validation study in methanol/water mixtures, *Int. J. Pharm.* **151**, 235–248 (1997).

167. Avdeef, A.; Zelazowski, A. J.; Garvey, J. S., Cadmium binding by biological ligands. 3. Five- and seven-cadmium binding in metallothionein: A detailed thermodynamic study, *Inorg. Chem.* **24**, 1928–1933 (1985).

168. Dunsmore, H. S.; Midgley, D., The calibration of glass electrodes in cells with liquid junction, *Anal. Chim. Acta* **61**, 115–122 (1972).

169. Bates, R. G., The modern meaning of pH, *CRC Crit. Rev. Anal. Chem.* **10**, 247–278 (1981).

170. Sprokholt, R.; Maas, A. H. J.; Rebelo, M. J.; Covington, A. K., Determination of the performance of glass electrodes in aqueous solutions in the physiological pH range and at the physiological sodium ion concentration, *Anal. Chim. Acta* **129**, 53–59 (1982).

171. Comer, J. E. A.; Hibbert, C., pH electrode performance under automatic management conditions, *J. Auto. Chem.* **19**, 213–224 (1997).

172. Avdeef, A.; Box, K. J.; Comer, J. E. A.; Gilges, M.; Hadley, M.; Hibbert, C.; Patterson, W.; Tam, K. Y., pH-metric logP. 11. pK_a determination of water-insoluble drugs in organic solvent-water mixtures, *J. Pharm. Biomed. Anal.*, **20**, 631–641 (1999).

173. Sweeton, F. H.; Mesmer, R. E.; Baes, C. F. Jr., Acidity measurements at elevated temperatures. 7. Dissociation of water, *J. Solut. Chem.* **3**, 191–214 (1974).

174. Good, N. E.; Wingert, G. D.; Winter, W.; Connolly, T. N.; Izawa, S.; Singh, R. M. M., Hydrogen ion buffers for biological research, *Biochemistry* **5**, 467–477 (1966).

175. Vega, C. A.; Bates, R. G., Buffers for the physiological pH range: Thermodynamic constants of four substituted aminoethanesulfonic acids from 5 to 50°C, *Anal. Chem.* **48**, 1293–1296 (1976).

176. Bates, R. G.; Vega, C. A.; White, D. R. Jr., Standards for pH measurements in isotonic saline media of ionic strength $I = 0.16$, *Anal. Chem.* **50**, 1295–1300 (1978).

177. Sankar, M.; Bates, R. G., Buffers for the physiological pH range: Thermodynamic constants of 3-(N-morpholino)propanesulfonic acid from 5 to 50°C, *Anal. Chem.* **50**, 1922–1924 (1978).

178. Roy, R. N.; Gibbons, J. J.; Padron, J. L.; Moeller, J., Second-stage dissociation constants of piperazine-N,N'-bis(2-ethanesulfonic acid) monosodium monohydrate and related thermodynamic functions in water from 5 to 55°C, *Anal. Chem.* **52**, 2409–2412 (1980).

179. Feng, D.; Koch, W. F.; Wu, Y. C., Second dissociation constant and pH of N-(2-hydroxyethyl)-piperazine-N'-2-ethanesulfonic acid from 0 to 50°C, *Anal. Chem.* **61**, 1400–1405 (1989).

180. Feng, D.; Koch, W. F.; Wu, Y. C., Investigation of the interaction of HCl and three amino acids, HEPES, MOPSO and glycine, by EMF measurements, *J. Solut. Chem.* **21**, 311–321 (1992).

181. Mizutani, M., *Z. Physik. Chem.* **116**, 350–358 (1925).

182. Mizutani, M., *Z. Physik. Chem.* **118**, 318–326 (1925).

183. Mizutani, M., *Z. Physik. Chem.* **118**, 327–341 (1925).

184. Hall, N. F.; Sprinkle, M. R., Relations between the structure and strength of certain organic bases in aqueous solution, *J. Am. Chem. Soc.* **54**, 3469–3485 (1932).

185. Kolthoff, I. M.; Lingane, J. J.; Larson, W. D., The relation between equilibrium constants in water and other solvents, *J. Am. Chem. Soc.* **60**, 2512–2515 (1938).

186. Kolthoff, I. M.; Guss, L. S., Ionization constants of acid-base indicators in methanol, *J. Am. Chem. Soc.* **60**, 2516–2522 (1938).

187. Albright, P. S.; Gosting, L. J., *J. Am. Chem. Soc.* **68**, 1061–1063 (1946).

188. Grunwald, E.; Berkowitz, B. J., The measurement and correlation of acid dissociation constants for carboxylic acids in the system ethanol-water. Activity coefficients and empirical activity functions, *J. Am. Chem. Soc.* **73**, 4939–4944 (1951).

189. Gutbezahl, B.; Grunwald, E., The acidity and basicity scale in the system ethanol-water. The evaluation of degenrate activity coefficients for single ions, *J. Am. Chem. Soc.* **75**, 565–574 (1953).

190. Marshall, P. B., Some chemical and physical properties associated with histamine antogonism, *Br. J. Pharmacol.* **10**, 270–278 (1955).

191. Bacarella, A. L.; Grunwald, E.; Marshall, H. P.; Purlee, E. L., The potentiometric measurement of acid dissociation constants and pH in the system methanol-water. pK_a values for carboxylic acids and anilinium ions, *J. Org. Chem.* **20**, 747–762 (1955).

192. Shedlovsky, T; Kay, R. L., The ionization constant of acetic acid in water-methanol mixtures at 25°C from conductance measurements, *J. Am. Chem. Soc.* **60**, 151–155 (1956).

193. Edmonson, T. D.; Goyan, J. E., The effect of hydrogen ion and alcohol concentration on the solubility of phenobarbital, *J. Am. Pharm. Assoc., Sci. Ed.* **47**, 810–812 (1958).

194. Yasuda, M., Dissociation constants of some carboxylic acids in mixed aqueous solvents, *Bull. Chem. Soc. Jpn.* **32**, 429–432 (1959).

195. de Ligny, C. L.; Rehbach, M., The liquid-liquid-junction potentials between some buffer solutions in methanol and methanol-water mixtures and a saturated KCl solution in water at 25°C, *Recl. Trav. Chim. Pays-Bas* **79**, 727–730 (1960).

196. de Ligny, C. L., The dissociation constants of some aliphatic amines in water and methanol-water mixtures at 25°C, *Recl. Trav. Chim. Pays-Bas* **79**, 731–736 (1960).

197. de Ligny, C. L.; Luykx, P. F. M.; Rehbach, M.; Wienecke, A. A., The pH of some standard solutions in methanol and methanol-water mixtures at 25°C. 1. Theoretical part, *Recl. Trav. Chim. Pays-Bas* **79**, 699–712 (1960).

198. de Ligny, C. L.; Luykx, P. F. M.; Rehbach, M.; Wienecke, A. A., The pH of some standard solutions in methanol and methanol-water mixtures at 25°C. 2. Experimental part, *Recl. Trav. Chim. Pays-Bas* **79**, 713–726 (1960).

199. Yasuda, M.; Yamsaki, K.; Ohtaki, H., *Bull. Chem. Soc. Jpn.* **23**, 1067 (1960).

200. de Ligny, C. L.; Loriaux, H.; Ruiter, A., The application of Hammett's acidity function Ho to solutions in methanol-water mixtures, *Recl. Trav. Chim. Pays-Bas* **80**, 725–739 (1961).

201. Shedlovsky, T., The behaviour of carboxylic acids in mixed solvents, in Pesce, B. (ed.), *Electrolytes*, Pergamon Press, New York, 1962, pp. 146–151.

202. Long, F. A.; Ballinger, P., Acid ionization constants of alcohols in the solvents water and deuterium oxide, in Pesce, B. (ed.), *Electrolytes*, Pergamon Press, New York, 1962, pp. 152–164.

203. Bates, R. G.; Paabo, M.; Robinson, R. A., Interpretation of pH measurements in alcohol-water solvents, *J. Phys. Chem.* **67**, 1833–1838 (1963).

204. Gelsema, W. J.; de Ligny, C. L.; Remijnse, A. G.; Blijleven, H. A., pH-measurements in alcohol-water mixtures, using aqueous standard buffer solutions for calibration, *Recl. Trav. Chim. Pays-Bas* **85**, 647–660 (1966).

205. Woolley, E. M.; Hurkot, D. G.; Hepler, L. G., Ionization constants for water in aqueous organic mixtures, *J. Phys. Chem.* **74**, 3908–3913 (1970).

206. Wooley, E. M.; Tomkins, J.; Hepler, L. G., Ionization constants for very weak organic acids in aqueous solution and apparent ionization constants for water in aqueous organic mixtures, *J. Solut. Chem.* **1**, 341–351 (1972).

207. Fisicaro, E.; Braibanti, A., Potentiometric titrations in methanol/water medium: Intertitration variability, *Talanta* **35**, 769–774 (1988).

208. Esteso, M. A.; Gonzalez-Diaz, O. M.; Hernandez-Luis, F. F.; Fernandez-Merida, L., Activity coefficients for NaCl in ethanol-water mixtures at 25°C, *J. Solut. Chem.* **18**, 277–288 (1989).

209. Papadopoulos, N.; Avranas, A., Dissociation of salicylic acid, 2,4-, 2,5- and 2,6-dihydroxybenzoic acids in 1-propanol-water mixtures at 25°C, *J. Solut. Chem.* **20**, 293–300 (1991).

210. Li, A.; Yalkowsky, S., Solubility of organic solutes in ethanol/water mixtures, *J. Pharm. Sci.* **83**, 1735–1740 (1994).

211. Halle, J.-C.; Garboriaud, R.; Schaal, R., Etude electrochimique des melanges d'eau et de dimethylsulfoxyde. Produit ionique apparent et niveau d'acidite, *Bull. Soc. Chim.* 1851–1857 (1969).

212. Halle, J.-C.; Garboriaud, R.; Schaal, R., Sur la realization de melanges tampons dans les milieux eau-DMSO, Application a l'etude de nitrodephenylamines, *Bull. Soc. Chim.* 2047–2053 (1970).

213. Woolley, E. M.; Hepler, L. G., Apparent ionization constants of water in aqueous organic mixtures and acid dissociation constants of protonated co-solvents in aqueous solution, *Anal. Chem.* **44**, 1520–1523 (1972).

214. Yakolev, Y. B.; Kul'ba, F. Y.; Zenchenko, D. A., Potentiometric measurement of the ionic products of water in water-dimetylsulphoxide, water-acetonitrile and water-dioxane mixtures, *Russ. J. Inorg. Chem.* **20**, 975–976 (1975).

215. Siow, K.-S.; Ang, K.-P., Thermodynamic of ionization of 2,4-dinitrophenol in water-dimethylsulfoxide solvents, *J. Solut. Chem.* **18**, 937–947 (1989).

216. Van Uitert, L. G.; Haas, C. G., Studies on coordination compounds. 1. A method for determining thermodynamic equilibrium constants in mixed solvents, *J. Am. Chem. Soc.* **75**, 451–455 (1953).

217. Grunwald, E., Solvation of electrolytes in dioxane-water mixtures, in Pesce, B. (ed.), *Electrolytes*, Pergamon Press, New York, 1962, pp. 62–76.

218. Marshall, W. L., Complete equilibrium constants, electrolyte equilibria, and reaction rates, *J. Phys. Chem.* **74**, 346–355 (1970).

219. Sigvartsen, T.; Songstadt, J.; Gestblom, B.; Noreland, E., Dielectric properties of solutions of tetra-isopentylammonium nitrate in dioxane-water mixtures, *J. Solut. Chem.* **20**, 565–582 (1991).

220. Casassas, E.; Fonrodona, G.; de Juan, A., Solvatochromic parameters for binary mixtures and a correlation with equilibrium constants. 1. Dioxane-water mixtures, *J. Solut. Chem.* **21**, 147–162 (1992).

221. Cavill, G. W. K.; Gibson, N. A.; Nyholm, R. S., *J. Chem. Soc.* 2466–2470 (1949).

222. Garrett, E. R., Variation of pK'_a-values of tetracyclines in dimethylformamide-water solvents, *J. Pharm. Sci.* **52**, 797–799 (1963).

223. Hawes, J. L.; Kay, R. L., Ionic association of potassium and cesium chloride in ethanol-water mixtures from conductance measurements at 25°, *J. Phys. Chem.* **69**, 2420–2431 (1965).

224. Takács-Novák, K.; Avdeef, A.; Podányi, B.; Szász, G., Determination of protonation macro- and microconstants and octanol/water partition coefficient of anti-inflammatory niflumic acid, *J. Pharm. Biomed. Anal.* **12**, 1369–1377 (1994).

225. Avdeef, A.; Takács-Novák, K.; Box, K. J., pH-metric logP. 6. Effects of sodium, potassium, and N-CH$_3$-D-glucamine on the octanol-water partitioning with prostaglandins E1 and E2, *J. Pharm. Sci.* **84**, 523–529 (1995).

226. Leggett, D. J., SQUAD: Stability quotients from absorbance data, in Leggett, D. J. (ed.), *Computational Methods for the Determination of Formation Constants*, Plenum Press, New York, 1985, pp. 158–217.

227. Lawson, C. L.; Hanson, R. J., *Solving Least Squares Problems*, Prentice-Hall, Englewood Cliffs, NJ, 1974.

228. Tam, K. Y. BUFMAKE: A computer program to calculate the compositions of buffers of defined buffer capacity and ionic strength for ultra-violet spectrophotometry, *Comp. Chem.* **23**, 415–419 (1999).

229. Comer, J.; Tam, K., Lipophilicity profiles: theory and measurement, in Testa, B.; van de Waterbeemd, H.; Folkers, G.; Guy, R. (eds.), *Pharmacokinetic Optimization in Drug Research*, Verlag Helvetica Chimica Acta, Zürich and Wiley-VCH, Weinheim, 2001, pp. 275–304.

230. Niebergall, P. J.; Schnaare, R. L.; Sugita, E. T., Spectral determination of microdissociation constants, *J. Pharm. Sci.* **61**, 232–234 (1972).

231. Streng, W. H., Microionization constants of commercial cephalosporins, *J. Pharm. Sci.* **67**, 666–669 (1978).

232. Noszál, B., Microspeciation of polypeptides, *J. Phys. Chem.* **90**, 6345–6349 (1986).

233. Noszál, B., Group constant: A measure of submolecular basicity, *J. Phys. Chem.* **90**, 4104–4110 (1986).

234. Noszál, B.; Osztás, E., Acid-base properties for each protonation site of six corticotropin fragments, *Int. J. Peptide Protein Res.* **33**, 162–166 (1988).

235. Noszál, B.; Sándor, P., Rota-microspeciation of aspartic acid and asparagine, *Anal. Chem.* **61**, 2631–2637 (1989).

236. Nyéki, O.; Osztás, E.; Noszál, B.; Burger, K., Acid-base properties and microspeciation of six angiotensin-type octapeptides, *Int. J. Peptide Protein Res.* **35**, 424–427 (1990).

237. Takács-Novák, K.; Noszál, B.; Hermecz, I.; Keresztúri, G.; Podányi, B.; Szász, G., Protonation equilibria of quinolone antibacterials, *J. Pharm. Sci.* **79**, 1023–1028 (1990).

238. Noszàl, B.; Guo, W.; Rabenstein, D. L., Rota-microspeciation of serine, cysteine and selenocysteine, *J. Phys. Chem.* **95**, 9609–9614 (1991).

239. Noszàl, B.; Rabenstein, D. L., Nitrogen-protonation microequilibria and C(2)-deprotonation microkinetics of histidine, histamine and related compounds, *J. Phys. Chem.* **95**, 4761–4765 (1991).

240. Noszàl, B.; Kassai-Tanczos, R., Microscopic acid-base equilibria of arginine, *Talanta* **38**, 1439–1444 (1991).

241. Takács-Novák, K.; Józan, M.; Hermecz, I.; Szász, G., Lipophilicity of antibacterial fluoroquinolones, *Int. J. Pharm.* **79**, 89–96 (1992).

242. Noszàl, B.; Guo, W.; Rabenstein, D. L., Characterization of the macroscopic and microscopic acid-base chemistry of the native disulfide and reduced dithiol forms of oxytocin, arginine-vasopressin and related peptides, *J. Org. Chem.* **57**, 2327–2334 (1992).

243. Takács-Novák, K.; Józan, M.; Szász, G., Lipophilicity of amphoteric molecules expressed by the true partition coefficient, *Int. J. Pharm.* **113**, 47 (1995).

244. Cannon, J. B.; Krill, S. L.; Porter, W. R., Physicochemical properties of A-75998, an antagonist of luteinizing hormone releasing hormone, *J. Pharm. Sci.* **84**, 953–958 (1995).

245. Dearden, J. C.; Bresnen, G. M., The measurement of partition coefficients, *Quant. Struct.-Act. Relat.* **7**, 133–144 (1988).

246. Hersey, A.; Hill, A. P.; Hyde, R. M.; Livingstone, D. J., Principles of method selection in partition studies, *Quant. Struct. Act. Relat.* **8**, 288–296 (1989).

247. Tsai, R.-S.; Fan, W.; El Tayar, N.; Carrupt, P.-A.; Testa, B.; Kier, L. B., Solute-water interactions in the oraganic phase of a biphasic system. 1. Structural influence of organic solutes on the "water-dragging" effect, *J. Am. Chem. Soc.* **115**, 9632–9639 (1993).

248. Fan, W.; Tsai, R. S.; El Tayar, N.; Carrupt, P.-A.; Testa, B, Solute-water interactions in the organic phase of a biphasic system. 2. Effects of organic phase and temperature on the "water-dragging" effect, *J. Phys. Chem.* **98**, 329–333 (1994).

249. Leahy, D. E.; Taylor, P. J.; Wait, A. R., Model solvent systems for QSAR. 1. Propylene glycol dipelargonate (PGDP). A new standard for use in partition coefficient determination, *Quant. Struct. Act. Relat.* **8**, 17–31 (1989).

250. Seiler, P., The simultaneous determination of partition coefficients and acidity constant of a substance, *Eur. J. Med. Chem. Chim. Therapeut.* **9**, 665–666 (1974).

251. van de Waterbeemd, H.; Kansy, M., Hydrogen-bonding capacity and brain penetration, *Chimia* **46**, 299–303 (1992).

252. von Geldern, T. W.; Hoffman, D. J.; Kester, J. A.; Nellans, H. N.; Dayton, B. D.; Calzadilla, S. V.; Marsh, K. C.; Hernandez, L.; Chiou, W.; Dixon, D. B.; Wu-wong, J. R.; Opgenorth, T. J., Azole endothelin antagonists. 3. Using ΔlogP as a tool to improve absorption, *J. Med. Chem.* **39**, 982–991 (1996).

253. Kubinyi, H., Hydrogen bonding: The last mystery in drug design? in Testa, B.; van de Waterbeemd, H.; Folkers, G.; Guy, R. (eds.), *Pharmacokinetic Optimization in Drug Research*, Verlag Helvetica Chimica Acta, Zürich and Wiley-VCH, Weinheim, 2001, pp. 513–524.

254. Raevsky, O.; Grigor'ev, V. Y.; Kireev; D.; Zefirov, N., Complete thermodynamic description of H-bonding in the framework of multiplicative approach, *Quant. Struct. Act. Relat.*, **11**, 49–63 (1992).

255. El Tayar, N.; Testa, B.; Carrupt, P.-A., Polar intermolecular interactions encoded in partition coefficients: An indirect estimation of hydrogen-bond parameters of polyfunctional solutes, *J. Phys. Chem.* **96**, 1455–1459 (1992).

256. Chikhale, E. G.; Ng, K. Y.; Burton, P. S.; Borchardt, R. T., Hydrogen bonding potential as a determinant of the in vitro and in situ blood-brain barrier permeability of peptides, *Pharm. Res.* **11**, 412–419 (1994).

257. Abraham, M.; Chadha, H.; Whiting, G.; Mitchell, R., Hydrogen bonding. 32. An analysis of water-octanol and water-alkane partitioning and the log P parameter of Seiler, *J. Pharm. Sci.* **83**, 1085–1100 (1994).

258. ter Laak, A. M.; Tsai, R.-S.; den Kelder, G. M., D.-O.; Carrupt, P.-A.; Testa, B., Lipophilicity and hydrogen-bonding capacity of H_1-antihistaminic agents in relation to their central sedative side effects, *Eur. J. Pharm. Sci.* **2**, 373–384 (1994).

259. Potts, R. O.; Guy, R. H., A predictive algorithm for skin permeability: The effects of molecular size and hydrogen bonding, *Pharm. Res.* **12**, 1628–1633 (1995).

260. Abraham, M. H.; Martins, F.; Mitchell, R. C., Algorithms for skin permeability using hydrogen bond descriptors: The problems of steroids, *J. Pharm. Pharmacol.* **49**, 858–865 (1997).

261. Reymond, F.; Steyaert, G.; Carrupt, P.-A.; Morin, D.; Tillement, J. P.; Girault, H.; Testa, B., The pH-partition profile of the anti-ischemic drug trimetazidine may explain its reduction of intracellular acidosis, *Pharm. Rese.* **16**, 616–624 (1999).

262. Carrupt, P.-A.; Testa, B.; Bechalany, A.; El Tayar, N.; Descas, P.; Perrissoud, D., Morphine 6-glucuronide and morphine 3-glucuronide as molecular chameleons with unexpectedly high lipophilicity, *J. Med. Chem.* **34**, 1272–1275 (1991).

263. Gaillard, P.; Carrupt, P.-A.; Testa, B., The conformation-dependent lipophilicity of morphine glucuronides as calculated from the molecular lipophilicity potential, *Bioorg. Med. Chem. Lett.* **4**, 737–742 (1994).

264. Poulin, P.; Schoenlein, K.; Theil, F.-P., Prediction of adipose tissue: Plasma partition coefficients for structurally unrelated drugs, *J. Pharm. Sci.* **90**, 436–447 (2001).

265. Chamberlain, K.; Evans, A. A.; Bromilow, R. H., 1-Octanol/water partition coefficients (K_{ow}) and pK_a for ionizable pesticides measured by a pH-metric method, *Pest. Sci.* **47**, 265–271 (1996).

266. Pliska, V.; Testa, B.; van de Waterbeemd, H. (eds.), *Methods and Principles in Medicinal Chemistry*, Vol. 4, VCH, Weinheim, Germany, 1996.

267. Testa, B.; van de Waterbeemd, H.; Folkers, G.; Guy, R. (eds.), *Pharmacokinetic Optimization in Drug Research*, Verlag Helvetica Chimica Acta, Zürich and Wiley-VCH: Weinheim, 2001.

268. Krämer, S. D.; Gautier, J.-C.; Saudemon, P., Considerations on the potentiometric log P determination, *Pharm. Res.* **15**, 1310–1313 (1998).

269. Bouchard, G.; Pagliara, A.; van Balen, G. P.; Carrupt, P.-A.; Testa, B.; Gobry, V.; Girault, H. H.; Caron, G.; Ermondi, G.; Fruttero, R., Ionic partition diagram of the zwitterionic antihistamine cetirizine, *Helv. Chim. Acta* **84**, 375–387 (2001).

270. Scherrer, R. A.; Howard, S. M., Use of distribution coefficients in quantitative structure-activity relationships, *J. Med. Chem.* **20**, 53–58 (1977).

271. Schaper, K.-J., Simultaneous determination of electronic and lipophilic properties [pK_a, P(ion), P(neutral)] for acids and bases by nonlinear regression analysis of pH-dependent partittion measurements, *J. Chem. Res.* (S) 357 (1979).

272. Taylor, P. J.; Cruickshank, J. M., Distribution coefficients of atenolol and sotalol, *J. Pharm. Pharmacol.* **36**, 118–119 (1984).

273. Taylor, P. J.; Cruickshank, J. M., Distribution coefficients of atenolol and sotalol, *J. Pharm. Pharmacol.* **37**, 143–144 (1985).

274. Manners, C. N.; Payling, D. W.; Smith, D. A., Distribution coefficient, a convenient term for the relation of predictable physico-chemical properties to metabolic processes, *Xenobiotica* **18**, 331–350 (1988).

275. Avdeef, A., Assessment of distribution-pH profiles, in Pliska, V.; Testa, B.; van de Waterbeemd, H. (eds.), *Methods and Principles in Medicinal Chemistry*, Vol. 4, VCH, Weinheim, Germany, 1996, pp. 109–139.

276. Scherrer, R. A., Biolipid pK_a values in the lipophilicity of ampholytes and ion pairs, in Testa, B.; van de Waterbeemd, H.; Folkers, G.; Guy, R. (eds.), *Pharmacokinetic Optimization in Drug Research*, Verlag Helvetica Chimica Acta, Zürich and Wiley-VCH, Weinheim, 2001, pp. 351–381.

277. Murthy. K. S.; Zografi, G., Oil-water partitioning of chlorpromazine and other phenothiazine derivatives using dodecane and n-octanol, *J. Pharm. Sci.* **59**, 1281–1285 (1970).

278. Tomlinson, E.; Davis, S. S., Interactions between large organic ions of opposite and unequal charge. II. Ion pair and ion triplet formation, *J. Colloid Interface Sci.* **74**, 349–357 (1980).

279. van der Giesen, W. F.; Janssen, L. H. M., *Int. J. Pharm.* **12**, 231–249 (1982).

280. Scherrer, R. A., The treatment of ionizable compounds in quantitative structure-activity studies with special consideration of ion partitioning, in Magee, P. S.; Kohn, G. K.; Menn, J. J. (eds.), *Pesticide Syntheses through Rational Approaches*, ACS Symposium Series 225, ACS (American Chemical Society), Washington, DC, 1984, pp. 225–246.

281. Scherrer, R. A.; Crooks, S. L., Titrations in water-saturated octanol: A guide to partition coefficients of ion pairs and receptor-site interactions, *Quant. Struct.-Act. Relat.* **8**, 59–62 (1989).

282. Akamatsu, M.; Yoshida, Y.; Nakamura, H.; Asao, M.; Iwamura, H.; Fujita, T., Hydrophobicity of di- and tripeptides having unionizable side chains and correlation with substituent and structural parameters, *Quant. Struc. -Act. Relat.* **8**, 195–203 (1989).

283. Manners, C. N.; Payling, D. W.; Smith, D. A., Lipophilicity of zwitterionic suphate conjugates of tiaramide, propanolol and 4′-hydroxypropranolol, *Xenobiotica* **19**, 1387–1397 (1989).

284. Westall, J. C.; Johnson, C. A.; Zhang, W., Distribution of LiCl, NaCl, KCl, HCl, $MgCl_2$, and $CaCl_2$ between octanol and water, *Environ. Sci. Technol.* **24**, 1803–1810 (1990).

285. Chmelík, J.; Hudeček, J.; Putyera, K.; Makovička, J.; Kalous, V.; Chmelíková, J., Characterization of the hydrophobic properties of amino acids on the basis of their partition and distribution coefficients in the 1-octanol-water system, *Collect. Czech. Chem. Commum.* **56**, 2030–2040 (1991).

286. Tsai, R.-S.; Testa, B.; El Tayar, N.; Carrupt, P.-A., Structure-lipophilicity relationships of zwitterionic amino acids, *J. Chem. Soc. Perkin Trans.* **2**, 1797–1802 (1991).

287. Akamatsu, M.; Fujita, T., Quantitative analysis of hydorphobicity of di- to pentapeptides having un-ionizable side chains with substituent and structural parameters, *J. Pharm. Sci.* **81**, 164–174 (1992).

288. Akamatsu, M.; Katayama, T.; Kishimoto, D.; Kurokawa, Y.; Shibata, H., Quantitative analysis of the structure-hydophibicity relationships for n-acetyl di- and tripeptide amides, *J. Pharm. Sci.* **83**, 1026–1033 (1994).

289. Abraham, M. H.; Takács-Novák, K.; Mitchell, R. C., On the partition of ampholytes: Application to blood-brain distribution, *J. Pharm. Sci.* **86**, 310–315 (1997).

290. Bierer, D. E.; Fort, D. M.; Mendez, C. D.; Luo, J.; Imbach, P. A.; Dubenko, L. G.; Jolad, S. D.; Gerber, R. E.; Litvak, J.; Lu, Q.; Zhang, P.; Reed, M. J.; Waldeck, N.; Bruening, R. C.; Noamesi, B. K.; Hector, R. F.; Carlson, T. J.; King, S. R., Ethnobotanical-directed discovery of the antihyperglycemic properties of cryptolepine: Its isolation from Cryptolepis sanguinolenta, synthesis, and in vitro and in vivo activities, *J. Med. Chem.* **41**, 894–901 (1998).

291. Takács-Novák, K.; Szász, G., Ion-pair partition of quaternary ammonium drugs: The influence of counter ions of different lipophilicity, size, and flexibility, *Pharm. Res.* **16**, 1633–1638 (1999).

292. Valkó, K.; Slégel, P., New chromotographic hydrophobicity index ($\varphi 0$) based on the slope and the intercept of the log k′ versus organic phase concentration plot, *J. Chromotogr.* **631**, 49–61 (1993).

293. Valkó, K.; Bevan, C.; Reynolds, D., Chromatographic hydrophobicity index by fast-gradient RP-HPLC: A high-thorughput alternative to logP/logD, *Anal. Chem.* **69**, 2022–2029 (1997).

294. Hitzel, L.; Watt, A. P.; Locker, K. L., An increased throughput method for the determination of partition coefficients, *Pharm. Res.* **17**, 1389–1395 (2000).

295. Schräder, W.; Andersson, J. T., Fast and direct method for measuring 1-octanol-water partition coefficients exemplified for six local anesthetics, *J. Pharm. Sci.* **90**, 1948–1954 (2001).

296. Lombardo, F.; Shalaeva, M. Y.; Tupper, K. A.; Gao, F.; Abraham, M. H., E logP_{oct}: A tool for lipophilicity determination in drug discovery, *J. Med. Chem.* **43**, 2922–2928 (2000).

297. Lombardo, F.; Shalaeva, M. Y.; Tupper, K. A.; Gao, F., E logD_{oct}: A tool for lipophilicity determination in drug discovery. 2. Basic and neutral compounds, *J. Med. Chem.* **43**, 2922–2928 (2000).

298. Valkó, K., Measurements of physical properties for drug design in industry, in Valkó, K. (ed.), *Separation Methods in Drug Synthesis and Purification*, Elsevier, Amsterdam, 2001, Ch. 12.

299. Pidgeon, C.; Venkataram, U. V., Immobilized artificial membrane chromatography: Supports composed of membrane lipids, *Anal. Biochem.* **176**, 36–47 (1989).

300. Pidgeon, C.; Ong, S.; Choi, H.; Liu, H., Preparation of mixed ligand immobilized artificial membranes for predicting drug binding to membranes, *Anal. Chem.* **66**, 2701–2709 (1994).

301. Pidgeon, C.; Ong, S.; Liu, H.; Qui, X.; Pidgeon, M.; Dantzig, A. H.; Munroe, J.; Hornback, W. J.; Kasher, J. S.; Glunz, L.; Szczerba, T., IAM chromatography: An in vitro screen for predicting drug membrane permeability, *J. Med. Chem.* **38**, 590–594 (1995).

302. Ong, S.; Liu, H.; Qiu, X.; Bhat, G.; Pidgeon, C., Membrane partition coefficients chromatographically measured using immobilized artificial membrane surfaces, *Anal. Chem.* **67**, 755–762 (1995).

303. Ong, S., Pidgeon, C., Thermodynamics of solute partitioning into immobilized artificial membranes, *Anal. Chem.* **67**, 2119–2128 (1995).

304. Ong, S.; Liu, H.; Pidgeon, C., Immobilized-artificial-membrane chromatography: Measurements of membrane partition coefficient and predicting drug membrane permeability, *J. Chromatogr. A* **728**, 113–128 (1996).

305. Barton, P.; Davis, A. M.; McCarthy, D. J.; Webborn, P. J. H., Drug-phospholipid interactions. 2. Predicting the sites of drug distribution using n-octanol/water and membrane/water distribution coefficients. *J. Pharm. Sci.* **86**, 1034–1039 (1997).

306. Barbato, F.; La Rotonda, M. I.; Quaglia, F., Interactions of nonsteroidal antiinflammatory drugs with phospholipids: comparison between octanol/buffer partition coefficients and chromatographic indexes on immobilized artificial membranes, *J. Pharm. Sci.* **86**, 225–229 (1997).

307. Yang, C. Y.; Cai, S. J.; Liu, H.; Pidgeon, C., Immobilized artificial membranes—screens for drug-membrane interactions, *Adv. Drug Deliv. Rev.* **23**, 229–256 (1997).

308. Stewart, B. H.; Chung, F. Y.; Tait, B.; John, C.; Chan, O. H., Hydrophobicity of HIV protease inhibitors by immobilized artificial membrane chromatography, *Pharm. Res.* **15**, 1401–1406 (1998).

309. Stewart, B. H.; Chan, O. H., Use of immobilized artificial membrane chromatography for drug transport applications, *J. Pharm. Sci.* **87**, 1471–1478 (1998).

310. Ottiger, C.; Wunderli-Allenspach, H., Immobilized artificial membrane (IAM)-HPLC for partition studies of neutral and ionized acids and bases in comparison with the liposomal partition system, *Pharm. Res.* **16**, 643–650 (1999).

311. Taillardat-Bertschinger, A.; Martinet, C. A. M.; Carrupt, P.-A.; Reist, M.; Caron, G.; Fruttero, R.; Testa, B., Molecular factors influencing retention on immobilized artificial membranes (IAM) compared to partitioning in liposomes and n-octanol, *Pharm. Res.* **19**, 729–737 (2002).

312. Genty, M.; González, G.; Clere, C.; Desangle-Gouty, V.; Legendre, J.-Y., Determination of the passive absorption through the rat intestine using chromatographic indices and molar volume, *Eur. J. Pharm. Sci.* **12**, 223–229 (2001).

313. Miyake, K.; Kitaura, F.; Mizuno, N., Phosphatidylcholine-coated silica as a useful stationary phase for high-performance liquid chromatographic determination of partition coefficients between octanol and water, *J. Chromatogr.* **389**, 47–56 (1987).

314. Yang, Q.; Lundahl, P., Binding of lysozyme on the surface of entrapped phosphati-dylserine-phosphatidylcholine vesicles and an example of high-performance lipid vesicle surface chromatography, *J. Chromatogr.* **512**, 377–386 (1990).

315. Zhang, Y.; Zeng, C.-M.; Li, Y.-M.; Stellan, H.; Lundahl, P., Immobilized liposome chromatography of drugs on calillary continuous beds for model analysis of drug-membrane interactions, *J. Chromatogr. A* **749**, 13–18 (1996).

316. Lundahl, P.; Beigi, F., Immobilized liposome chromatography of drugs for model analysis of drug-membrane interactions, *Adv. Drug Deliv. Rev.* **23**, 221–227 (1997).

317. Österberg, T.; Svensson, M.; Lundahl, P., Chromatographic retention of drug molecules on immobilized liposomes prepared from egg phospholipids and from chemically pure phospholipids, *Eur. J. Pharm. Sci.* **12**, 427–439 (2001).

318. Danelian, E.; Karlén, A.; Karlsson, R.; Winiwarter, S.; Hansson, A.; Löfås, S.; Lennernäs, H.; Hämäläinen, M. D., SPR biosensor studies of the direct interaction between 27 drugs and a liposome surface: Correlation with fraction absorbed in humans, *J. Med. Chem.* **43**, 2083–2086 (2000).

319. Loidl-Stahlhofen, A.; Hartmann, T.; Schöttner, M.; Röhring, C.; Brodowsky, H.; Schmitt, J.; Keldenich, J., Multilamellar liposome and solid-supported lipid membranes (TRAN-SIL): Screening of lipid-water partitioning toward a high-throughput scale, *Pharm. Res.* **18**, 1782–1788 (2001).

320. Ishihama, Y.; Oda, Y.; Uchikawa, K.; Asakawa, N., Evaluation of solute hydrophobi-city by microemulsion electrokinetic chromatography, *Anal. Chem.* **67**, 1588–1595 (1995).

321. Razak, J. L.; Cutak, B. J.; Larive, C. K.; Lunte, C. E., Correlation of the capacity factor in vesicular electrokinetic chromatography with the octanol:water partition coefficient for charged and neutral analytes, *Pharm. Res.* **18**, 104–111 (2001).

322. Burns, S. T.; Khaledi, M. G., Rapid determination of liposome-water partition coefficient (K_{lw}) using liposome electrokinetic chromatography (LEKC), *J. Pharm. Sci.* **91**, 1601–1612 (2002).

323. Schräder, W.; Andersson, J. T., Fast and direct method for measuring 1-octanol-water partition coefficients exemplified for six local anesthetics, *J. Pharm. Sci.* **90**, 1948–1954 (2001).

324. Dyrssen, D., Studies on the extraction of metal complexes. IV. The dissociation constants and partition coefficients of 8-quinolinol (oxine) and N-nitroso-N-phenylhydroxylamine (cupferron), *Sv. Kem. Tidsks.* **64**, 213–224 (1952).

325. Rydberg, J., Studies on the extraction of metal complexes. IX. The distribution of acetylacetone between chloroform or hexone and water, *Sv. Kem. Tidskr.* **65**, 37–43 (1953).

326. Hök, B., Studies on the extraction of metal complexes. XV. The dissociation constants of salicylic acid, 3,5-dinitrobenzoic acid, and cinnamic acid and the distribution between chloroform-water and methyl isobutyl ketone (hexone)-water, *Sv. Kem. Tidskr.* **65**, 182–194 (1953).

327. Dyrssen, D., Studies on the extraction of metal complexes. XVIII. The dissociation constant and partition coefficient of tropolone, *Acta Chem. Scand.* **8**, 1394–1397 (1954).

328. Dyrssen, D.; Dyrssen, M.; Johansson, E., Studies on the extraction of metal complexes. XXXI. Investigation with some 5,7-dihalogen derivatives of 8-quinolinol, *Acta Chem. Scand.* **10**, 341–352 (1956).

329. Hök-Bernström, B., Studies on the extraction of metal complexes. XXIII. On the complex formation of thorium with salicylic acid, methoxybenzoic acid and cinnamic acid, *Acta Chem. Scand.* **10**, 174–186 (1956).

330. Dyrssen, D., Studies on the extraction of metal complexes. XXXII. N-phenylbenzohydroxamic acid, *Acta Chem. Scand.* **10**, 353–359 (1957).

331. Dyrssen, D., Studies on the extraction of metal complexes. XXX. The dissociation, distribution, and dimerization of di-n-buty phosphate (DBP), *Acta Chem. Scand.* **11**, 1771–1786 (1957).

332. Brändström, A., A rapid method for the determination of distribution coefficinet of bases for biological purposes, *Acta Chem. Scand.* **17**, 1218–1224 (1963).

333. Korenman, I. M.; Gur'ev, I. A., Acid-base titration in the presence of extractants, *J. Anal. Chem. USSR* **30**, 1601–1604 (Engl.) (1975).

334. Kaufman, J. J.; Semo, N. M.; Koski, W. S., Microelectrometric titration measurement of the pK_a's and partition and drug distribution coefficients of narcotics and narcotic antagonists and their pH and temperature dependence, *J. Med. Chem.* **18**, 647–655 (1975).

335. Johansson, P.-A.; Gustavii, K., Potentiometric titration of ionizable compounds in two phase systems. 2. Determination of partition coefficients of organic acids and bases, *Acta Pharm. Suecica* **13**, 407–420 (1976).

336. Johansson, P.-A.; Gustavii, K., Potentiometric titration of ionizable compounds in two phase systems. 3. Determination of extraction constants, *Acta Pharm. Suecica* **14**, 1–20 (1977).

337. Gur'ev, I. A.; Korenman, I. M.; Aleksandrova, T. G.; Kutsovskaya, V. V., Dual-phase titration of strong acids, *J. Anal. Chem. USSR* **32**, 192–195 (Engl.) (1977).

338. Gur'ev, I. A.; Gur'eva, Z. M., Dual-phase titration of aromatic acids, *J. Anal. Chem. USSR* **32**, 1933–1935 (Engl.) (1977).

339. Gur'ev, I. A.; Kiseleva, L. V., Dual-phase titration of aliphatic acids in the presence of neutral salts, J. Anal. Chem. USSR **33**, 427–430 (Engl.) (1978).

340. Gur'ev, I. A.; Zinina, O. B., Improving conditions for titrating acids by using dual-phase systems, *J. Anal. Chem. USSR* **33**, 1100–1102 (Engl.) (1978).

341. Gur'ev, I. A.; Gushchina, E. A.; Mitina, E. N., Dual-phase titration of phenols with an ion selective electrode, *J. Anal. Chem. USSR* **34**, 913–916 (Engl.) (1979).

342. Gur'ev, I. A.; Gushchina, E. A.; Gadashevich, M. Z., Dual-phase potentiometric titration of aromatic carboxylic acids with liquid ion selective electrodes, *J. Anal. Chem. USSR* **36**, 803–810 (Engl.) (1981).

343. Gur'ev, I. A.; Gushchina, E. A., Use of extraction parameters to predict the results of the dual-phase titration of phenolate ions with a liquid ion-slective electrode, *J. Anal. Chem. USSR* **37**, 1297–1301 (Engl.) (1982).

344. Clarke, F. H., Ionization constants by curve fitting. Application to the determination of partition coefficients, *J. Pharm. Sci.* **73**, 226–230 (1984).

345. Clarke, F. H.; Cahoon, N. M., Ionization constants by curve fitting: Determination of partition and distribution coefficients of acids and bases and their ions, *J. Pharm. Sci.* **76**, 611–620 (1987).

346. Avdeef, A., Fast simultaneous determination of logP and pK_a by potentiometry: Para-alkoxyphenol series (methoxy to pentoxy), in Silipo, C.; Vittoria, A. (eds.), *QSAR: Rational Approaches to the Design of Bioactive Compounds*, Elsevier, Amsterdam, 1991, pp. 119–122.

347. Avdeef, A., pH-metric log P. 1. Difference plots for determining ion-pair octanol-water partition coefficients of multiprotic substances, *Quant Struct. Act. Relat.* **11**, 510–517 (1992).

348. Avdeef, A.; Comer, J. E. A., Measurement of pK_a and logP of water-insoluble substances by potentiometric titration, in Wermuth, C. G. (ed.), *QSAR and Molecular Modelling*, Escom, Leiden, 1993, pp. 386–387.

349. Comer, J. E. A., The acid test; ionization and lipophilicity of drugs; influence on biological activity, *Chem. Britain* **30**, 983–986 (1994).

350. Clarke, F. H.; Cahoon, N. M., Ionization constants by curve fitting: determination of partition and distribution coefficients of acids and bases and their ions, *J. Pharm. Sci.* **83**, 1524 (1994).

351. Clarke, F. H.; Cahoon, N. M., Potentiometric determination of the partition and distribution coefficients of dianionic compounds, *J. Pharm. Sci.* **84**, 53–54 (1995).

352. Comer, J. E. A.; Avdeef, A.; Box, K. J., Limits for successful measurement of pK_a and log P by pH-metric titration, *Am. Lab.* **4**, 36c–36i (1995).

353. Herbette, L. G.; Vecchiarelli, M.; Trumlitz, G., *NSAID Mechanism of Action: Membrane Interactions in the Role of Intracellular Pharmacokinetics*, Vane, Botting, Botting, pp. 85–102 (1995).

354. Karajiannis, H.; van de Waterbeemd, H., The prediction of the lipophilicity of peptidomimetics. A comparison between experimental and theoretical lipophilicity values of renin inhibitors and the building blocks, *Pharm. Acta Helv.* **70**, 67–77 (1995).

355. Danielsson, L.-G.; Zhang, Y.-H., Methods for determining n-octanol-water partition constants, *Trends Anal. Chem.* **15**, 188–196 (1996).

356. Caron, G.; Pagliara, A.; Gaillard, P.; Carrupt, P-A.; Testa, B., Ionization and partitioning profiles of zwitterions: The case of the anti-inflammatory drug azapropazone, *Helv. Chim. Acta.* **79**, 1683–1695 (1996).

357. Takács-Novák, K.; Avdeef, A., Interlaboratory study of log P determination by shake-flask and potentiometric methods, *J. Pharm. Biomed. Anal.* **14**, 1405–1413 (1996).

358. McFarland, J. W.; Berger, C. M.; Froshauer, S. A.; Hayashi, S. F.; Hecker, S. J.; Jaynes, B. H.; Jefson, M. R.; Kamicker, B. J.; Lipinski, C. A.; Lundy, K. M.; Reese, C. P.; Vu, C. B., Quantitative structure-activity relationships among macrolide antibacterial agents: In vitro and in vivo potency against pasteurella multocida, *J. Med. Chem.* **40**, 1340–1346 (1997).

359. Caron, G.; P. Gaillard, P.;.Carrupt, P.-A.; Testa, B., Lipophilicity behavior of model and medicinal compounds containing a sulfide, sulfoxide, or sulfone moiety, *Helv. Chim. Acta.* **80**, 449–462 (1997).

360. Herbette, L. G.; Vecchiarelli, M.; Leonardi, A., Lercanidipine: Short plasma half-life, long duration of action, *J. Cardiovasc. Pharmacol.*, **29** (Suppl 1), S19–S24 (1997).

361. Morgan, M. E.; Liu, K.; Anderson, B. D., Microscale titrimetric and spectrophotometric methods for determination of ionization constants and partition coefficients of new drug candidates, *J. Pharm. Sci.* **87**, 238–245 (1998).

362. Caron, G.; Steyaert, G.; Pagliara, A.; Reymond, F.; Crivori, P.; Gaillard, P.; Carrupt, P.-A.; Avdeef, A.; Comer, J.; Box, K. J.; Girault, H. H.; Testa, B., Structure-lipophilicity relationships of neutral and protonated β-blockers. Part I: Intra and intermolecular effects in isotropic solvent systems, *Helv. Chim. Acta.* **82**, 1211–1222 (1999).

363. Franke, U.; Munk, A.; Wiese, M., Ionization constants and distribution coefficients of phenothiazines and calcium channel antagonists determined by a pH-metric method and correlation with calculated partition coefficients, *J. Pharm. Sci.* **88**, 89–95 (1999).

364. Leo, A.; Hansch, C.; Elkins, D., Partition coefficients and their uses, *Chem. Rev.* **71**, 525–616 (1971).

365. Sangster, J., Log KOW Databank, Montreal, Quebec, Canada, Sangster Research Laboratories, 1994.

366. Howard, P. H.; Meylan, W., *PHYSPROP Database*, Syracuse Research Corp., Syracuse, NY, 2000.

367. *Physicochemical Parameter Database*, Medicinal Chemisry Project, Pomona College, Claremont, CA, USA (www.biobyte.com/bb/prod/cqsar.html).

368. Krämer, S. D.; Jakits-Deiser, C.; Wunderli-Allenspach, H., Free-fatty acids cause pH-dependent changes in the drug-lipid membrane interations around the physiological pH, *Pharm. Sci.* **14**, 827–832 (1997).

369. Jonkman, J. H. G.; Hunt, C. A., Ion pair absorption of ionized drugs—fact or fiction? *Pharm. Weekblad Sci. Ed.* **5**, 41–47 (1983).

370. Dollery, C. (ed.)., *Therapeutic Drugs*, 2nd ed., Vols. 1 and 2, Churchill Livingstone, Edinburgh, 1999.

371. El Tayar, N.; Tsai, R.-S.; Carrupt, P.-A.; Testa, B., Octan-1-ol-water partition coefficinets of zwitterionic α-amino acids. Determination by centrifugal partition chromatography and factorization into steric/hydrophobic and polar components, *J. Chem. Soc. Perkin Trans.* **2**, 79–84 (1992).

372. Wilson, C. G.; Tomlinson, E.; Davis, S. S.; Olejnik, O., Altered ocular absorption and disposition of sodium cromoglycate upon ion-pair and complex coacervate formation with dodecylbenzyldimethyl-ammonium chloride, *J. Pharm. Pharmacol.* **31**, 749–753 (1981).

373. Megwa, S. A.; Cross, S. E.; Benson, H. A. E.; Roberts, M. S., Ion-pair formation as a strategy to enhance topical delivery of salicylic acid, *J. Pharm. Pharmacol.* **52**, 919–928 (2000).

374. M. S. Mirrlees, M. S.; S. J. Moulton, S. J.; C. T. Murphy; C. T.; P. J. Taylor, P. J., Direct measurement of octanol-water partiton coefficients by high-pressure liquid chromatography, *J. Med. Chem.* **19**, 615–619 (1976).

375. Unger, S. H.; Cook, J. R.; Hollenberg, J. S., Simple procedure for determining octanol-aqueous partition, distribution, and ionization coefficients by reverse-phase high-pressure liquid chromatography, *J. Pharm. Sci.* **67**, 1364–1366 (1978).

376. Loidl-Stahlhofen, A.; Eckert, A.; Hartmann, T.; Schöttner, M., Solid-supported lipid membranes as a tool for determination of membrane affinitity: High-throughput screening of a physicochemical parameter, *J. Pharm. Sci.* **90**, 599–606 (2001).

377. Terada, H.; Murayama, W.; Nakaya, N.; Nunogaki, Y.; Nunogaki, K.-I., Correlation of hydrophobic parameters of organic compounds determined by centrifugal partition chromatography with partiotion coefficients between octanol and water, *J. Chromatog.* **400**, 343–351 (1987).

378. El Tayar, N.; Tsai, R.-S.; Vallat, P.; Altomare, C.; Testa, B., Measurement of partition coefficients by various centrifugal partition chromatographic techniques. A comparative evaluation, *J. Chromatogr.* **556**, 181–194 (1991).

379. Tsai, R.-S.; El Tayar; N.; Carrupt, P.-A.; Testa, B., Physicochemical properties and transport behavior of piribedil: Considerations on its membrane-crossing potential, *Int. J. Pharm.* **80**, 39–49 (1992).

380. Cevc, G. (ed.), *Phospholipid Handbook*, Marcel Dekker, New York, 1993.

381. Miyoshi, H.; Maeda, H.; Tokutake, N.; Fujita, T., Quantitative analysis of partition behavior of substituted phenols from aqueous phase into liposomes made of lecithin and various lipids, *Bull. Chem. Soc. Jpn.* **60**, 4357–4362 (1987).

382. Escher, B. I.; Schwarzenbach, R. P., Partitioning of substituted phenols in liposome-water, biomembrane-water, and octanol-water systems, *Environ. Sci. Tech.* **30**, 260–270 (1996).

383. Escher, B. I.; Snozzi, M.; Schwarzenbach, R. P., Uptake, speciation, and uncoupling activity of substituted phenols in energy transducing membranes, *Environ. Sci. Technol.* **30**, 3071–3079 (1996).

384. Escher, B. I.; Hunziker, R.; Schwarzenbach, R. P.; Westall, J. C., Kinetic model to describe the intrinsic uncoupling activity of substituted phenols in energy transducing membranes, *Environ. Sci. Technol.* **33**, 560–570 (1999).

385. Balon, K.; Mueller, B. W.; Riebesehl, B. U., Determination of liposome partitioning of ionizable drugs by titration, *Pharm. Res.* **16**, 802–806 (1999).

386. Balon, K.; Mueller, B. W.; Riebesehl, B. U., Drug liposome partitioning as a tool for the prediction of human passive intestinal absorption, *Pharm. Res.* **16**, 882–888 (1999).

387. Fruttero, R.; Caron, G.; Fornatto, E.; Boschi, D.; Ermondi, G.; Gasco, A.; Carrupt, P.-A.; Testa, B., Mechanism of liposome/water partitioning of (p-methylbenzyl)alkylamines, *Pharm. Res.* **15**, 1407–1413 (1998).

388. Allan, D., Mapping the lipid distribution in the membranes of BHK cells, *Molec. Membr. Biol.* **13**, 81–84 (1996).

389. Hope, M. J.; Bally, M. B.; Webb, G.; Cullis, P. R., Production of large unilamellar vesicles by a rapid extrusion procedure. Characterization of size distribution, trapped volume and ability to maintain a membrane potential, *Biochim. Biophys. Acta* **812**, 55–65 (1985).

390. Mayer, L. D.; Hope, M. J.; Cullis, P. R.; Janoff, A. S., Solute distributions and trapping efficiencies observed in freeze-thawed multilamellar vesicles, *Biochim. Biophys. Acta* **817**, 193–196 (1985).

391. Mayer, L. D.; Bally, M. B.; Hope, M. J.; Cullis, P. R., Techniques for encapsulating bioactive agents into liposomes, *Chem. Phys. Lipids* **40**, 333–345 (1986).

392. Perkins, W. R.; Minchey, S. R.; Ahl, P. H.; Janoff, A. S., The determination of liposome captured volume, *Chem. Phys. Lipids* **64**, 197–217 (1993).

393. Lasic, D. D.; Needham, D., The "stealth" liposome. A prototypical biomaterial, *Chem. Rev.* **95**, 2601–2628 (1995).

394. Davis, S. S.; James, M. J.; Anderson, N. H., The distribution of substituted phenols in lipid vescles, *Faraday Discuss. Chem. Soc.* **81**, 313–327 (1986).

395. Bäuerle, H.-D.; Seelig, J., Interaction of charged and uncharged calcium channel antagoninsts with phospholipid membranes. Binding equilibrium, binding enthalpy, and membrane location, *Biochemistry* **30**, 7203–7211 (1991).

396. Miyazaki, J.; Hideg, K.; Marsh, D., Inerfacial ionization and partitioning of membrane-bound local anesthetics, *Biochim. Biophys. Acta* **1103**, 62–68 (1992).

397. Thomas, P. G.; Seelig, J., Binding of the calcium antagonist flunarizine to phosphatidyl-choline bilayers: Charge effects and thermodynamics, *Biochem. J.* **291**, 397–402 (1993).

398. Wenk, M. R.; Fahr, A.; Reszka, R.; Seelig, J., Paclitaxel partitioning into lipid bilayers, *J. Pharm. Sci.* **85**, 228–231 (1996).

399. Yeagle, P. L.; Hutton, W. C.; Huang, C.-H.; Martin, R. B., Headgroup conformation and lipid-cholesterol association in phosphatidylcholine vesicles: A $^{31}P\{^1H\}$ nuclear Overhauser effect study, *Proc. Natl. Acad. Sci.* **72**, 3477–3481 (1975).

400. Gur, Y.; Ravina, I.; Babchin, A. J., On the electrical double layer theory. II. The Poisson-Boltzman equation including hydration forces, *J. Colloid Inter. Sci.* **64**, 333–341 (1978).

401. Brown, M. F.; Seelig, J., Influence of cholesterol on the polar region of phosphatidylcholine and phosphatidylethanolamine bilayers, *Biochemistry* **17**, 381–384 (1978).

402. Eisenberg, M.; Gresalfi, T.; Riccio, T.; McLaughlin, S., Adsorption of monovalent cations to bilayer membranes containing negative phospholipids, *Biochemistry* **18**, 5213–5223 (1979).

403. Cevc, G.; Watts, A.; Marsh, D., Titration of the phase transition of phosphatidylserine bilayer membranes. Effects of pH, surface electrostatics, ion binding, and head group hydration. *Biochemistry* **20**, 4955–4965 (1981).

404. Rooney, E. K.; Lee, A. G., Binding of hydrophobic drugs to lipid bilayers and to the $(Ca^{2+} + Mg^{2+})$-ATPase, *Biochim. Biophys. Acta* **732**, 428–440 (1983).

405. Cevc, G.; Marsh, D. Properties of the electrical double layer near the interface between a charged bilayer membrane and electrolyte solution: Experiment vs. theory, *J. Phys. Chem.* **87**, 376–379 (1983).

406. Cevc, G., Membrane electrostatics, *Biochim. Biophys. Acta* **1031**, 311–382 (1990).

407. McLaughlin, S., Electrostatic potentials at membrane-solution interfaces, *Curr. Topics. Membr. Transport* **9**, 71–144 (1977).

408. Shinitzky, M. (ed.), *Biomembranes: Physical Aspects*, VCH, Weinheim, 1993.

409. Biegel, C. M.; Gould, J. M., Kinetics of hydrogen ion diffusion across phospholipid vesicle membranes, *Biochemistry* **20**, 3474–3479 (1981).

410. Perkins, W. R.; Cafiso, D. S., An electrical and structural characterization of H^+/OH^- currents in phospholipid vesicles, *Biochemistry*, **25**, 2270–2276 (1986).

411. Meier, E. M.; Schummer, D.; Sandhoff, K., Evidence for the presence of water within the hydrophobic core of membranes, *Chem. Phys. Lipids* **55**, 103–113 (1990).

412. Nagle, J. F., Theory of passive proton conductance in lipid bilayers, *J. Bioenerg. Biomem.* **19**, 413–426 (1987).

413. Gutknecht, J., Proton conductance through phospholipid bilayers: Water wires or weak acids, *J. Bioenerg. Biomem.* **19**, 427–442 (1987).

414. Redelmeier, T. E.; Mayer, L. D.; Wong, K. F.; Bally, M. B.; Cullis, P. R., Proton flux in large unilamellar vesicles in response to membrane potentials and pH gradients, *Biophys. J.* **56**, 385–393 (1989).

415. Norris, F. A.; Powell, G. L., The apparent permeability coefficient for proton flux through phosphatidylcholine vesicles is dependent on the direction of flux, *Biochim. Biophys. Acta* **1030**, 165–171 (1990).

416. Madden, T. D.; Harrigan, P. R.; Tai, L. C. L.; Bally, M. B.; Mayer, L. D.; Redelmeier, T. E.; Loughrey, H. C.; Tilcock, C. P. S.; Reinish, L. W.; Cullis, P. R., The accumulation of drugs within large unilamellar vesicles exhibiting a proton gradient: A survey, *Chem. Phys. Lipids* **53**, 37–46 (1990).

417. Chakraborti, A. C.; Deamer, D. W., Permeability of lipid bilayers to amino acids and phosphate, *Biochim. Biophys. Acta* **1111**, 171–177 (1992).

418. Romanowski, M.; Zhu, X.; Kim, K.; Hruby, V.; O'Brien, D. F., Interactions of enkaphalin peptides with anionic model membraines, *Biochim. Biophys. Acta* **1558**, 45–53 (2002).

419. Lee, R. J.; Wang, S.; Low, P. S., Measurement of endosome pH following folate receptor-mediated endocytosis, *Biochim. Biophys. Acta* **1312**, 237–242 (1996).

420. Boulanger, Y.; Schreier, S.; Leitch, L. C.; Smith, I. C. P., Multiple binding sites for local anesthetics in membranes: Characterization of the sites and their equilibria by deuterium NMR of specifically deuterated procaine and tetracaine, *Can. J. Biochem.* **58**, 986–995 (1980).

421. Boulanger, Y.; Schreier, S.; Smith, I. C. P., Molecular details of anesthetic-lipid interaction as seen by deuterium and phosphorus-31 nuclear magnetic resonance, *Biochemistry* **20**, 6824–6830 (1981).

422. Westman, J.; Boulanger, Y.; Ehrenberg, A.; Smith, I. C. P., Charge and pH dependent drug binding to model membranes: A ^{2}H-NMR and light absorption study, *Biochim. Biophys. Acta* **685**, 315–328 (1982).

423. Kelusky, E. C.; Smith, I. C. P., Anethetic-membrane interaction: A ^{2}H nuclear magnetic resonance study of the binding of specifically deuterated tetracaine and procaine to phosphatidylcholine, *Can. J. Biochem. Cell Biol.* **62**, 178–184 (1984).

424. Schreier, S.; Frezzatti, W. A., Jr.; Araujo, P. S.; Chaimovich, H.; Cuccovia, I. M., Effect of lipid membranes on the apparent pK of the local anesthetic tetracaine: Spin label and titration studies, *Biochim. Biophys. Acta* **769**, 231–237 (1984).

425. Trumbore, M.; Chester, D. W.; Moring, J.; Rhodes, D.; Herbette, L. G., Structure and location of amiodarone in a membrane bilayer as determined by molecular mechanics and quantitative X-ray diffraction, *Biophys. J.* **54**, 535–543 (1988).

426. Mason, R. P.; Campbell, S. F.; Wang, S.-D.; Herbette, L. G., Comparison of location and binding for the positively charged 1,4-dihydropyridine calcium channel antagonist amlopidine with uncharged drugs of this class in cardiac membranes, *Molec. Pharmacol.* **36**, 634–640 (1989).

427. Mason, R. P.; Rhodes, D. G.; Herbette, L. G., Reevaluating equilibrium and kinetic binding parameters for lipophilic drugs based on a structural model for drug interaction with bilogical membranes, *J. Med. Chem.* **34**, 869–877 (1991).

428. Herbette, L. G.; Rhodes, D. R.; Mason, R. P., New approaches to drug design and delivery based on drug-membrane interactions, *Drug Design Deliv.* **7**, 75–118 (1991).

429. Chatelain, P.; Laruel, R., Amiodarone partitioning with phospholipid bilayers and erythrocyte membranes, *J. Pharm. Sci.* **74**, 783–784 (1985).

430. Betageri, G. V.; Rogers, J. A., Thermodynamics of partitioning of β-blockers in the n-octanol buffer and liposome systems, *Int. J. Pharm.* **36**, 165–173 (1987).

431. Choi, Y. W.; Rogers, J. A., The liposome as a model membrane in correlations of partitioning with α-adrenoreceptor against activities, *Pharm. Res.* **7**, 508–512 (1990).

432. Rogers, J. A.; Choi, Y. W., The liposome partitioning system for correlating biological activities for imidazolidine derivatives, *Pharm. Res.* **10**, 913–917 (1993).

433. Alcorn, C. J.; Simpson, R. J.; Leahy, D. E.; Peters, T. J., Partition and distribution coefficients of solutes and drugs in Brush Border membrane vesicles, *Biochem. Pharmacol.* **45**, 1775–1782 (1993).

434. Smejtek, P.; Wang, S., Distribution of hydrophobic ionizable xenobiotics between water and lipid membrane: Pentachlorophenol and pentachlorophenate. A comparison with octanol-water partition, *Arch. Environ. Contam. Toxicol.* **25**, 394–404 (1993).

435. Pauletti, G. M.; Wunderli-Allenspach, H., Partition coefficients in vitro: Artificial membranes as a standardized distribution model, *Eur. J. Pharm. Sci.* **1**, 273–281 (1994).

436. Krämer, S. D.; Jakits-Dieser, C.; Wunderli-Allenspach, H., Free fatty acids cause pH-dependent changes in drug-lipid membrane interactions around physiological pH, *Pharm. Res.* **14**, 827–832 (1997).

437. Ottiger, C.; Wunderli-Allenspach, H., Partition behavior of acids and bases in a phosphatidylcholine liposome-buffer equilibrium dialysis system, *Eur. J. Pharm. Sci.* **5**, 223–231 (1997).

438. Krämer, S. D.; Braun, A.; Jakits-Dieser, C.; Wunderli-Allenspach, H., Towards the predictability of drug-lipid membrane interactions: The pH-dependent affinity of propranolol to phosphatidyliniositol containing liposomes, *Pharm. Rev.* **15**, 739–744 (1998).

439. Xiang, T.-X.; Anderson, B. D., Phospholipid surface density determines the partitioning and permeability of acetic acid in DMPC: cholesterol bilayers, *J. Membrane Biol.* **148**, 157–167 (1995).

440. Xiang, T.-X.; Anderson, B. D., Development of a combined NMR paramagnetic ion-induced line-broadening dynamic light scattering method for permeability measurements across lipid bilayer membranes, *J. Pharm. Sci.* **84**, 1308–1315 (1995).

441. Austin, R. P.; Davis, A. M.; Manners, C. N., Partitioning of ionized molecules between aqueous buffers and phospholipid vesicles, *J. Pharm. Sci.* **54**, 1180–1183 (1995).

442. Austin, R. P.; Barton, P.; Davis, A. D.; Manners, C. N.; Stansfield, M. C., The effect of ionic strength on liposome-buffer and 1-octanol-buffer distribution coefficients, *J. Pharm. Sci.* **87**, 599–607 (1998).

443. Tatulian, S. A., Ionization and ion binding, in Cevc, G. (ed.), *Phospholipid Handbook*, Marcel Dekker, New York, 1993, pp. 511–550.

444. Miller, K. W.; Yu, S.-C. T., The dependence of the lipid bilayer membrane: Buffer partition coefficient of pentobarbitone on pH and lipid composition, *Br. J. Pharm.* **61**, 57–63 (1977).

445. Herbette, L. G., "Pharmacokinetic" and "pharmacodynamic" design of lipophilic drugs based on a structural model for drug interactions with biological membranes, *Pest. Sci.* **35**, 363–368 (1992).

446. Bellemare, F.; Fragata, M., *J. Colloid Interface Sci.* **77**, 243 (1980).

447. Colbow, K.; Chong, C. S., in Colbow, K. (ed.), *Biological Membranes*, Simon Fraser Univ. Burnaby, Canada, 1975, p. 145.

448. Fernández, M. S.; Fromherz, P., Lipoid pH indicators as probes of electrical potential and polarity in micelles, *J. Phys. Chem.* **81**, 1755–1761 (1977).

449. Thomas, J. K., *Chem. Rev.* **80**, 283 (1980).

450. Iwamoto, K.; Sunamoto, J., *Bull. Chem. Soc. Jpn.* **54**, 399 (1981).

451. Laver, D. R.; Smith, J. R.; Coster, H. G. L., *Biochim. Biophys. Acta* **772**, 1 (1984).

452. Kanashina, S.; Kamaya, H.; Ueda, I., *Biochim. Biophys. Acta* **777**, 75 (1984).

453. Lessard, J. G.; Fragata, M., Micropolarities of lipid bilayers and micelles. 3. Effect of monovalent ions on the dielectric constant of the water-membrane interface of unilamellar phosphatidylcholine vesicles, *J. Phys. Chem.* **90**, 811–817 (1986).

454. Kariv, I.; Cao, H.; Oldenburg, K. R., Development of a high-throughput equilibrium dialysis method, *J. Pharm. Sci.* **90**, 580–587 (2001).

455. Lee, A. G., Effects of charged drugs on the phase transition temperature of phospholipid bilayers, *Biochim. Biophys. Acta* **514**, 95–104 (1978).

456. Davis, M. G.; Manners, C. N.; Payling, D. W.; Smith, D. A.; Wilson, C. A., GI absorption for the strongly acidic drug proxicromil, *J. Pharm. Sci.* **73**, 949–953 (1984).

457. Bunton, C. A.; Ohmenzetter, K.; Sepulvida, L., Binding of hydrogen ions to anionic micelles, *J. Phys. Chem.* **81**, 2000–2004 (1977).

458. Garcia-Soto, J.; Fernández, M. S., The effect of neutral and charged micelles on the acid-base dissociation of the local anesthetic tetracaine, *Biochim. Biophys. Acta* **731**, 275–281 (1983).

459. Saad, H. Y.; Higuchi, T., Water solubility of cholesterol, *J. Pharm. Sci.* **54**, 1205–1206 (1965).

460. Mosharraf, M.; Nyström, C., Solubility characterization of practically insoluble drugs using the coulter counter principle, *Int. J. Pharm.* **122**, 57–67 (1995).

461. Quartermain, C. P.; Bonham, N. M.; Irwin, A. K., Improving the odds—high-throughput techniques in new drug selection, *Eur. Pharm. Rev.* **18**, 27–32 (1998).

462. Bevan, C. D.; Lloyd, R. S., A high-throughput screening method for the determination of aqueous drug solubility using laser nephelometry in microtitre plates, *Anal. Chem.* **72**, 1781–1787 (2000).

463. Pan, L.; Ho, Q.; Tsutsui, K.; Takahashi, L., Comparison of chromatographic and spectroscopic methods used to rank compounds for aqueous solubility, *J. Pharm. Sci.* **90**, 521–529 (2001).

464. Hancock, B. C.; Zografi, G., Characterization and significance of the amorphous state in pharmaceutical systems, *J. Pharm. Sci.* **86**, 1–12 (1997).

465. Feitknecht, W.; Schindler, P., *Solubility Constant of Metal Oxides, Metal Hydroxides and Metal Hydroxide Salts in Aqueous Solution*, Butterworths, London, 1963.

466. Brittain, H. G., Spectral methods for the characterization of polymorphs and solvates, *J. Pharm. Sci.* **86**, 405–411 (1997).

467. Levy, R. H.; Rowland, M., Dissociation constants of sparingly soluble substances: Nonlogarithmic linear titration curves, *J. Pharm. Sci.* **60**, 1155–1159 (1971).

468. Charykov, A. K.; Tal'nikova, T.V., pH-metric method of determining the solubility and distribution ratios of some organic compounds in extraction systems, *J. Anal. Chem. USSR* **29**, 818–822 (Engl.) (1974).

469. Kaufman, J. J.; Semo, N. M.; Koski, W. S., Microelectrometric titration measurement of the pK_as and partition and drug distribution coefficients of narcotics and narcotic antagonists and their pH and temperature dependence, *J. Med. Chem.* **18**, 647–655 (1975).

470. Streng, W. H.; Zoglio, M. A., Determination of the ionization constants of compounds which precipitate during potentiometric titration using extrapolation techniques, *J. Pharm. Sci.* **73**, 1410–1414 (1984).

471. Todd, D.; Winnike, R. A., A rapid method for generating pH-solubility profiles for new chemical entities, Abstr. 9th Ann. Mtng., Amer. Assoc. Pharm. Sci., San Diego, 1994.

472. Avdeef, A., pH-metric solubility. 1. Solubility-pH profiles from Bjerrum plots. Gibbs buffer and pK_a in the solid state, *Pharm. Pharmacol. Commun.* **4**, 165–178 (1998).

473. Avdeef, A.; Berger, C. M.; Brownell, C., pH-metric solubility. 2. Correlation between the acid-base titration and the saturation shake-flask solubility-pH methods, *Pharm. Res.* **17**, 85–89 (2000).

474. Avdeef, A.; Berger, C. M., pH-metric solubility. 3. Dissolution titration template method for solubility determination, *Eur. J. Pharm. Sci.* **14**, 281–291 (2001).

475. Chowhan, Z. T., pH-solubility profiles of organic carboxylic acids and their salts, *J. Pharm. Sci.* **67**, 1257–1260 (1978).

476. Bogardus, J. B.; Blackwood, R. K. Jr., Solubility of doxycycline in aqueous solution, *J. Pharm. Sci.* **68**, 188–194 (1979).

477. Ahmed, B. M.; Jee, R. D., The acidity and solubility constants of tetracyclines in aqueous solution, *Anal. Chim. Acta* **166**, 329–333 (1984).

478. Chiarini, A.; Tartarini, A.; Fini, A., pH-solubility relationship and partition coefficients for some anti-inflammatory arylaliphatic acids, *Arch. Pharm.* **317**, 268–273 (1984).

479. Anderson, B. D., Conradi, R. A., Predictive relationships in the water solubility of salts of a nonsteroidal anti-inflammatory drug, *J. Pharm. Sci.* **74**, 815–820 (1985).

480. Streng, W. H.; Tan, H. G. H., General treatment of pH solubility profiles of weak acids and bases. II. Evaluation of thermodynamic parameters from the temperature dependence of solubility profiles applied to a zwitterionic compound, *Int. J. Pharm.* **25**, 135–145 (1985).

481. Zimmermann, I., Determination of overlapping pK_a values from solubility data, *Int. J. Pharm.* **31**, 69–74 (1986).

482. Garren, K. W., Pyter, R. A., Aqueous solubility properties of a dibasic peptide-like compound, *Int. J. Pharm.* **63**, 167–172 (1990).

483. Islam, M. H.; Narurkar, M. M., Solubility, stability and ionization behavior of famotidine, *J. Pharm. Pharmacol.* **45**, 682–686 (1993).

484. Streng, W. H.; Yu, D. H.-S. Precision tests of a pH-solubility profile computer program, *Int. J. Pharm.* **164**, 139–145 (1998).

485. Roseman, T. J.; Yalkowsky, S. H., Physical properties of prostaglandin $F_{2\alpha}$ (tromethamine salt): Solubility behavior, surface properties, and ionization constants, *J. Pharm. Sci.* **62**, 1680–1685 (1973).

486. Attwood, D.; Gibson, J., Aggregation of antidepressant drugs in aqueous solution, *J. Pharm. Pharmacol.* **30**, 176–180 (1978).

487. Streng, W. H.; Yu, D. H.-S.; Zhu, C., Determination of solution aggregation using solubility, conductivity, calorimetry, and ph measurements, *Int. J. Pharm.* **135**, 43–52 (1996).

488. Zhu, C.; Streng, W. H., Investigation of drug self-association in aqueous solution using calorimetry, conductivity, and osmometry, *Int. J. Pharm.* **130**, 159–168 (1996).

489. Ritschel, W. A.; Alcorn, G. C.; Streng, W. H.; Zoglio, M. A., Cimetidine-theophylline complex formation, *Meth. Find. Exp. Clin. Pharmacol.* **5**, 55–58 (1983).

490. Li, P.; Tabibi, S. E.; Yalkowsky, S. H., Combined effect of complexation and pH on solubilization, *J. Pharm. Sci.* **87**, 1535–1537 (1998).

491. Nuñez, F. A. A.; Yalkowsky, S. H., Solubilization of diazepam, *J. Pharm. Sci. Tech.* 33–36 (1997).

492. Meyer, J. D.; Manning, M. C., Hydrophobic ion pairing: Altering the solubility properties of biomolecules, *Pharm. Res.* **15**, 188–193 (1998).

493. Narisawa, S.; Stella, V. J., Increased shelf-life of fosphenytoin: solubilization of a degradant, phenytoin, through complexation with (SBE)$_{7m}$-β-CD, *J. Pharm. Sci.* **87**, 926–930 (1998).

494. Higuchi, T.; Shih, F.-M. L.; Kimura, T.; Rytting, J. H., Solubility determination of barely aqueous soluble organic solids, *J. Pharm. Sci.* **68**, 1267–1272 (1979).

495. Venkatesh, S.; Li, J.; Xu, Y.; Vishnuvajjala, R.; Anderson, B. D., Intrinsic solubility estimation and pH-solubility behavior of cosalane (NSC 658586), and extremely hydrophobic diprotic acid, *Pharm. Res.* **13**, 1453–1459 (1996).

496. Badwan, A. A.; Alkaysi, H. N.; Owais, L. B.; Salem, M. S.; Arafat, T. A., Terfenadine, *Anal. Profiles Drug Subst.* **19**, 627–662 (1990).

497. Streng, W. H., Hsi, S. K.; Helms, P. E.; Tan, H. G. H., General treatment of pH-solubility profiles of weak acids and bases and the effect of different acids on the solubility of a weak base, *J. Pharm. Sci.* **73**, 1679–1684 (1984).

498. Miyazaki, S.; Oshiba, M.; Nadai, T., Precaution on use of hydrochloride salts in pharmaceutical formulation, *J. Pharm. Sci.* **70**, 594–596 (1981).

499. Ledwidge, M. T.; Corrigan, O. I., Effects of surface active characteristics and solid state forms on the pH solubility profiles of drug-salt systems, *Int. J. Pharm.* **174**, 187–200 (1998).

500. Jinno, J.; Oh, D.-M.; Crison, J. R.; Amidon, G. L., Dissolution of ionizable water-insoluble drugs: The combined effect of pH and surfactant, *J. Pharm. Sci.* **89**, 268–274 (2000).

501. Mithani, S. D.; Bakatselou, V.; TenHoor, C. N.; Dressman, J. B., Estimation of the increase in solubility of drugs as a function of bile salt concentration, *Pharm. Res.* **13**, 163–167 (1996).

502. Anderson, B. D.; Flora, K. P., Preparation of water-soluble compounds through salt formation, in Wermuth, C. G. (ed.), *The Practice of Medicinal Chemistry*, Academic Press, London, 1996, pp. 739–754.

503. Engel, G. L.; Farid, N. A.; Faul, M. M.; Richardson, L. A.; Winneroski, L. L., Salt selection and characterization of LY333531 mesylate monohydrate, *Int. J. Pharm.* **198**, 239–247 (2000).

504. McFarland, J. W.; Avdeef, A.; Berger, C. M.; Raevsky, O. A., Estimating the water solubilities of crystalline compounds from their chemical structure alone, *J. Chem. Inf. Comput. Sci.* **41**, 1355–1359 (2001).

505. McFarland, J. W.; Du, C. M.; Avdeef, A., Factors influencing the water solubility of crystalline drugs, in van de Waterbeemd, H.; Lennernäs, H.; Artursson, P. (eds.), *Drug Bioavailability. Estimation of Solubility, Permeability, Absorption and Bioavailability*, Wiley-VCH, Weinheim, 2003 (in press).

506. Bergström, C. A. S.; Strafford, M.; Lazarova, L.; Avdeef, A.; Luthman, K.; Artursson, P., Absorption classification of oral drugs based on molecular surface properties, *J. Med. Chem.* **46**, 558–570 (2003).

507. Yalkowsky, S. H.; Dannenfelser, R.-M. (eds.), *AQUASOL dATAbASE of Aqueous Solubility*, 5th ed., College of Pharmacy, Univ. Arizona, Tucson, 1998.

508. Abraham, M. H.; Le, J., The correlation and prediction of the solubility of compounds in water using an amended solvation energy relationship, *J. Pharm. Sci.* **88**, 868–880 (1999).

509. Faller, B.; Wohnsland, F., Physicochemical parameters as tools in drug discovery and lead optimization, in Testa, B.; van de Waterbeemd, H.; Folkers, G.; Guy, R. (eds.), *Pharmacokinetic Optimization in Drug Research*, Verlag Helvetica Chimica Acta, Zürich and Wiley-VCH, Weinheim, 2001, pp. 257–274.

510. Hidalgo, I. J.; Kato, A.; Borchardt, R. T., Binding of epidermal growth factor by human colon carcinoma cell (Caco-2) monolayers, *Biochem. Biophys. Res. Commun.* **160**, 317–324 (1989).

511. Artursson, P., Epithelial transport of drugs in cell culture. I: A model for studying the passive diffusion of drugs over intestinal absorptive (Caco-2) cells, *J. Pharm. Sci.* **79**, 476–482 (1990).

512. Karlsson, J. P.; Artursson, P., A method for the determination of cellular permeability coefficients and aqueous boundary layer thickness in monolayers of intestinal epithelial (Caco-2) cells grown in permeable filter chambers, *Int. J. Pharm.* **7**, 55–64 (1991).

513. Hilgers, A. R.; Conradi, R. A.; Burton, P. S., Caco-2 cell monolayers as a model for drug transport across the intestinal mucosa, *Pharm. Res.* **7**, 902–910 (1990).

514. Ho, N. F. H.; Raub, T. J.; Burton, P. S.; Barsuhn, C. L.; Adson, A.; Audus, K. L.; Borchardt, R. T., Quantitative approaches to delineate passive transport mechanisms in cell culture monolayers, in Amidon, G. L.; Lee, P. I.; Topp, E. M. (eds.), *Transport Processes in Pharmaceutical Systems*, Marcel Dekker, New York, 2000, pp. 219–317.

515. Hidalgo, I. J.; Hillgren, K. M.; Grass, G. M.; Borchardt, R. T., A new side-by-side diffusion cell for studying transport across epithelial cell monolayers, *in Vitro Cell Dev. Biol.* **28A**, 578–580 (1992).

516. Mueller, P.; Rudin, D. O.; Tien, H. T.; Westcott, W. C., Reconstitution of cell membrane structure in vitro and its transformation into an excitable system, *Nature* **194**, 979–980 (1962).

517. Tanford, C., *Ben Franklin Stilled the Waves: An Informal History of Pouring Oil on Water with Reflections on the Ups and Downs of Scientific Life in General*, Duke Univ. Press: Durham, NC, 1989.

518. Tien, T. H.; Ottova, A. L., The lipid bilayer concept and its experimental realization: From soap bubbles, kitchen sink, to bilayer lipid membranes, *J. Membr. Sci.* **189**, 83–117 (2001).

519. Ottova, A.; Tien, T. H., The 40th anniversary of bilayer lipid membrane research, *Bioelectrochemistry* **5693** (2002).

520. Hooke, R., Royal Society Meeting, in Birch, T. (ed.). *The History of the Royal Society of London*, Vol. 3, No. 29, A., Miller, London, 1672, p. (1757).

521. Newton, I., *Optics*, Dover, New York, 1952 (1704), pp. 215–232 (reprinted).

522. Franklin, B., Of the stilling of waves by means of oil, *Phil. Trans. (Roy. Soc.)* **64**, 445–460 (1974).

523. Tien, H. T.; Ottova, A. L., *Membrane Biophysics: As Viewed from Experimental Bilayer Lipid Membranes (Planar Lipid Bilayers and Spherical Liposomes)*, Elsevier: Amsterdam, 2000.

524. Overton, E., *Vjschr. Naturforsch. Ges. Zurich* **44**, 88 (1899).

525. Collander, R.; Bärlund, H., Permeabilititätsstudien an Chara eratophylla, *Acta Bot. Fenn.* **11**, 72–114 (1932).

526. Langmuir, I., The constitution and fundamental properties of solids and liquids. II. Liquids, *J. Am. Chem. Soc.* **39**, 1848–1906 (1917).

527. Gorter, E.; Grendel, F., On biomolecular layers of lipids on the chromocytes of the blood, *J. Exp. Med.* **41**, 439–443 (1925).

528. Danielli, J. F.; Davson, H., A contribution to the theory of permeability of thin films, *J. Cell Comp. Physiol.* **5**, 495–508 (1935).

529. Singer, S. J.; Nicolson, G. L., The fluid mosaic model of the structure of cell membranes, *Science* **175**, 720–731 (1972).

530. Singer, S. J., The molecular organization of membranes, *Ann. Rev. Biochem.* **43**, 805–834 (1974).

531. Fishman, P. (ed.), in *Proc. Symp. Plasma Membrane*, Am. Heart Assoc. and NY Heart Assoc., New York City, Dec 8–9, 1961, Circulation **26**(5) (1962) (suppl.).

532. Mueller, P.; Rudin, D. O.; Tien, H. T.; Wescott, W. C., Reconstitution of cell membrane structure in vitro and its transformation into an excitable system, *J. Phys. Chem.* **67**, 534 (1963).

533. Bangham, A. D., Surrogate cells or Trojan horses, *BioEssays* 1081–1088 (1995).

534. Barry, P. H.; Diamond, J. M., Effects of the unstirred layers on membrane phenomena, *Physiol. Rev.* **64**, 763–872 (1984).

535. Gutknecht, J.; Tosteson, D. C., Diffusion of weak acids across lipid membranes: Effects of chemical reactions in the unstirred layers, *Science* **182**, 1258–1261 (1973).

536. Gutknecht, J.; Bisson, M. A.; Tosteson, F. C., Diffusion of carbon dioxide through lipid bilayer membranes. Effects of carbonic anhydrase, bicarbonate, and unstirred layers, *J. Gen. Physiol.* **69**, 779–794 (1977).

537. Walter, A.; Gutknecht, J., Monocarboxylic acid permeation through lipid bilayer membranes, *J. Membr. Biol.* **77**, 255–264 (1984).

538. Xiang, T.-X.; Anderson, B. D., Substituent contributions to the transport of substituted p-toluic acids across lipid bilayer membranes, *J. Pharm. Sci.* **83**, 1511–1518 (1994).

539. Pohl, P.; Saparov, S. M.; Antonenko, Y. N., The size of the unstirred water layer as a function of the solute diffusion coefficient, *Biphys. J.* **75**, 1403–1409 (1998).

540. Antonenko, Y. N.; Denisov, G. A.; Pohl, P., Weak acid transport across bilayer lipid membrane in the presence of buffers, *Biophys. J.* **64**, 1701–1710 (1993).

541. Cotton, C. U.; Reuss, L., Measurement of the effective thickness of the mucosal unstirred layer in Necturus gallbladder epithelium, *J. Gen. Physiol.* **93**, 631–647 (1989).

542. Mountz, J. M.; Tien, H. T., Photoeffects of pigmented lipid membranes in a micropourous filter, *Photochem. Photobiol.* **28**, 395–400 (1978).

543. Thompson, M.; Lennox, R. B.; McClelland, R. A., Structure and electrochemical properties of microfiltration filter-lipid membrane systems, *Anal. Chem.* **54**, 76–81 (1982).

544. O'Connell, A. M.; Koeppe, R. E., II; Andersen, O. S., Kinetics of gramicidin channel formation in lipid bilayers: Transmembrane monomer association, *Science* **250**, 1256–1259 (1990).

545. Cools, A. A.; Janssen, L. H. M., Influence of sodium ion-pair formation on transport kinetics of warfarin through octanol-impregnated membranes, *J. Pharm. Pharmacol.* **35**, 689–691 (1983).

546. Camenisch, G.; Folkers, G.; van de Waterbeemd, H., Comparison of passive drug transport through Caco-2 cells and artificial membranes, *Int. J. Pharm.* **147**, 61–70 (1994).

547. Kansy, M.; Senner, F.; Gubernator, K., Physicochemical high throughput screening: Parallel artificial membrane permeability assay in the description of passive absorption processes, *J. Med. Chem.* **41**, 1070–1110 (1998).

548. Ghosh, R., Novel membranes for simulating biological barrier transport, *J. Membr. Sci.* **192**, 145–154 (2001).

549. Zhu, C.; Chen, T.-M.; Hwang, K., A comparative study of parallel artificial membrane permeability assay for passive absorption screening, in CPSA2000: The Symposium on Chemical and Pharmaceutical Structure Analysis. Milestone Development Services. Princeton, NJ, Sept. 26–28, 2000.

550. Kansy, M.; Fischer, H.; Kratzat, K.; Senner, F.; Wagner, B.; Parrilla, I., High-throughput aritificial membrane permeability studies in early lead discovery and development, in Testa, B.; van de Waterbeemd, H.; Folkers, G.; Guy, R. (eds.), *Pharmacokinetic Optimization in Drug Research*, Verlag Helvetica Chimica Acta, Zürich and Wiley-VCH: Weinheim, 2001, pp. 447–464.

551. Krishna, G.; Chen, K.-J.; Lin, C.-C.; Nomeir, A. A., Permeability of lipophilic compounds in drug discovery using in-vitro human absorption model, Caco-2, *Int. J. Pharm.* **222**, 77–89 (2001).

552. Adson, A.; Burton, P. S.; Raub, T. J.; Barsuhn, C. L.; Audus, K. L.; Ho, N. F. H., Passive diffusion of weak organic electrolytes across Caco-2 cell monolayers: Uncoupling the contributions of hydrodynamic, transcellular, and paracellular barriers, *J. Pharm. Sci.* **84**, 1197–1204 (1995).

553. Pontier, C.; Pachot, J.; Botham, R.; Lefant, B.; Arnaud, P., HT29-MTX and Caco-2/TC7 monolayers as predictive models for human intestinal absorption: Role of mucus layer, *J. Pharm. Sci.* **90**, 1608–1619 (2001).

554. Wohnsland, F.; Faller, B., High-throughput permeability pH profile and high-throughput alkane/water log P with artificial membranes, *J. Med. Chem.* **44**, 923–930 (2001).

555. Hwang, K., Predictive artificial membrane technology for high throughput screening, in *New Technologies to Increase Drug Candidate Survivability Conf.*, SR Institute, May 17–18, 2001, Somerset, NJ.

556. Avdeef, A.; Strafford, M.; Block, E.; Balogh, M. P.; Chambliss, W.; Khan, I., Drug absorption in vitro model: Filter-immobilized artificial membranes. 2. Studies of the permeability properties of lactones in piper methysticum forst, *Eur. J. Pharm. Sci.* **14**, 271–280 (2001).

557. Ruell, J. A., Membrane-based drug assays. *Modern Drug Disc.* Jan., 28–30 (2003).

558. Ruell, J. A.; Tsinman, K. L.; Avdeef, A., PAMPA—a drug absorption in vitro model. 4. Unstirred water layer in iso-pH mapping assays in pK_a flux–optimized design (pOD-PAMPA) (submitted).

559. Avdeef, A., PAMPA—a drug absorption in vitro model. 16. Gradient-pH permeability (in preparation).

560. Avdeef, A.; Takács-Novák, K., PAMPA—a drug absorption in vitro model. 23. Permeability of quaternary ammonium drugs and other charged substances (in preparation).

561. Sugano, K.; Hamada, H.; Machida, M.; Ushio, H., High throughput prediction of oral absorption: Improvement of the composition of the lipid solution used in parallel artificial membrane permeability assay, *J. Biomolec. Screen.* **6**, 189–196 (2001).

562. Sugano, K.; Hamada, H.; Machida, M.; Ushio, H.; Saitoh, K.; Terada, K., Optimized conditions of biomimetic artificial membrane permeability assay, *Int. J. Pharm.* **228**, 181–188 (2001).

563. Zhu, C.; Jiang, L.; Chen, T.-M.; Hwang, K.-K., A comparative study of artificial membrane permeability assay for high-throughput profiling of drug absorption potential, *Eur. J. Med. Chem.* **37**, 399–407 (2002).

564. Veber, D. F.; Johnson, S. R.; Cheng, H.-Y; Smith, B. R.; Ward, K. W.; Kopple, K. D., Molecular properties that influence the oral bioavailability of drug candidates, *J. Med. Chem.* **45**, 2615–2623 (2002).

565. www.pampa2002.com.

566. Haase, R. et al., The phospholipid analogue hexadecylphosphatidylcholine inhibits phosphatidyl biosynthesis in Madin-Darby Canine Kidney Cells, *FEBS Lett.* **288**, 129–132 (1991).

567. Rawn, J. D., *Biochemistry*, Niel Patterson Publishers, Burlington, NC, 1989, pp. 223–224.

568. Hennesthal, C.; Steinem, C., Pore-spanning lipid bilayers visualized by scanning force microscopy. *J. Am. Chem. Soc.* **122**, 8085–8086 (2000).

569. Lichtenberg, D., Micelles and liposomes, in Shinitzky, M. (ed.), *Biomembranes: Physical Aspects*, VCH, Weinheim, 1993, pp. 63–96.

570. Lamson, M. J.; Herbette, L. G.; Peters, K. R.; Carson, J. H.; Morgan, F.; Chester, D. C.; Kramer, P. A., Effects of hexagonal phase induction by dolichol on phospholipid membrane permeability and morphology, *Int. J. Pharm.* **105**, 259–272 (1994).

571. Proulx, P., Structure-function relationships in intestinal brush border membranes, *Biochim. Biophys. Acta* **1071**, 255–271 (1991).

572. Krämer, S. D.; Begley, D. J.; Abbott, N. J., Relevance of cell membrane lipid composition to blood-brain barrier function: Lipids and fatty acids of different BBB models, *Am. Assoc. Pharm. Sci. Ann. Mtg.*, 1999.

573. Krämer, S. D.; Hurley, J. A.; Abbott, N. J.; Begley, D. J., Lipids in blood-brain barrier models in vitro I: TLC and HPLC for the analysis of lipid classes and long polyunsaturated fatty acids, *J. Lipid Res.* (2003), in press.

574. Sawada, G. A.; Ho, N. F. H.; Williams, L. R.; Barsuhn, C. L.; Raub, T. J., Transcellular permeability of chlorpromazine demonstrating the roles of protein binding and membrane partitioning, *Pharm. Res.* **11**, 665–673 (1994).

575. Sawada, G. A.; Barsuhn, C. L.; Lutzke, B. S.; Houghton, M. E.; Padbury, G. E.; Ho, N. F. H.; Raub, T. J., Increased lipophilicity and subsequent cell partitioning decrease passive transcellular diffusion of novel, highly lipophilic antioxidants, *J. Pharmacol. Exp. Ther.* **288**, 1317–1326 (1999).

576. Sawada, G. A., Williams, L. R.; Lutzke, B. S., Raub, T. J., Novel, highly lipophilic antioxidants readily diffuse across the blood-brain barrier and access intracellular sites, *J. Pharmacol. Exp. Ther.* **288**, 1327–1333 (1999).

577. Helenius, A.; Simons, K., Solubilization of membranes by detergents, *Biochim. Biophys. Acta* **413**, 29–79 (1975).

578. Palm, K.; Luthman, K.; Ros, J.; Gråsjö, J.; Artursson, P., Effect of molecular charge on intestinal epithelial drug transport: pH-dependent transport of cationic drugs, *J. Pharmacol. Exp. Ther.* **291**, 435–443 (1999).

579. Kakemi, K.; Arita, T.; Hori, R.; Konishi, R., Absorption and excretion of drugs. XXX. Absorption of barbituric acid derivatives from rat stomach, *Chem. Pharm. Bull.* **17**, 1534–1539 (1967).

580. Wagner, J. G.; Sedman, A. J., Quantitation of rate of gastrointestinal and buccal absorption of acidic and basic drugs based on extraction theory, *J. Pharmacokinet. Biopharm.* **1**, 23–50 (1973).

581. Garrigues, T. M.; Pérez-Varona, A. T.; Climent, E.; Bermejo, M. V.; Martin-Villodre, A.; Plá-Delfina, J. M., Gastric absorption of acidic xenobiotics in the rat: Biophysical interpretation of an apparently atypical behaviour, *Int. J. Pharm.* **64**, 127–138 (1990).

582. Lee, A. J.; King, J. R.; Barrett, D. A., Percutaneous absorption: A multiple pathway model, *J. Control. Release* **45**, 141–151 (1997).

583. Ho, N. F. H.; Park, J. Y.; Morozowich, W.; Higuchi, W. I., Physical model approach to the design of drugs with improved intestinal absorption, in Roche, E. B. (ed.), *Design of Biopharmaceutical Properties through Prodrugs and Analogs* APhA/APS, Washington, DC, pp. 136–227.

584. Ho, N. F. H.; Park, J. Y.; Ni, P. F.; Higuchi, W. I., Advancing quantitative and mechanistic approaches in interfacing gastrointestinal drug absorption studies in animals and man, in Crouthamel, W. G.; Sarapu, A. (eds.), *Animal Models for Oral Drug Delivery in Man: In Situ and In Vivo Approaches*, APhA/APS, Washington, DC, pp. 27–106.

585. Hansch, C., A quantitative approach to biochemical structure-activity relationships, *Acc. Chem. Res.* **2**, 232–239 (1969).

586. Kubinyi, H., Quantitative structure-activity relationships. IV. Nonlinear dependence of biological activity on hydrophobic character: a new model, *Arzneim. Forsch. (Drug Res.)* **26**, 1991–1997 (1976).

587. Collander, R., The partition of organic compounds between higher alcohols and water, *Acta Chem. Scand.* **5**, 774–780 (1951).

588. van de Waterbeemd, H.; van Boeckel, S.; Jansen, A. C. A.; Gerritsma, K., Transport in QSAR. II: Rate-equilibrium relationships and the interfacial transfer of drugs, *Eur. J. Med. Chem.* **3**, 279–282 (1980).

589. van de Waterbeemd, H.; van Boeckel, S.; de Sevaux, R.; Jansen, A.; Gerritsma, K., Transport in QSAR. IV: The interfacial transfer model. Relationships between partition coefficients and rate constants of drug partitioning, *Pharm. Weekbl. Sci. Ed.* **3**, 224–237 (1981).

590. van de Waterbeemd, H.; Jansen, A., Transport in QSAR. V: Application of the interfacial drug transfer model, *Pharm. Weekbl. Sci. Ed.* **3**, 587–594 (1981).

591. Wils, P.; Warnery, A.; Phung-Ba, V.; Legrain, S.; Scherman, D., High lipophilicity decreases drug transport across intestinal epithelial cells, *J. Pharmacol. Exp. Ther.* **269**, 654–658 (1994).

592. Poulin, P.; Thiel, F.-P., Prediction of pharmacokinetics prior to in vivo studies. 1. Mechanism-based prediction of volume of distribution, *J. Pharm. Sci.* **91**, 129–156 (2002).

593. Vilà, J. I.; Calpena, A. C.; Obach, R.; Domenech, J., Gastric, intestinal and colonic absorption of a series of β-blockers in the rat, *J. Clin. Pharmacol. Ther. Toxicol.* **30**, 280–286 (1992).

594. Cussler, E. L. *Diffusion*, 2nd ed., Cambridge Univ. Press, Cambridge, UK, 1997, p. 112.

595. Chen, Y.; Pant, A. C.; Simon, S. M., P-glycoprotein does not reduce substrate concentration from the extracellular leaflet of the plasma membrane in living cells, *Cancer Res.* **61**, 7763–7768 (2001).

596. Higgins, C. F.; Callaghan, R.; Linton, K. J.; Rosenberg, M. F.; Ford, R. C., Structure of the multidrug resistance P-glycoprotein, *Cancer Biol.* **8**, 135–142 (1997).

597. Regev, R.; Eytan, G. D., Flip-flop of doxorubicin across erythrocyte and lipid membranes, *Biochem. Pharmacol.* **54**, 1151–1158 (1997).

598. Langguth, P.; Kubis, A.; Krumbiegel, G.; Lang, W.; Merkle, H. P.; Wächter, W.; Spahn-Langguth, H.; Weyhenmeyer, R., Intestinal absorption of the quaternary trospium chloride: Permeability-lowering factors and bioavailabilities for oral dosage forms, *Eur. J. Pharm. Biopharm.* **43**, 265–272 (1997).

599. Young, R. C.; Mitchell, R. C.; Brown, T. H.; Ganellin, C. R.; Griffiths, R.; Jones, M.; Rana, K. K.; Saunders, D.; Smith, I. R.; Sore, N. E.; Wilks, T. J., Development of a new physicochemical model for brain penetration and its application to the design of centrally acting H_2 receptor histamine antagonists, *J. Med. Chem.* **31**, 565–671 (1988).

600. Garrone, A.; Marengo, E.; Fornatto, E.; Gasco, A., A study on pK_a^{app} and partition coefficient of substituted benzoic acids in SDS anionic micellar system, *Quant. Struct.-Act. Relat.* **11**, 171–175 (1992).

601. Henis, Y. I., Lateral and rotational diffusion in biological membranes, in Shinitzky, M. (ed.), *Biomembranes: Physical Aspects*, VCH, Weinheim, 1993, pp. 279–340.

602. Yazdanian, M.; Glynn, S. L.; Wright, J. L.; Hawi, A., Correlating partitioning and Caco-2 cell permeability of stucturally diverse small molecular weight compounds, *Pharm. Res.* **15**, 1490–1494 (1998).

603. Karlsson, J.; Ungell, A.-L.; Grasjo, J.; Artursson, P., Paracellular drug transport across intestinal epithelia: Influence of charge and induced water flux, *Eur. J. Pharm. Sci.* **9**, 47–56 (1999).

604. Irvine, J. D.; Takahashi, L.; Lockhart, K.; Cheong, J.; Tolan, J. W.; Selick, H. E.; Grove, J. R., MDCK (Madin-Darby canine kidney) cells: A tool for membrane permeability screening, *J. Pharm. Sci.* **88**, 28–33 (1999).

605. Adson, A.; Raub, T. J.; Burton, P. S.; Barsuhn, C. L.; Hilgers, A. R.; Audus, K. L.; Ho, N. F. H., Quantitative approaches to delineate paracellular diffusion in cultured epithelial cell monolayers, *J. Pharm. Sci.* **83**, 1529–1536 (1994).

606. Conradi, R. A., Pharmacia & Upjohn, personal correspondence, Feb. 2000.

607. Stenberg, P.; Norinder, U.; Luthman, K.; Artursson, P., Experimental and computational screening models for the prediction of intestinal drug absorption, *J. Med. Chem.* **44**, 1927–1937 (2001).

608. Walter, E.; Janich, S.; Roessler, B. J.; Hilfinger, J. M.; Amidon, G. L., HT29-MTX/Caco-2 cocultures as an in vitro model for the intestinal epithelium: In vitro-in vivo correlation with permeability data from rats and humans, *J. Pharm. Sci.* **85**, 1070–1076 (1996).

609. Hilgendorf, C.; Spahn-Langguth, H.; Regardh, C. G.; Lipka, E.; Amidon, G. L.; Langguth, P., Caco-2 vs Caco-2/HT29-MTX co-cultured cell lines: Permeabilities via diffusion, inside- and outside-directed carrier-mediated transport, *J. Pharm. Sci.* **89**, 63–75 (2000).

610. Aungst, B. J.; Nguyen, N. H.; Bulgarelli, J. P.; Oates-Lenz, K. The influence of donor and reservoir additives on Caco-2 permeability and secretory transport of HIV protease inhibitors and other lipophilic compounds, *Pharm. Res.* **17**, 1175–1180 (2000).

611. Collett, A.; Sims, E.; Walker, D.; He, Y.-L.; Ayrton, J.; Rowland, M.; Warhurst, G., Comparison of HT29-18-C1 and Caco-2 cell lines as models for studying intestinal paracellular drug absorption, *Pharm. Res.* **13**, 216–221 (1996).

612. Partridge, W. M., *Peptide Drug Delivery to the Brain*, Raven Press, New York, 1991, pp. 52–88.

INDEX

Absorption, 242–246
Absorption-distribution-metabolism-excretion (ADME)
  medicinal chemists, 4
  multimechanism, 3, 6
  the 'A' in, 5
Aggregates, effect on solubility
  charged, 112
  uncharged, 112
Amphotericin B, 125
Apparent partition coefficient, log D, 45–53
Aqueous dilution solubility method, 107

Bile acids, effects on permeability, 135, 226–228
Bile salts, 136, 228
Biological membranes, lipid compositions, 132
Biopharmaceutics Classification System, 20
Bjerrum plot, 25–28, 57–58, 103–106, 198
Black lipid membrane (BLM), 123

Caco-2, 3, 131, 134, 136, 138, 142, 170, 221, 238, 242
Charged species transport
  discrimination, 17
  evidence of transport, 215, 218–224
Chemical sink, 139

Cosolvent, effect on permeability, 135, 226–228
Cosolvent solubility method, 108–111
Cosolvent titrations, 29
Cyclodextrin
  effect on permeability, 228
  effect on solubility, 113

Danielli and Davson, 121
Δ log P_e concept, 222–226
Δ shift method, 113
diff 1-2, liposome-water rule, 81–83
diff 3-4, octanol-water rule, 53
Diffusivity, D_{aq}, 9, 207
Dissolution template titration method, 101
DMSO effect on solubility, 111
Double-sink PAMPA, 151
Double-sink Σ-P_m PAMPA GIT model, 244–246, 249

Effects of cosolvents, bile acids, and surfactants on permeability, 135, 226–228
Egg lecithin from different sources, 183
Epithelial cells, 13, 14

Fick's law, 7–11
Focused libraries, 2
Franklin, Benjamin, experiment, 119

Absorption and Drug Development: Solubility, Permeability, and Charge State. By Alex Avdeef
ISBN 0-471-423653. Copyright © 2003 John Wiley & Sons, Inc.

Gastrointestinal tract (GIT), 11–17
  food effect, 12, 13
  pH, 12
  surface area, 12, 13
  transit time, 13
Glycocalyx, 14, 15
Gorter and Grendel, 120
Gradient-pH
  equilibrium concentrations, 151
  permeability, 211–215, 241

Henderson-Hasselbalch equation, 23
Human absorption prediction, 242–246
Human jejunal permeability, 237–239, 246

Intrinsic permeability, 11, 201
Intrinsic permeability vs. alkane partition
    coefficient, 124
Ion-pair absorption
  liposome, experimental methods, 75
  octanol, fact or fiction, 53
Ion-pair partition, 45

Langmuir, I., 120
Lateral diffusion, 16, 17
Lipid bilayer concept, history, 118–121
Lipophilicity profiles
  liposome-water, 80
  octanol-water, 47–52
Liposomes
  charged, 85
  partition coefficients, 67–90. *See also* Partition
      coefficients
log *D*, apparent partition coefficient, 45–53
log *S* vs. pH plots, 92–97

Maximum chemical diversity, 1
Membrane retention, %*R*, 117, 142, 169–171,
    196
Membrane permeability, 199, 205
Metallothionein, 28
Microfilters as artificial membrane supports,
    124
Microvilli, 15
Mucus layer, 14, 15

Octanol
  partition coefficients, 42–66. *See also* Partition
      coefficients
  structure, 19
Octanol-impregnated filters, 126–128
Olive oil, 43, 119, 167
Overton, 119

Parallel artificial membrane permeability assay
    (PAMPA), 116–249
  brush-border lipid (Chugai model), 130, 181
  dodecane, 168
  DOPC, 166
  effect of buffer, 229–231
  egg lecithin (Roche model), 128
  five-component lipid model, 181
  hexadecane (Novartis model), 129
  hydrophilic filters (Aventis model), 131
  intra- and inter-plate reproducibility, 232–234
  membrane retention %*R*, 117, 142, 169–171, 196
  models for gastrointestinal tract, 156–160
  neutral lipid models, 160–169
  olive oil, 167
  optimized double-sink model, 236–238
  retention-gradient-sink (*p*ION model), 131
  two-component anionic lipid models, 171–181
  UV spectral data quality, 233–235
PAMPA lipid models table, 157, 240
Partition coefficients
  charged liposome, 85–87
  entropy-/enthalpy-driven, 70
  immobilized artificial membrane (IAM)
      chromatography, 54, 83
  liposomes
    charged compounds, 79–83, 85
    databases, 69
    dielectric medium, 71–73
    drug location in bilayer, 69
    Gouy-Chapman, 81–82
    one-pH measurement, 84
    preparation, 74
    prediction of absorption, 90
    prediction from octanol, 76–79
    table of values, 87–89
    water wires, 73
  liposome-water, 67–90
  octanol
    Bjerrum plot, 57, 58
    charged-species partitioning, 53
    chromatographic, 54
    common-ion effect, 52
    conditional constants, 45
    cyclic voltammetry, 55
    databases, 45
    high-throughput, 59
    log *D*, 45–50
    microconstants, 54
    peptides, 50
    potentiometric, 55–58
    table of values, 59–66
  octanol-water, 42–66

Permeability, 9–11, 116–249
  effect of stirring, 205, 231
  gastrointestinal tract and blood-brain barrier, 116–118
  mass balance, 143
Permeability equation
  Caco-2, 142
  gradient pH, 148–150
  iso-pH with membrane retention, 142–147
  precipitate in donor compartment, 147
  sink condition in acceptor well, 147
  without retention, 139–142
Permeability-lipophilicity, nonlinear relations, 153–156
Permeability-pH, 132–134
Permeability-solubility product, 10, 11
pH electrode calibration, 27
pH microclimate, 17, 133, 249
pH partition hypothesis, 7
pH, intracellular, 18
pK_a (ionization constant)
  Bjerrum plot, 25–26, 28
  capillary electrophoresis, 32
  chromatographic, 33
  databases, 24
  equilibria and quotients, 23
  ionic strength dependence, 24
  microconstants, 33
  poorly-soluble compounds, 29
  potentiometric, 25
  spectroscopic, 31
  table of values, 35–41
pK_a^{flux}, 202, 206, 216
pK_a^{gibbs}, 93, 98
pK_a^{mem}, 67, 83–84
pK_a^{oct}, 44, 83–84
Pockels, Agnes, 119

Salt solubility, 97–98
Screening, target vs. ADME, 2

sdiff 3–4 rule, 97, 99
Serum proteins, role in permeability assays, 135
Sialic acid, 15
Sink condition, 9, 138–139, 147, 150, 171, 177, 196, 228, 235
Solubility, 10–11, 91–115
  databases, 100
  equilibria and quotients, 92–93
  fast UV plate spectrophotometer method, 107
  high-throughput, 107–111
  HPLC method, 101
  potentiometric method, 101–107
  saturation shake-flask method, 101
  table of values, dissolution template method, 114–115
  turbidity method, 101
  use of Bjerrum plots, 103–107
Solubility product, 97
Soy lecithin, 187–196
Stirring in PAMPA, 231–232
Structure of phospholipid membranes, 131
Summary of simple prediction rules, 247–249
Surface ion-pair (SIP), 82
Surfactants, effect on permeability, 135, 226–228

Tetrad of equilibria
  liposome partition coefficients, 67
  octanol partition coefficients, 43
  solubility, 99
Tight junction pores 15, 18, 248

Unstirred water layer (UWL), 15, 135, 177, 179, 199–208, 216, 236, 241
  thickness, 199, 205, 207–208, 236
UV spectra, 235

Vancomycin, 28
Villi, 13

Yasuda-Shedlovsky procedure, 29